APPLIED CODING AND INFORMATION THEORY FOR ENGINEERS

PRENTICE HALL INFORMATION AND SYSTEM SCIENCES SERIES

Thomas Kailath, Editor

ANDERSON & MOORE	*Optimal Control: Linear Quadratic Methods*
ANDERSON & MOORE	*Optimal Filtering*
ASTROM & WITTENMARK	*Computer-Controlled Systems: Theory and Design, 2/E*
BASSEVILLE & NIKIROV	*Detection of Abrupt Changes: Theory & Application*
BOYD & BARRATT	*Linear Controller Design: Limits of Performance*
DICKINSON	*Systems: Analysis, Design and Computation*
FRIEDLAND	*Advanced Control System Design*
GARDNER	*Statistical Spectral Analysis: A Nonprobabilistic Theory*
GRAY & DAVISSON	*Random Processes: A Mathematical Approach for Engineers*
GREEN & LIMEBEER	*Linear Robust Control*
HAYKIN	*Adaptive Filter Theory*
HAYKIN	*Blind Deconvolution*
JAIN	*Fundamentals of Digital Image Processing*
JOHANSSON	*Modeling and System Identification*
JOHNSON	*Lectures on Adaptive Parameter Estimation*
KAILATH	*Linear Systems*
KUNG	*VLSI Array Processors*
KUNG, WHITEHOUSE, & KAILATH, EDS.	*VLSI and Modern Signal Processing*
KWAKERNAAK & SIVAN	*Signals and Systems*
LANDAU	*System Identification and Control Design Using P.I.M. + Software*
LJUNG	*System Identification: Theory for the User*
LJUNG & GLAD	*Modeling of Dynamic Systems*
MACOVSKI	*Medical Imaging Systems*
MOSCA	*Stochastic and Predictive Adaptive Control*
NARENDRA & ANNASWAMY	*Stable Adaptive Systems*
RUGH	*Linear System Theory*
RUGH	*Linear System Theory, Second Edition*
SASTRY & BODSON	*Adaptive Control: Stability, Convergence, and Robustness*
SOLIMAN & SRINATH	*Continuous and Discrete-Time Signals and Systems*
SOLO & KONG	*Adaptive Signal Processing Algorithms: Stability & Performance*
SRINATH, RAJASEKARAN, & VISWANATHAN	*Introduction to Statistical Signal Processing with Applications*
VISWANADHAM & NARAHARI	*Performance Modeling of Automated Manufacturing Systems*
WELLS	*Applied Coding and Information Theory for Engineers*
WILLIAMS	*Designing Digital Filters*

APPLIED CODING AND INFORMATION THEORY FOR ENGINEERS

Richard B. Wells

University of Idaho
Department of Electrical Engineering
MRC Institute

Prentice Hall
Upper Saddle River, New Jersey 07458

Library of Congress Cataloging-in-Publication Data

Wells, Richard B.
   Applied coding and information theory for engineers / Richard B. Wells.
      p.   cm. — (Prentice-Hall information and system sciences series)
   Includes bibliographical references and index.
   ISBN: 0-13-961327-7
   1. Coding theory. 2. Information theory. I. Title. II. Series
TK5102.92.W45 1999
003′ .54—dc21   98-24946
                                                        CIP

Publisher: **TOM ROBBINS**
Associate editor: **ALICE DWORKIN**
Production editor: **CAROLE SURACI**
Editor-in-chief: **MARCIA HORTON**
Managing editor: **EILEEN CLARK**
Assistant vice president of production and manufacturing: **DAVID W. RICCARDI**
Cover designer: **BRUCE KENSELAAR**
Copy editor: **ETHAN BAKER**
Manufacturing buyer: **PAT BROWN**
Editorial assistant: **DAN DEPASQUALE**

This book is dedicated to John Stedman, my friend, who made many things possible.

©1999 by Prentice-Hall, Inc.
Simon & Schuster / A Viacom Company
Upper Saddle River, New Jersey 07458

All rights reserved. No part of this book may be
reproduced, in any form or by any means,
without permission in writing from the publisher.

The author and publisher of this book have used their best efforts in preparing this book. These efforts include the development, research, and testing of the theories and programs to determine their effectiveness. The author and publisher make no warranty of any kind, expressed or implied, with regard to these programs or the documentation contained in this book. The author and publisher shall not be liable in any event for incidental or consequential damages in connection with, or arising out of, the furnishing, performance, or use of these programs.

Printed in the United States of America

10 9 8 7 6 5 4 3 2 1

ISBN: 0-13-961327-7

Prentice-Hall International (UK) Limited, *London*
Prentice-Hall of Australia Pty. Limited, *Sydney*
Prentice-Hall Canada Inc., *Toronto*
Prentice-Hall Hispanoamericana, S.A., *Mexico*
Prentice-Hall of India Private Limited, *New Delhi*
Prentice-Hall of Japan, Inc., *Tokyo*
Simon & Schuster Asia Pte. Ltd., *Singapore*
Editora Prentice-Hall do Brasil, Ltda., *Rio de Janeiro*

Contents

PREFACE   xi

1. DISCRETE SOURCES AND ENTROPY   1

1.1   Overview of Digital Communication and Storage Systems   1

1.2   Discrete Information Sources and Entropy   2
- 1.2.1   Source alphabets and entropy, 2
- 1.2.2   Joint and conditional entropy, 6
- 1.2.3   Entropy of symbol blocks and the chain rule, 8

1.3   Source Coding   10
- 1.3.1   Mapping functions and efficiency, 10
- 1.3.2   Mutual information, 12
- 1.3.3   A brief digression on encryption, 14
- 1.3.4   Summary of section 1.3, 15

1.4   Huffman Coding   16
- 1.4.1   Prefix codes and instantaneous decoding, 16
- 1.4.2   Construction of Huffman codes, 17
- 1.4.3   Hardware implementation approaches, 19
- 1.4.4   Robustness of Huffman coding efficiency, 20

1.5   Dictionary Codes and Lempel–Ziv Coding   21
- 1.5.1   The rationale behind dynamic dictionary coding, 21
- 1.5.2   A linked-list LZ algorithm, 22
- 1.5.3   The decoding process, 25
- 1.5.4   Large-block requirement of LZ compression, 26

1.6   Arithmetic Coding   28
- 1.6.1   Code-word length and the asymptotic equipartition property, 28
- 1.6.2   The arithmetic coding method, 30
- 1.6.3   Decoding arithmetic codes, 32
- 1.6.4   Other issues in arithmetic coding, 33

1.7   Source Models and Adaptive Source Coding   34

1.8 Chapter Summary   35

References   36

Exercises   37

2. CHANNELS AND CHANNEL CAPACITY   39

2.1 The Discrete Memoryless Channel Model   39
  2.1.1 The transition probability matrix, 39
  2.1.2 Output entropy and mutual information, 41

2.2 Channel Capacity and the Binary Symmetric Channel   43
  2.2.1 Maximization of mutual information and channel capacity, 43
  2.2.2 Symmetric channels, 45

2.3 Block Coding and Shannon's Second Theorem   48
  2.3.1 Equivocation, 48
  2.3.2 Entropy rate and the channel-coding theorem, 49

2.4 Markov Processes and Sources with Memory   51
  2.4.1 Markov processes, 51
  2.4.2 Steady-state probability and the entropy rate, 54

2.5 Markov Chains and Data Processing   56

2.6 Constrained Channels   58
  2.6.1 Modulation theory and channel constraints, 58
  2.6.2 Linear and time-invariant channels, 60

2.7 Autocorrelation and Power Spectrum of Sequences   62
  2.7.1 Statistics of time sequences, 62
  2.7.2 The power spectrum, 64

2.8 Data Translation Codes   68
  2.8.1 Constraints on data sequences, 68
  2.8.2 State space and trellis descriptions of codes, 70
  2.8.3 Capacity of a data translation code, 73

2.9 (d,k) Sequences   75
  2.9.1 Run-length-limited codes and maxentropic sequences, 75
  2.9.2 Power spectrum of maxentropic sequences, 77

2.10 Chapter Summary   82

References   83

Exercises   83

3. RUN-LENGTH-LIMITED CODES   89

3.1 General Considerations for Data Translation Coding   89

Contents

- 3.2 Prefix Codes and Block Codes 91
  - 3.2.1 Fixed-length block codes, 91
  - 3.2.2 Variable-length block codes, 92
  - 3.2.3 Prefix codes and the Kraft inequality, 94
- 3.3 State-Dependent Fixed-Length Block Codes 96
- 3.4 Variable-Length Fixed-Rate Codes 98
- 3.5 Look-Ahead Codes 102
  - 3.5.1 Code-word concatenation, 102
  - 3.5.2 The k constraint, 104
  - 3.5.3 Informal and formal design methods, 105
- 3.6 DC-Free Codes 107
  - 3.6.1 The running digital sum and the digital sum variation, 107
  - 3.6.2 State-splitting and matched spectral null codes, 109
- 3.7 Chapter Summary 114

References 115

Exercises 115

4. LINEAR BLOCK ERROR-CORRECTING CODES 117

- 4.1 General Considerations 117
  - 4.1.1 Channel coding for error correction, 117
  - 4.1.2 Error rates and error distribution for the binary symmetric channel, 118
  - 4.1.3 Error detection and correction, 121
  - 4.1.4 The maximum likelihood decoding principle, 123
  - 4.1.5 Hamming distance code capability, 124
- 4.2 Binary Fields and Binary Vector Spaces 126
  - 4.2.1 The binary field, 126
  - 4.2.2 Representing linear codes in a vector space, 130
- 4.3 Linear Block Codes 131
  - 4.3.1 Elementary properties of vector spaces, 131
  - 4.3.2 Hamming weight, Hamming distance, and the Hamming cube, 133
  - 4.3.3 The Hamming sphere and bounds on redundancy requirements, 135
- 4.4 Decoding Linear Block Codes 136
  - 4.4.1 Complete decoders and bounded-distance decoders, 136
  - 4.4.2 Syndrome decoders and the parity-check theorem, 138
- 4.5 Hamming Codes 140
  - 4.5.1 The design of Hamming codes, 140
  - 4.5.2 The dual code of a Hamming code, 143
  - 4.5.3 The expanded Hamming code, 144

4.6 Error Rate Performance Bounds for Linear Block Error-Correcting Codes 147
　4.6.1 *Block error rates, 147*
　4.6.2 *Bit error rate, 148*

4.7 Performance of Bounded-Distance Decoders with Repeat Requests 149
　4.7.1 *Approximate error performance, 152*
　4.7.2 *Effective code rate of ARQ systems, 154*
　4.7.3 *ARQ protocols, 156*

4.8 Chapter Summary 157

References 158

Exercises 158

5. CYCLIC CODES　　　　160

5.1 Definition and Properties of Cyclic Codes 160

5.2 Polynomial Representation of Cyclic Codes 162

5.3 Polynomial Modulo Arithmetic 164
　5.3.1 *Polynomial rings, 164*
　5.3.2 *Some important algebraic identities, 166*

5.4 Generation and Decoding of Cyclic Codes 169
　5.4.1 *Generator, parity-check, and syndrome polynomials, 169*
　5.4.2 *Systematic cyclic codes, 169*
　5.4.3 *Hardware implementation of encoders for systematic cyclic codes, 171*
　5.4.4 *Hardware implementation of decoders for cyclic codes, 174*
　5.4.5 *The Meggitt decoder, 175*

5.5 Error-Trapping Decoders 178
　5.5.1 *Updating the syndrome during correction, 178*
　5.5.2 *Burst error patterns and error trapping, 180*

5.6 Some Standard Cyclic Block Codes 184
　5.6.1 *The Hamming codes, 184*
　5.6.2 *BCH codes, 185*
　5.6.3 *Burst-correcting codes, 186*
　5.6.4 *Cyclic redundancy check codes, 187*

5.7 Simple Modifications to Cyclic Codes 189
　5.7.1 *Expanding a code, 189*
　5.7.2 *Shortening a code, 190*
　5.7.3 *Noncyclicity of shortened codes, 193*
　5.7.4 *Interleaving, 194*

5.8 Chapter Summary 197

References 197

Exercises 198

6. CONVOLUTIONAL CODES   207

- 6.1 Definition of Convolutional Codes   201
- 6.2 Structural Properties of Convolutional Codes   205
  - 6.2.1 The state diagram and trellis representations, 205
  - 6.2.2 Transfer functions of convolutional codes, 207
- 6.3 The Viterbi Algorithm   210
- 6.4 Why the Viterbi Algorithm Works I: Hard-Decision Decoding   215
  - 6.4.1 Maximum likelihood under hard-decision decoding, 215
  - 6.4.2 Error event probability, 217
  - 6.4.3 Bounds on bit error rate, 219
- 6.5 Some Known Good Convolutional Codes   221
- 6.6 Why the Viterbi Algorithm Works II: Soft-Decision Decoding   223
  - 6.6.1 Euclidean distance and maximum likelihood, 223
  - 6.6.2 Elimination of ties and information loss, 226
  - 6.6.3 Calculation of the likelihood metric, 228
- 6.7 The Traceback Method of Viterbi Decoding   229
- 6.8 Punctured Convolutional Codes   234
  - 6.8.1 Puncturing, 234
  - 6.8.2 Good punctured convolutional codes, 236
- 6.9 Chapter Summary   238

  References   239

  Exercises   239

7. TRELLIS-CODED MODULATION   242

- 7.1 Multiamplitude/Multiphase Discrete Memoryless Channels   242
  - 7.1.1 I—Q modulation, 242
  - 7.1.2 The n-ary PSK signal constellation, 243
  - 7.1.3 PSK error rate, 245
  - 7.1.4 Quadrature amplitude modulation, 247
- 7.2 Systematic Recursive Convolutional Encoders   248
- 7.3 Signal Mapping and Set Partitioning   251
- 7.4 Known Good Trellis Codes for PSK and QAM   254
- 7.5 Chapter Summary   257

  References   257

  Exercises   258

8. INFORMATION THEORY AND CRYPTOGRAPHY 259

- 8.1 Cryptosystems 259
  - 8.1.1 Basic elements of ciphersystems, 259
  - 8.1.2 Some simple ciphersystems, 261
- 8.2 Attacks on Cryptosystems 265
- 8.3 Perfect Secrecy 266
- 8.4 Language Entropy and Successful Ciphertext Attacks 269
  - 8.4.1 The key-equivocation theorem, 269
  - 8.4.2 Spurious keys and key equivocation, 270
  - 8.4.3 Language redundancy and unicity distance, 271
- 8.5 Computational Security 272
- 8.6 Diffusion and Confusion 274
- 8.7 Product Cipher Systems 275
  - 8.7.1 Commuting, noncommuting, and idempotent product ciphers, 276
  - 8.7.2 Mixing transformations and good product ciphers, 278
- 8.8 Codes 279
- 8.9 Public-Key Cryptosystems 280
- 8.10 Other Issues 281
- 8.11 Chapter Summary 281

  References 282

  Exercises 283

9. SHANNON'S CODING THEOREMS 285

- 9.1 Random Coding 285
- 9.2 The Average Random Code 287
- 9.3 A Discussion of Shannon's Second Theorem 289
- 9.4 Shannon–Fano Coding 290
- 9.5 Shannon's Noiseless-Coding Theorem 292
- 9.6 A Few Final Words 293

  References 294

  Answers to Selected Exercises 295

INDEX 299

Preface

Welcome to the study of information and coding theory. We are living, as the saying goes, in the dawn of the information era, a time which many have likened to the next industrial revolution. Although this observation has been made so often as to become a cliche, the underlying significance of this new age for business, industry, and society in general is nonetheless difficult to overstate. When the industrial revolution occurred, it brought with it a new need for people to become skilled in the technical sciences, arts, and crafts. In a similar fashion, the information revolution brings with it the need for a greater number of people with understanding of and skills in the crafting of information for a variety of uses.

This book has been written for the beginner. It is the result of class notes from an undergraduate-level course I teach to students in electrical and computer engineering, computer science, and mathematics. The level of exposition in this book has been aimed at undergraduate students in their junior or senior years of study and for the practicing engineer who has little or no previous exposure to this subject. The goal of this book is to help you get started in the *practice* of information engineering.

In recent years, introductory textbooks on this subject have become virtually extinct. Some very good graduate-level textbooks do exist, but these texts are often a bit too theory laden and a bit light on practice for the eager new student motivated by the need to develop a marketable skill or for the busy practitioner looking for an introductory-level treatment so she can get started on that hot new project. With these readers in mind, I have deliberately abandoned the traditional "theorem-proof" format found in most books on this subject. The material in this text comes equipped with theory but I have tried to structure the book in such a way that the required mathematical developments immediately precede the material on "how to" apply the methods and theory. Thus, there is no grand chapter where all of the mathematical theorems are condensed. Topics are introduced as they are needed and in a "just in time" fashion. The text is liberally sprinkled with examples (with numbers) to illustrate the concepts.

The material in this textbook is adequate for a one-semester course in the junior or senior year. The text assumes the reader has previously acquired a background in elementary linear algebra and in introductory probability. A previous course in digital logic design or in introductory communication systems is helpful but is not vital.

Chapter 1 begins with an overview of digital communication systems and an introduction to the concept of information. I have found that students are frequently surprised to learn that "information" and "data" are not the same thing. Chapter 1 introduces discrete information sources and the fundamental concepts of entropy and joint entropy. This leads to an introduction to source coding for data compression where we apply the theory to Huffman codes, Lempel–Ziv codes, and arithmetic codes. The topics of source modeling and adaptive coding are lightly touched upon and references are provided for the reader wishing to delve deeper into this important topic.

Our study of information theory continues in Chapter 2 with the introduction of discrete memoryless channels. After describing and defining these channels, we introduce mutual information and channel capacity. The Arimoto—Blahut algorithm for calculating channel capacity for discrete memoryless channels is described. We also are introduced to the all-important binary symmetric channel, which is described in some detail. This leads us to the idea of block coding and Shannon's famous second theorem. This chapter also introduces Markov processes and channels with memory which leads us to a number of important concepts. Constrained channels are then introduced, along with the important notions of the autocorrelation and power spectrum of a sequence. We close the chapter with applications of the theory to data translation codes and introduce run-length-limited (d,k) codes.

Chapter 3 is all application and the instructor may skip this chapter without loss of continuity if he feels the pinch of the clock and calendar. This chapter is about the particular class of data translation codes known variously as line codes, modulation codes, or run-length limited codes. It is a survey of prefix block coding techniques including state-independent fixed-rate/fixed-block codes, state-dependent coding for fixed-length block codes, variable-length/fixed-rate block codes, look-ahead codes, and concludes with a few words about dc-free codes.

Chapter 4 introduces the general theory of linear-block error correcting codes. It begins with a discussion of the coding problem and the calculation of error probabilities for noisy channels. Error correction using binary repetition codes is next discussed along with some important bounds and constraints which any linear block code must obey. We then provide some brief background on binary fields and binary vector spaces in preparation for the more theoretical development of algebraic codes. The fundamental ideas of Hamming distance, Hamming weight, and the Hamming cube are introduced along with some other important mathematical definitions and concepts. Decoding is introduced using the standard array, and systematic block codes are defined. This leads us to a very in-depth discussion of the Hamming codes, our first "important" practical code. Along with the basic Hamming codes, we also discuss some useful "variations on a theme" including the dual codes and the expanded Hamming code. Codes for correction and detection are discussed. The chapter concludes with a discussion of error rates for linear-block codes and code performance for error-correcting codes and for automatic repeat-request systems.

Chapter 5 continues the discussion of linear-block codes with the introduction of cyclic block codes. Following basic definitions and properties, the polynomial representation of cyclic codes is introduced and we discuss polynomial modulo arithmetic for polynomials constructed from the binary field. Attention is then turned to efficient methods for the generation and decoding of cyclic codes. A number of practical circuits for implementing encoders and decoders are given. We briefly discuss error trapping and pipelined error-trapping decoders. We finish up this chapter by providing several useful and important standard codes including the Hamming codes (again!), some of the simpler BCH codes, and some good burst-correcting codes. Error detection using cyclic redundancy check (CRC) codes is also discussed along with some useful "variations on a theme," including interleaving and shortened codes.

In Chapter 6, we turn away from block codes and introduce linear convolutional codes. After discussing the basic encoder, we examine some structural properties of convolutional codes and the representation of these codes using state diagrams and trellis diagrams. The notions of the transfer function of a code and its uses are discussed. We discuss the Viterbi algorithm in depth. The presentation here differs from that of most texts and papers in that we first describe what the algorithm is and how it works before describing why it works. This reversal of the usual presentation is reported by my students to be easier to follow than the traditional pedagogy. We discuss both hard- and soft-decision Viterbi decoding and compare and discuss the performance differences between these two methods. Some of the known good convolutional codes are then presented in tabular form. We return the Viterbi algorithm to discuss some practical implementation matters including the traceback method of decoding and the use of punctured convolutional codes to obtain higher coding rates.

Chapter 7 is a brief introduction to trellis-coded modulation. We introduce two-dimensional *I-Q* channels and transmitters and receivers for these channels. The error rate properties of encoded channels is discussed for phase modulation and for quadrature amplitude modulation systems. This is followed by an introduction to systematic recursive convolutional encoders and their representation using trellis diagrams. Ungerboeck's canonical encoder is presented and octal description of TCM codes using parity check polynomials is given. This is followed by a discussion of set partitioning and how this is used to construct TCM codes using Ungerboeck encoders. The chapter concludes with a summary of some known good codes for phase modulation and quadrature amplitude modulation.

Chapter 8 is a brief introduction to the application of information theory to cryptography. Some simple cryptosystems, based on ciphers, are introduced along with a brief description of some of the methods by which cryptosystems may be attacked. Shannon's definition of perfect secrecy is then introduced and conditions for attaining perfect secrecy are derived. The entropy rate of a natural language and how the redundancy of natural languages can be exploited in cryptanalysis is then discussed. This leads us to the important notions of spurious keys and unicity distance. Following this, the issue of computational security is discussed. Shannon's techniques of diffusion and confusion are described, leading to the important technique of product cipher systems. Finally, the chapter concludes with brief descriptions of codes, public-key cryptosystems, and certain other issues. Public-key cryptosystems, while important, are not described in depth since, the theory of public key cryptosystems is more involved with number theory than with information theory, but the reader is provided with several good references for following up in more detail on the theory and practice of public key cryption.

The text concludes with Chapter 9. This brief chapter provides a proof of Shannon's second theorem for the special case of the binary symmetric channel. It introduces the notion of the random-coding argument and provides a discussion of what the theorem does and does *not* tell us. We then turn to the derivation of Shannon's noiseless coding theorem and the existence of prefix codes for performing source compression. The text then concludes with a few final words on information theory and where the reader may go from here.

Although a certain amount of formalism is necessary in presenting coding-and-information theory, I have tried to keep the presentation a little informal whenever possible. The reader will run across, from time to time, some light-hearted commentary from the author who feels, "If the writing of a textbook has its lighter moments, shouldn't the reading of it have some too?" One of the more important aspects of coding and information theory that is often lost in the classical graduate-level textbooks on this subject is that it is *fun*. I have tried not to lose sight of this aspect during the swirl of mathematical presentation.

While it is true that an author is a character of some importance in the writing of a textbook, the merit of a textbook lies not with its author but, rather, with the audience that book is meant to serve. Furthermore, any textbook (and this one in particular) owes its existence to a number of people who make important contributions. With this in mind, I would like to acknowledge my appreciation to my students who provided me with plenty of feedback about the manuscript, the homework exercises, and the solution manual. I would also like to thank Mr. Aaron Brennan for his assistance in the design of some of the homework exercises. I would especially like to thank Dr. George Freeman of the University of Waterloo for his insightful comments and suggestions which did much to improve the quality of this book. Finally, although they are seldom mentioned by an author, I would like to thank Alice, Tom, and the rest of the merry band at Prentice-Hall whose craftsmanship makes the difference between a manuscript and a book.

Richard B. Wells
Moscow, Idaho

CHAPTER 1

Discrete Sources and Entropy

1.1 OVERVIEW OF DIGITAL COMMUNICATION AND STORAGE SYSTEMS

Systems dedicated to the communication or storage of information are commonplace in everyday life. Broadly speaking, a communication system is a system which sends information from one place to another. Examples include, but are not limited to, the telephone network, radio, television, cellular telephones, local-area computer networks, and so on. Storage systems are systems for the storage and later retrieval of information. In a sense, such systems may be regarded as communication systems which transmit information from *now* (the present) to *then* (the future). Examples include magnetic and optical disk drives, magnetic tape recorders, video tape players, and so on.

Both of these types of systems may be represented abstractly by the block diagram given in Figure 1.1.1. In all cases, there is a *source* from which the information originates. Information from the source is processed by a system for encoding and modulating the information. This encoder/modulator processes the information into some form of signal, which is designed to facilitate the transmission (or storage) of the information in physical form. In communication systems, this function is often referred to as a transmitter while in storage systems it is often called a recorder or a writer.

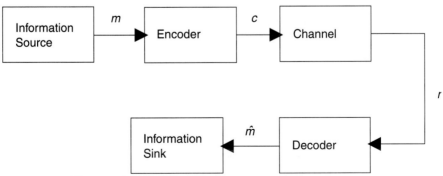

Figure 1.1.1: Basic Information Processing System

The output of the encoding system is then transmitted through some physical communication channel (in the case of a communication system) or stored in some physical storage medium (in the case of a storage system). Examples of the former include wireless transmission using electromagnetic waves and wire transmission using copper telephone cables or fiber optic cables. Examples of the latter case include magnetic disks, such as those used by a floppy disk drive, magnetic tape, and optical disks, such as those used by a CD-ROM or a compact-disk player. Regardless of the explicit form of the medium, we shall refer to it as the "channel."

Information conveyed through (or stored in) the channel must be recovered at the destination and processed to restore the original representation of the information. This is the task of the decoder/demodulator. In the case of a communication system, this device is often referred to as the receiver. In the case of a storage system, this block is often called the playback system or the reader. The signal processing performed by the decoder can be viewed as the inverse of the function performed by the encoder. The output of the decoder is then presented to the final user or destination, which we call the information sink.

The physical channel typically produces a received signal r which differs from the original input signal c. This is because of signal distortion and noise introduced by the channel. Consequently, the decoder can only produce an estimate $\hat{m}$ of the original information message m. The goal of all well-designed system is to attempt to reproduce m as reliably as possible while, at the same time, sending as much information as possible per unit time (communication system) or per unit storage (storage system).

In the typical case, the source message m consists of a time sequence of *symbols* emitted by the information source. The source is said to be a continuous-time source if this sequence is continuous in time. Otherwise, the source is said to be discrete-time. An example of a continuous-time source would be a speech waveform. Examples of discrete-time sources include data sequences from a computer or the printed text on this page (which is an example of a data *storage* system).

The symbols emitted from the source can also be characterized as continuous-amplitude or as discrete-amplitude. Speech is an example of a continuous-amplitude source, since a model of a speech waveform consists of real-valued signal amplitudes. This text is again an example of a discrete-amplitude source since it draws its characters from a finite symbol alphabet.

In this introductory text, we will be primarily concerned with discrete-time/discrete-amplitude sources since these sources have the simplest mathematical treatment and since practically all new communication or storage systems currently fall into this category. Extension of the theory presented here to the continuous case is more appropriately dealt with in an advanced course.

1.2 DISCRETE INFORMATION SOURCES AND ENTROPY

1.2.1 Source Alphabets and Entropy

Information theory is heavily based on the concepts and mathematics of probability theory. This is because the term *information* carries with it a connotation of unpredictability in the transmitted messages. The information content of a message is directly related to the amount of "surprise" conveyed by the message. For example, suppose someone were to say to you, "The capitol of the United States of America is Washington, D.C." Once, early in your life, you did not know this, and so the first time you heard of it, it was *informative*. Seeing it just now, in the previous sentence, this message was completely uninformative (at least for any resident of the United States!). From the point of view of information theory, the statement above held zero information for you after the word "is."

There is a distinction to be made between *information* and *knowledge*. They are not the same thing although one gives rise to the other. Knowledge has the following definitions according to Webster's dictionary: (1) The fact or condition of knowing something with familiarity gained by experience or association; (2) the fact or condition of being aware of something; (3) the sum of what is known. That which adds to our knowledge may be said to be *informative*. Information is therefore distinguished by the property that it adds to our knowledge. That which is not informative delivers no information. Consequently, that which is informative carries with it this element of surprise or uncertainty. It also follows from this character of information that data and information are not the same thing. Were I to recite the English alphabet to you, I would be supplying you with data but, assuming you can read this text, I would be providing no information to you.

An information source is defined by the set of output symbols it is capable of producing and the probability rules which govern the emission of these symbols. A finite discrete source is one for which there is a finite number of unique symbols. The symbol set is frequently called the *source alphabet*. For a source alphabet with M possible symbols, we represent the symbol alphabet as a set

$$A = \{a_0, a_1, \ldots, a_{M-1}\}. \tag{1.2.1}$$

The number of elements in a set is called its *cardinality* and is written

$$M = |A|.$$

The source outputs symbols in a time sequence represented by the notation

$$\bar{a} = (s_0 s_1 \ldots s_t \ldots), \tag{1.2.2}$$

where $s_t \in A$ is the symbol emitted by the source at time t. In this text, we shall take t to be an integer time index unless stated otherwise. At any given time index, the probability that the source emits symbol a_m is written as $p_m = \Pr(a_m)$. If the set of probabilities

$$P_A = \{p_0, p_1, \cdots, p_{M-1}\} \tag{1.2.3}$$

is not a function of time, the source is said to be *stationary*. Since it is certain that the source emits only members of its alphabet A, we have

$$\sum_{m=0}^{M-1} p_m = 1. \tag{1.2.4}$$

Mathematically, the simplest sources to treat are *synchronous* sources which emit a new symbol at a fixed time interval T_s. An *asynchronous* source is one in which the time interval between emitted symbols is not fixed. Such a source can be modeled in an approximate fashion by defining one of its symbols, say a_0, to be a *null character*. If the source does *not* emit a character at time index t, we say that the source produces a null character at time t. A null character, therefore, is a kind of "virtual" symbol denoting the absence of a "real" symbol at t. This convention allows many discrete-time asynchronous sources to be approximated as synchronous sources.

The symbols emitted by a source in a physical communication system must be represented somehow. In digital communication systems, this is typically done using binary representation. For example, a source with $M = 4$ might have its symbols represented by a pair of binary digits. In this case, a_0 might be represented as 00, a_1 as 01, and so on. It is common to refer to symbols represented in this fashion as *source data*.

Information theory makes an important distinction between *data* and *information*. The two concepts are *not* the same and, generally, data is *not* equivalent to information. To see this, consider an information source that has an alphabet with only one symbol. The source can emit only this symbol and nothing else. The representation of this symbol is "data" but, clearly, this data is completely uninformative. There is no "surprise" whatsoever tied to this data. Since information carries the connotation of uncertainty in what will come next, the *information content* of this source is zero.

The information content of a source is an important attribute and can be measured. In his original paper (which founded the science of information theory), Shannon gave a precise mathematical definition of the *average amount of information conveyed per source symbol*. This measure is called the *entropy* of the source and is defined as

$$H(A) = \sum_{m=0}^{M-1} p_m \log_2(1/p_m). \qquad 1.2.5$$

Equation 1.2.5 tells us that the information content of a source is determined by the individual symbol probabilities. Note that for the case of our uninformative $M = 1$ source, the probability of its symbol is unity and $H = 0$.

In the definition given above for entropy, we take the logarithm to the base 2 of the reciprocal of the probabilities. It is also possible to define entropy using any other base logarithm. For entropy defined as in Equation 1.2.5, the unit of measure of entropy is called a "bit." If we were to replace the logarithm to the base 2 in Equation 1.2.5 by the natural logarithm (log to the base e), the entropy is said to be measured in "natural units" or "nats." The most common practice is to measure entropy in bits.

EXAMPLE 1.2.1

What is the entropy of a 4-ary source having symbol probabilities

$$P_A = \{0.5, 0.3, 0.15, 0.05\}?$$

Solution:

$$H(A) = .5\log_2(2) + .3\log_2(10/3) + .15\log_2(100/15) + .05\log_2(20) = 1.6477 \text{ bits}.$$

(Recall that $\log_2(x) = \ln(x)/\ln(2)$.)

A binary data representation of the source in Example 1.2.1 would require two binary digits to represent the four symbols in A. Yet the information content of A is less than two bits. As we shall see later, when the entropy in bits is less than the number of binary digits used to represent the source data it is possible to devise a more efficient scheme for encoding the source information using, on the average, fewer binary digits. This is called *data compression* and an encoder used for this purpose is called a *source encoder*. The coded representation itself is called *source code*.

The ratio of the entropy of a source to the (average) number of binary digits used to represent the source data is a measurement of the information *efficiency* of the source. For instance, the source in Example 1.2.1 has an efficiency of $1.6477/2 = 82.387\%$. This means that approximately 17.6% of the binary digits emitted by the source are "wasted" in the sense that the actual information carried by the sequence of source symbols is less than the number of binary digits transmitted (or stored) by this percentage.

EXAMPLE 1.2.2

Consider an M-ary source. What distribution of probabilities P_A maximizes the information content of A?

Solution: To maximize the information content of A, we must maximize the entropy with respect to the individual probabilities p_m subject to the constraint that the probabilities sum to one. To solve this problem, we express the entropy in the equivalent form

$$f(p_0, \cdots, p_{M-1}) = \frac{1}{\ln(2)} \sum_{m=0}^{M-1} p_m \ln\left(\frac{1}{p_m}\right) + \lambda \left(\sum_{m=0}^{M-1} p_m - 1 \right),$$

where λ is an arbitrary constant. Notice that λ is multiplied by zero in the expression above (we've just written zero in a funny way; this approach is called the method of Lagrange multipliers in basic calculus).

We maximize the entropy by maximizing the function f above for each of the individual probabilities. For probability p_j (with j in the range from 0 to $M-1$), we have

$$\frac{\partial f}{\partial p_j} = \frac{1}{\ln(2)}\left[\ln\left(\frac{1}{p_j}\right) - 1 \right] + \lambda = 0$$

or

$$\ln(1/p_j) = 1 - \lambda \ln(2).$$

Since the right-hand side of the above expression is a constant independent of j, this tells us that *all* of the probabilities must be equal to each other. Since the probabilities must sum to unity, this means that $p_m = 1/M$.

The result of Example 1.2.2 makes sense intuitively. If every symbol in A is equally probable, an observer would have no idea what symbol will be emitted next by the source. Each symbol thus carries maximum "surprise value" and the average amount of information conveyed is similarly maximized. On the other hand, if any one of the symbols in A occurs with greater likelihood than all other symbols combined, an enterprising gambler could confidently make money by betting on the outcome of the next symbol produced by the source.

1.2.2 Joint and Conditional Entropy

Most communication systems are designed to be used by a large number of users. The designers of such a system are concerned with maximizing the *total* information-carrying capacity of the system. Likewise, many computer systems support multiple users and the designers of the system are equally concerned with the total data storage requirements of the system. We can look at the implications of this by considering a situation where we have two information sources, A and B. Let's assume $|A| = M_A$ and $|B| = M_B$. If sources A and B are statistically independent, the total entropy of this system will turn out to be simply $H(A, B) = H(A) + H(B)$. On the other hand, if the information sent by B is statistically *dependent* on the information sent by A, the situation is less obvious. (Such a situation could arrive if A and B were cooperating in some way).

Let the *joint probability* that A sends symbol a_i and B sends symbol b_j be written as

$$p_{i,j} = \Pr(a_i, b_j).$$

If the two sources are statistically independent, then $p_{i,j} = \Pr(a_i) \cdot \Pr(b_j) = p_i p_j$. This is not true, however, if the two sources are statistically *dependent*. Let us regard the combined emission of symbol a_i and b_j as a *compound symbol* $c_{i,j} \equiv \langle a_i, b_j \rangle$ having probability $p_{i,j}$. (In mathematics, this notation denotes what is called an *ordered pair* and is a convenient method of representing functions of two variables; this idea also extends easily to the situation where we have n variables; in that case, the compound symbol is called an ordered n-tuple).

If C is the set of all compound symbols $c_{i,j}$, the entropy of C is calculated by applying Equation 1.2.5 to all of the elements of C

$$H(C) = \sum_{c_{i,j} \in C} p_{i,j} \log_2(1/p_{i,j}) = \sum_{i=0}^{M_A-1} \sum_{j=0}^{M_B-1} p_{i,j} \log_2(1/p_{i,j}). \qquad 1.2.6$$

Now, a joint probability $p_{i,j}$ may be written in terms of a *conditional* probability

$$p_{j|i} \equiv \Pr(b_j | a_i)$$

as

$$p_{i,j} = p_{j|i} \cdot p_i.$$

Using this and the property $\log(ab) = \log(a) + \log(b)$, equation 1.2.6 becomes

$$H(C) = \sum_{i=0}^{M_A-1} p_i \log_2(1/p_i) + \sum_{i=0}^{M_A-1} \sum_{j=0}^{M_B-1} p_{i,j} \log_2(1/p_{j|i}), \qquad 1.2.7$$

where we have used the relationship

$$\sum_i \sum_j p_i p_{j|i} \log_2(1/p_i) = \sum_i p_i \log_2(1/p_i) \sum_j p_{j|i}$$

and the fact that, for any given i, the sum over all j of $p_{j|i}$ must equal 1. (Whatever i may be, *some* symbol b_j must occur so the sum over all of the possibilities must add to 1.)

The first term on the right-hand side of Equation 1.2.7 is simply $H(A)$. The second term is called a *conditional entropy*. It is the uncertainty (entropy) of B given A and is written $H(B|A)$. Thus, we may write

$$H(C) = H(A,B) = H(A) + H(B|A). \qquad 1.2.8$$

(Note the similarity of Equation 1.2.8 with the logarithm of $p_{i,j} = p_{j|i} \cdot p_i$). It is a simple matter to prove that $H(A, B) = H(B) + H(A|B)$ as well.

Now, if B is statistically independent of A,

$$H(B|A) = \sum_i p_i \sum_j p_{j|i} \log_2(1/p_{j|i}) = \sum_i p_i \sum_j p_j \log_2(1/p_j) = H(B).$$

In this case, the total entropy is simply the sum of the entropies of the two sources. On the other hand, if B is dependent on A, it can be shown that $H(B|A) < H(B)$. (The proof of this gives rise to a standard theorem in information theory that says "side information never increases entropy.") Therefore,

$$H(A,B) \leq H(A) + H(B), \qquad 1.2.9$$

with equality if and only if A and B are statistically independent.

EXAMPLE 1.2.3

Many computer backplanes and memory systems employ a *parity bit* as a simple means of error detection. Let A be an information source with alphabet $A = \{0, 1, 2, 3\}$. Let each symbol a be equally probable and let $B = \{0, 1\}$ be a parity generator with

$$b_j = \begin{cases} 0 & \text{if} \quad a = 0 \text{ or } a = 3 \\ 1 & \text{if} \quad a = 1 \text{ or } a = 2 \end{cases}.$$

What are $H(A), H(B)$, and $H(A, B)$?

Solution: From the definition of entropy, $H(A) = 4 \cdot \frac{1}{4} \cdot \log_2(4) = 2$. Likewise, the two symbols in B each have probability 0.5, so $H(B) = 1$. However, the conditional probabilities $\Pr(b|a)$ are

$$\Pr(0|0) = 1, \quad \Pr(1|0) = 0,$$
$$\Pr(0|1) = 0, \quad \Pr(1|1) = 1,$$
$$\Pr(0|2) = 0, \quad \Pr(1|2) = 1,$$
$$\Pr(0|3) = 1, \quad \Pr(1|3) = 0.$$

Therefore,

$$H(B|A) = \sum_{i=0}^{3} p_i \sum_{j=0}^{1} p_{j|i} \log_2(1/p_{j|i}).$$

Since $\lim_{x \to 0} x \log(x) = 0$, this expression evaluates to $H(B|A) = 4 \cdot 0.25 \cdot (1 \cdot \log_2(1) - 0 \cdot \log_2(0)) = 0$ which simply says that B is *completely determined* by A. Therefore,

$$H(A,B) = H(A) + H(B|A) = 2 + 0 = 2.$$

Source B contributes no information to the compound signal. (Source B is said to be "redundant.")

1.2.3 Entropy of Symbol Blocks and the Chain Rule

Another useful application for entropy is in finding the information content of a *block* of symbols. Suppose a source A produces a sequence of n symbols $(s_0, s_1, \cdots, s_{n-1})$ with each s_t drawn from alphabet A. The entropy of this block is denoted as $H(A_0, A_1, \cdots, A_{n-1})$ where the arguments A_t indicate that the symbol at index t was drawn from alphabet A. This H is a joint entropy since its value depends on the joint probability of the symbols in the block. If we regard the subsequence $(s_1, s_2, \cdots, s_{n-1})$ as a compound symbol, similarly to the preceding discussion, we can use our previous result to express the entropy as

$$H(A_0, A_1, \cdots, A_{n-1}) = H(A_0) + H(A_1, A_2, \cdots, A_{n-1} | A_0).$$

Now, the second term on the right-hand side is a joint entropy conditioned on A_0. It can be shown that this term can be expressed as

$$H(A_1, A_2, \cdots, A_{n-1} | A_0) = H(A_1 | A_0) + H(A_2, \cdots, A_{n-1} | A_0, A_1).$$

Repeating this argument inductively and plugging the results back into the first equation gives us

$$H(A_0, A_1, \cdots, A_{n-1}) = H(A_0) + H(A_1 | A_0) + H(A_2 | A_0, A_1) \\ + \cdots H(A_{n-1} | A_0, \cdots, A_{n-2}). \quad 1.2.10$$

This result is known as the *chain rule for entropy*.

Since $H(B|A) \leq H(B)$, one immediate consequence of Equation 1.2.10 is

$$H(A_0, \cdots, A_{n-1}) \leq \sum_{i=0}^{n-1} H(A_i), \quad 1.2.11$$

with equality if and only if all of the symbols in the sequence are statistically independent. (A source with this property is called a *memoryless source*; it doesn't "remember" from one time to the next what symbols it has previously emitted.) While we have developed 1.2.10 and 1.2.11 assuming the same source A emitted all of the symbols, these equations are also true if the symbols are emitted by *different* sources.

When all of the symbols are emitted from the same information source, and if the probabilities of the source do not change over time, then all of the H terms under the summation in 1.2.11 are equal to the entropy of source A, and we have

$$H(A_0, \cdots, A_{n-1}) \leq n \cdot H(A), \quad 1.2.12$$

with equality if and only if the symbols are statistically independent. Equality in 1.2.12 signifies a memoryless source.

EXAMPLE 1.2.4

Suppose a memoryless source with $A = \{0,1\}$ having equal symbol probabilities emits a sequence of six symbols. Following the sixth symbol, suppose a seventh symbol is transmitted which is the sum modulo 2 of the six previous symbols (this is just the exclusive-or of the symbols emitted by A). What is the entropy of the seven-symbol sequence?

Solution: Let $b = \sum_{t=0}^{5} \oplus s_t$ where $\sum \oplus$ denotes a summation modulo 2 and $s_t \in A$. The entropy of the sequence is $H(A_0, A_1, \cdots, A_5, b)$. By the chain rule for entropies, this is equivalent to

$$H(A_0, A_1, \cdots, A_5, b) = H(A_0) + H(A_1|A_0) + H(A_2|A_0, A_1) + \cdots H(b|A_0, \cdots, A_5).$$

However, $H(A_i|A_j) = H(A_i)$ if the symbols are statistically independent. (Statistical independence means that A_j tells us *nothing* about A_i if $i \neq j$). Since the first six symbols are drawn from the same source, A, the first six terms on the right-hand side of the foregoing expression are all equal, and since $p_0 = p_1 = 0.5$, these terms sum to 6. In the last term on the right-hand side, the six symbols $A_0, \cdots, A_5$ completely determine b so knowledge of these six symbols leaves no uncertainty in b and $H(b|A_0, \cdots, A_5) = 0$. Therefore, $H(A_0, A_1, \cdots, A_5, b) = 6$.

EXAMPLE 1.2.5

Rigorously show that $H(b|A_0, \cdots, A_5) = 0$ in example 1.2.4.

Solution: From the definition of entropy, we can write

$$H(b|A_0,\cdots,A_5) = \sum_{b=0}^{1} p_{b|A_0,\cdots,A_5} \log_2(1/p_{b|A_0,\cdots,A_5}),$$

where $p_{b|A_0,\cdots,A_5}$ is the conditional probability of b given $A_0,\cdots,A_5$. Now, for any given $A_0,\cdots,A_5$, b is either 0 or 1 (depending on $A_0, \cdots, A_5$), so we have either $p_{b|A_0,\cdots,A_5} = 0$ or $p_{b|A_0,\cdots,A_5} = 1$, depending on the value of b in the summation. Therefore,

$$H(b|A_0,\cdots,A_5) = -0 \cdot \log_2(0) + 1 \cdot \log_2(1) = 0 + 0 = 0.$$

EXAMPLE 1.2.6

For an information source having alphabet A with $|A|$ symbols, what is the range of entropies possible?

Solution: The definition of entropy is given by Equation 1.2.5. Since the symbol probabilities must lie in the range $0 \leq p_m \leq 1$ for every symbol, $\log_2(1/p_m) \geq 0$ for every symbol. Therefore, each term in the summation in Equation 1.2.5 is non-negative. Therefore, $H(A) \geq 0$. In Example 1.2.2, we saw that the entropy is maximized when every symbol in A is equally probable with $p_m = 1/|A|$. Consequently,

$$H(A) \leq \sum_{m=0}^{|A|-1} \frac{1}{|A|} \log_2(|A|) = \log_2(|A|).$$

Therefore,

$$0 \leq H(A) \leq \log_2(|A|). \qquad 1.2.13$$

1.3 SOURCE CODING

1.3.1 Mapping Functions and Efficiency

Example 1.2.6 showed that arbitrary information sources can have a considerable range of possible entropies. The entropy of a source is the average information carried per symbol. Since each symbol will either be transmitted (in the case of a communication system) or stored (in the case of a storage system), and since each use of the channel (or each unit of storage) has some associated cost, it is clearly desirable to obtain the most information possible per symbol (on the average). If we have an inefficient source, *i.e.*, $H(A) < \log_2(|A|)$, our system can be made more cost effective through the use of a *source encoder*.

A source encoder can be looked at as a data processing element which takes an input sequence of $s_0, s_1, \cdots$ symbols $s_t \in A$ from the information source and produces an output sequence $s'_0, s'_1, \cdots$ using symbols s'_t drawn from a *code alphabet* B. These symbols are called *code words*. The objective of the encoder is to process the input in such a way that the average information transmitted (or stored) per channel use closely approaches $H(A)$. (If A is a compound symbol, such as in example 1.2.3, $H(A)$ would actually be a joint entropy.)

In its simplest form, the encoder can be viewed as a mapping of the source alphabet A to a code alphabet B. Mathematically, this is represented as

$$C: A \rightarrow B.$$

Since the encoded sequence must eventually be *decoded*, the function C must be invertible. This means there must exist another function C^{-1} such that if $C(a) \mapsto b$ then $C^{-1}(b) \mapsto a$. This is possible only if $C(a) \mapsto b$ is unique, *i.e.*, for every $b \in B$, there is exactly one $a \in A$ such that $C(a) \mapsto b$, and for every $a \in A$, there is exactly one $b \in B$ such that $C^{-1}(b) \mapsto a$. (We are assuming that every $a \in A$ has a non-zero probability of occurrence.)

The source encoder acts as a data-processing unit. In this section, we will discuss some of the details how data processing effects information. In particular, we must be concerned with the relationships between information carried by code words b and information carried by source symbols a. Before diving into these details, however, it is worthwhile to illustrate the function of a source encoder by means of the following example.

EXAMPLE 1.3.1

Let A be a 4-ary source with symbol probabilities as given in Example 1.2.1. Let C be an encoder which maps the symbols in A into strings of binary digits, as follows:

$$p_0 = 0.5 \qquad C(a_0) \mapsto 0$$
$$p_1 = 0.3 \qquad C(a_1) \mapsto 10$$
$$p_2 = 0.15 \qquad C(a_2) \mapsto 110$$
$$p_3 = 0.05 \qquad C(a_3) \mapsto 111$$

Let L_m be the number of binary digits in code word b_m. If the code words are transmitted one binary digit at a time, the average number of transmitted binary digits per code word is given by

$$\overline{L} = \sum_{m=0}^{3} p_m L_m = .5(1) + .3(2) + .15(3) + .05(3) = 1.70.$$

From Example 1.2.1, $H(A) = 1.6477$ bits. The *efficiency* of this encoder is $H(A)/\overline{L} = 0.96924$. If the source encoder were not used, we would need two binary digits to represent each source symbol. The efficiency of the uncoded source would be $H(A)/2 = 0.82385$. Put another way, almost 18% of the transmitted binary digits for the uncoded source would be redundant, whereas only about 3% of the transmitted binary digits from B are redundant.

More sophisticated data processing by the encoder can also be carried out. Suppose that symbols emitted by source A were first grouped into ordered pairs $\langle a_i, a_j \rangle$ and the code words produced by the encoder were based on these pairs. The set of all possible pairs $\langle a_i, a_j \rangle$ is called the *Cartesian product* of set A with itself and is denoted by $A \times A$. The encoding process then becomes a function of two variables and can be denoted by a mapping $C: A \times A \to B$ or by a function $C(a_i, a_j) \mapsto b$.

EXAMPLE 1.3.2

Let A be a 4-ary memoryless source with the symbol probabilities given in Example 1.3.1. Since A is a memoryless source, the probability of any given pair of symbols is given by $\Pr(a_i, a_j) = \Pr(a_i) \cdot \Pr(a_j)$. Let the encoder map pairs of symbols into the code words shown in the following table:

$\langle a_i, a_j \rangle$	$\Pr(a_i, a_j)$	b_m	$\langle a_i, a_j \rangle$	$\Pr(a_i, a_j)$	b_m
a_0, a_0	.25	00	a_2, a_0	.075	1101
a_0, a_1	.15	100	a_2, a_1	.045	0111
a_0, a_2	.075	1100	a_2, a_2	.0225	111110
a_0, a_3	.025	11100	a_2, a_3	.0075	1111110
a_1, a_0	.15	101	a_3, a_0	.025	11101
a_1, a_1	.09	010	a_3, a_1	.015	111101
a_1, a_2	.045	0110	a_3, a_2	.0075	11111110
a_1, a_3	.015	111100	a_3, a_3	.0025	11111111

Since the symbols from A are independent, we know from Equation 1.2.12 that the entropy $H(A \times A) = 2H(A) = 3.2954$. Since $p_m \equiv \Pr(b_m) = \Pr(a_i, a_j)$ for each symbol, the average number of bits per transmitted code word is

$$\overline{L} = \sum_{m=0}^{15} p_m L_m = 3.3275.$$

The efficiency of the encoder is therefore $H(A \times A)/\overline{L} = .99035$. Less than 1% of the transmitted bits are redundant.

1.3.2 Mutual Information

It is worth noting in both of the previous examples that $\overline{L} \neq H(B)$. The entropy of the set B is defined by Equation 1.2.5 and has nothing to do with the binary representation of the symbols b. Since the symbol probabilities in B are precisely the same as A (in Example 1.3.1) or $A \times A$ (in Example 1.3.2), $H(B) = H(A)$ in Example 1.3.1 and $H(B) = H(A \times A)$ in Example 1.3.2.

Is this true for arbitrary relations between source sets A and coded sets B? In general, the answer is "no." In both previous examples, the relation between the source alphabet and the code set was a one-to-one and onto *function* with source and code sets of equal size. Each a specified a unique b and for each b there was a unique a. Another way to put this is in terms of probabilities. In all cases from the previous examples,

$$\Pr(b|a) = \begin{cases} 1, & C(a) \mapsto b \\ 0, & C(a) \mapsto b' \neq b \end{cases} \qquad 1.3.1$$

and similarly for $\Pr(a|b)$. This is a sufficient condition for $H(B) = H(A)$ as we will shortly see.

To further clarify this point, suppose we have an information source A composed of $|A|$ symbols and having entropy $H(A) < \log_2(|A|)$. Can we construct an encoder with a code-word alphabet B such that $|B| < |A|$? This is certainly possible. However, can we do this in such a way that no information from A is *lost* in the encoding process? Our intuition tells us the answer to this question is "no," but how do we *know* our intuition is correct?

Suppose we represent the information source and the encoder as "black boxes" and station two perfect observers at the scene to watch what happens. The first observer observes the symbols output from the source A, while the second observer watches the code symbols output from the encoder B. We assume that the first observer has perfect knowledge of alphabet A and symbol probabilities P_A and the second observer has equally perfect knowledge of code alphabet B and code word probabilities P_B. Neither observer, however, has any knowledge whatsoever of the other observer's black box.

Now suppose each time observer B observes a code word he asks observer A what symbol had been sent by the information source. How much information does observer B obtain from observer A? If the answer to this is "none," then all of the information presented to the encoder passed through it to reach observer B and the encoder was information *lossless*. On the other hand, if observer A's report occasionally *surprises* observer B, then some information was lost in the encoding process. A's report then serves to decrease the uncertainty observer B has concerning the symbols being emitted by black box B. The reduction in uncertainty about B conveyed by the observations A is called the *mutual information, $I(B; A)$*.

Let us make this idea quantitative. The information presented to observer B by his observations is merely the entropy $H(B)$. If the observer observes symbol b and then learns from his partner that the source symbol was a, observer A's report conveys information

$$H(B|A = a) = \sum_{b \in B} p_{b|a} \log_2(1/p_{b|a})$$

and, averaged over the course of all observations, the average information conveyed by A's reports will be

$$H(B|A) \equiv \sum_{a \in A} p_a H(B|A = a) = \sum_{a \in A} \sum_{b \in B} p_a p_{b|a} \log_2(1/p_{b|a}). \qquad 1.3.2$$

The amount by which B's uncertainty is therefore reduced is

$$I(B; A) = H(B) - H(B|A). \qquad 1.3.3$$

Combining Equations 1.2.5 and 1.3.2 and recalling the relationships

$$p_{a,\,b} = p_{b,\,a} = p_a p_{b|a} = p_b p_{a|b}$$

gives us

$$I(B; A) = \sum_{b \in B} \sum_{a \in A} p_{b,a} \log_2\left(\frac{p_{b,a}}{p_b p_a}\right). \qquad 1.3.4$$

From Equation 1.3.4, we can immediately observe that $I(A; B) = I(B; A)$ and, consequently, it makes no difference if observer A reports to observer B or *vice versa*.

Suppose $p_{b|a}$ is given by Equation 1.3.1. Then $-p_{b|a} \log_2(p_{b|a}) = 0$ and $H(B|A) = 0$. Put into words, observer B learns nothing new from observer A's reports. In this case, $I(B; A) = H(B)$, which means the encoder is information lossless.

Now suppose $|B| < |A|$. We can assume without loss of generality that the encoder maps every $a \in A$ into some $b \in B$, since, if the encoder does not encode some a, this is equivalent to encoding to a null symbol in B. However, the encoding can no longer be *unique*, since there must be some $b \in B$ for which $C(a_i) \mapsto b$ and for at least two symbols $a_i \neq a_j$ in A.
Therefore, $0 < p_{a_i|b} < 1$ and $0 < p_{a_j|b} < 1$. Consequently, $H(A|B) \neq 0$. Since

$$I(A; B) = H(A) - H(A|B),$$

we have $I(A; B) < H(A)$. But $I(A; B) = I(B; A) \Rightarrow I(B; A) < H(A)$. It follows from Equation 1.3.3 that $H(B) < H(A) + H(B|A)$. However, $p_{b|a} = 1$ if $C(a) \mapsto b$ and $p_{b|a} = 0$ otherwise. Therefore, $H(B|A) = 0$, so $H(B) < H(A)$. This proves the encoder is information *lossy*. We cannot build a lossless encoder with $|B| < |A|$.

What about the converse, $|B| > |A|$? Can we build an encoder with $|B| > |A|$ and somehow obtain *more* information, *i.e.*, $H(B) > H(A)$, than was emitted by the source? Our intuition might argue that the answer to this is "no," but anyone who has had experience with the town gossip might be a little less certain about this. Suppose we construct an encoder with $|B| > |A|$ such that for every $b \in B$, there is *exactly* one $a \in A$ such that $C(a) \mapsto b$. Since $|B| > |A|$, at least one of the elements a must specify more than one b. How is this possible? One possibility is that the selection could be made *at random*. If a can be coded into either of say b_i or b_j, the encoder could "toss a coin" to decide which selection would be made each time.

Since each b is associated with a unique a, $p_{a|b}$ is either 0 or 1 so $H(A|B) = 0$ and $I(A; B) = H(A)$. On the other hand, there are at least two elements of B such that, for some a, $0 < p_{b|a} < 1$ so $H(B|A) \neq 0$. It therefore seems that

$$I(A; B) = I(B; A) \Rightarrow H(A) = H(B) - H(B|A) \Rightarrow H(B) > H(A).$$

If this result seems a bit unsettling, it should. It appears we have gotten "something for nothing" out of the encoder. However, let's think about what the result means. $H(B) > H(A)$ says there is more uncertainty in B than there is in A. Where did this uncertainty come from? In the discussion above, we said we could allow the encoder to pick from a selection of b symbols associated with a at random. If the encoder is "tossing a coin," the extra uncertainty in B comes from this random process and *not* from A. There is indeed more information in B but the excess $H(B) - H(A)$ is *useless* in terms of getting more information out of A. The encoder is like the town gossip: Along with the facts come other unrelated or speculative tidbits.

1.3.3 A Brief Digression on Encryption

All of this is not to say that such an encoder is useless. The difficulty and expense of building a *decoder* is directly related to the entropy of the signal being decoded. Suppose you wanted to make it more difficult for a snoopy third party to decode a private message between yourself and a friend. One way to do so would be to construct an encoder such as that in the example above. Encoders constructed for this purpose are often called *encryption encoders*. Their function, however, is to add more uncertainty to the transmitted message, *i.e.*, raise the entropy, without adding to the *knowledge* conveyed by the message. The excess information, $H(B) - H(A)$, is useless to your friend but could be terribly frustrating to eavesdroppers (and, so, of value to you).

EXAMPLE 1.3.3

Let the information source have alphabet $A = \{0, 1\}$ with $p_0 = p_1 = 0.5$. Then $H(A) = 1$. Let encoder C have alphabet $B = \{0, 1, \cdots, 7\}$ and let the elements of B have binary representation $\overline{b} = (b_2\, b_1\, b_0)^T$. The encoder is shown in Figure 1.3.1 below where the "addition" blocks are modulo-2 adders (*i.e.*, exclusive-or gates). Find the entropy of the coded output and the output sequence if the input sequence is

$$a(t) = \{1010010110000011001111011\}$$

and the initial contents of the registers are $\overline{b}(t=0) = 5\ (=101)$.

Solution: The encoder is a finite state machine. By direct examination, the output sequence can be expressed as

$$\overline{b}(t+1) = \begin{bmatrix} 0 & 1 & 0 \\ 1 & 0 & 1 \\ 1 & 0 & 0 \end{bmatrix} \overline{b}(t) + \begin{bmatrix} 0 \\ 0 \\ 1 \end{bmatrix} a(t),$$

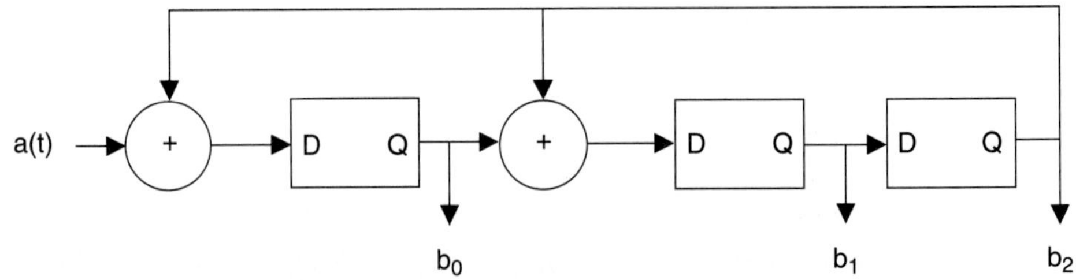

Figure 1.3.1: Encoder for Example 1.3.3

where all additions are modulo 2. Let $\pi_0 = \Pr(\bar{b} = 0)$, $\pi_1 = \Pr(\bar{b} = 1)$, etc., and define the *probability state* as the vector

$$P_B = [\pi_0 \, \pi_1 \cdots \pi_7]^T.$$

At any time t, P_B gives the probability of the encoder being in each of the eight possible states. By drawing a state diagram for the encoder of Figure 1.3.1, we can show that the probability state as a function of time is given by

$$P_B(t+1) = \begin{bmatrix} .5 & 0 & 0 & 0 & 0 & .5 & 0 & 0 \\ .5 & 0 & 0 & 0 & 0 & .5 & 0 & 0 \\ 0 & .5 & 0 & 0 & .5 & 0 & 0 & 0 \\ 0 & .5 & 0 & 0 & .5 & 0 & 0 & 0 \\ 0 & 0 & .5 & 0 & 0 & 0 & 0 & .5 \\ 0 & 0 & .5 & 0 & 0 & 0 & 0 & .5 \\ 0 & 0 & 0 & .5 & 0 & 0 & .5 & 0 \\ 0 & 0 & 0 & .5 & 0 & 0 & .5 & 0 \end{bmatrix} P_B(t),$$

subject to the constraint

$$\sum_{j=0}^{7} \pi_j = 1.$$

In the steady state, $P_B(t+1) = P_B(t)$. By direct substitution, we see that $\pi_j = 1/8$, for all elements of B satisfies this condition. Therefore,

$$H(B) = -\sum_{j=0}^{7} \pi_j \log_2(\pi_j) = 3 > H(A).$$

The encoder of Figure 1.3.1 is a pseudorandom number generator which generates all elements of B from 1 through 7 if input $a = 0$. With equiprobable $a = 0$ and $a = 1$, the encoder generates all elements of B with equal probability. The increase in entropy is due to the *memory* of the pseudorandom number generator. For the input sequence given above, the encoder's output sequence is found from the next-state equation above to be

$$\{\bar{b}\} = \{0, 0, 1, 2, 4, 2, 4, 2, 5, 1, 2, 4, 3, 6, 6, 6, 7, 5, 0, 1, 3, 6, 6, 6\}.$$

This example would not make a very good encryption code because the encoder is too simple. Note, for instance, that 0 is always followed by either 0 or 1, 2 is always followed by either 4 or 5, and so on. An expert in encryption would have no difficulty "cracking" this code by compiling a list of what code words follow each other and reverse-engineering the encoder (try it!). However, the basic theme outlined in this example can be expanded on to create an encryption system which is rather formidable. We shall say more on the subject of encryption in Chapter 8.

1.3.4 Summary of Section 1.3

Let's summarize what we've learned in this section. Source coding consists of transforming the information source's alphabet to a code word alphabet. One use for this is to improve the transmission efficiency by data compression. In this application, redundancy in the source message is removed so that the average number of bits actually transmitted (or stored) per code word more closely approaches the entropy of the information source. Data compression can either be lossless ($H(B) = H(A)$) or lossy ($H(B) < H(A)$). In the lossy case, there is no way to recover all of the information transmitted by the source.

Source coding can also be used to obtain $H(B) > H(A)$. The usual motivation for this is data encryption. In this case, all of the original information is still contained in the transmitted (or stored) message and the increase in code word entropy is due to "noise" information supplied by the encoder. (The use of the term "noise" in this context refers to the extra apparent randomness added by the encoder; as far as the receiver of the message is concerned, the "information" added by the encoder is useless information.)

This section also introduced two new important concepts: Mutual information $I(B; A)$ and conditional entropy $H(B|A)$. These two concepts will be used later when we discuss the information capacity of a communication channel.

1.4 HUFFMAN CODING

1.4.1 Prefix Codes and Instantaneous Decoding

Huffman codes are lossless data compression codes. They play an important role in data communications, speech coding, and video or graphical image compression. Huffman codes are optimal in the sense that they can deliver code word sequences which asymptotically approach the source entropy. They are also have practical implementations for both the encoder and decoder.

Huffman codes generally have variable-length code words. For instance, Examples 1.3.1 and 1.3.2 are Huffman codes. The Huffman codes belong to a class of data compression codes known as *prefix codes*. A prefix code is a code that has the property of being self-punctuating. To understand what this means, look at the set of code words in Example 1.3.2. These code words vary in length from 2 bits to 8 bits. If the code words are transmitted bit-serially from left to right, how does the decoder at the receiver know when one code word has ended and another has begun?

When one divides a string of symbols into words, one is said to *punctuate* the string. For example, consider the string

   IGOTINAROUNDNOONEAGERANDINFORMED.

One might punctuate this string as

   I GOT IN AROUND NOON EAGER AND INFORMED.

However, one could equally well punctuate this as

   I GO TIN A ROUND NO ONE AGE RAN DIN FORM ED.

The first punctuation makes more sense if one expects to see a meaningful English sentence. However, every individual word in the second punctuation is a valid English word or proper name. What if the message is in a secret code and "I", "GO", and so on are code words hiding a message I don't want YOU to see?

Even more to the point, how would you program a computer to punctuate the original character string? If you merely scanned the message until you found legal words in a dictionary, your computer would probably deliver the second punctuation above. You would have to add some rules governing sentence structure to delineate between the two punctuations given above. Suppose you did this. Consider the sentence

TIME FLIES LIKE AN ARROW

(noun) (verb) (predicate).

If your program could parse TIMEFLIESLIKEANARROW using the example sentence structure rule given above, what would it do with

FRUIT FLIES LIKE A BANANA (Wouldn't you if you were a fruit fly?)

The English language is not generally self-punctuating because it contains sufficient ambiguities that some messages could be correctly parsed in two different ways

IF I WANTED TO PICK ONE vs. IF I WANT ED TO PICK ONE

and, in other cases, the rules connecting possible words to valid sentences are complicated. It is for this reason that we use spaces, commas, and other punctuation *marks* in written language.

A prefix code is a code having punctuation built in to the structure (rather than added in using special punctuation symbols). This is accomplished by designing the code such that no code word is a prefix of some other (longer) code word. Such codes are said to satisfy the *prefix condition*. They are also said to be *instantaneously decodable*.

EXAMPLE 1.4.1

For the code in Example 1.3.2, decode the sequence

10000011111111011101

assuming the code words are transmitted bit-serially from left to right.

Solution: Scan the sequence from left to right, and place a comma after every string of bits that is a valid code word. Observe that neither 1 nor 10 is a valid code word, but 100 is a valid code word. Parsing the sequence in this way, we get

100, 00, 0111, 111110, 11101,

which decodes as

$a_0 \, a_1 \, a_0 \, a_0 \, a_2 \, a_1 \, a_2 \, a_2 \, a_3 \, a_0$.

Notice that this parsing is unique. For example, there are no code words such as 1000 or 01111. Any code which satisfies the prefix condition is structured so that whenever a string of n bits is recognizable as a code word, there is no possibility of some other (longer) code word which begins in the same way as the first recognizable code word over its first n bits.

1.4.2 Construction of Huffman Codes

Huffman codes are constructed using a tree-building algorithm. The algorithm is defined in a manner which guarantees the code will satisfy the prefix condition. The first step in the procedure is to list the source symbols in a column in descending order of probability. This is illustrated in Example 1.4.2. The source symbols form the leaves of the tree. In Example 1.4.2, symbol a_0 with probability $p_0 = 0.5$ is at the top of the column, and symbol a_3 with $p_3 = 0.05$ is at the bottom.

Step 2 of the algorithm is to combine symbols, starting with the two *lowest probability* symbols, to form a new compound symbol. In Example 1.4.2, symbols a_2 and a_3 are combined to form compound symbol $a_2 a_3$ with probability $p_{23} = p_2 + p_3 = 0.20$. The combining of two symbols forms a branch in the tree with stems going from the compound symbol to the two symbols which formed it. The top stem is labeled with a "0" and the bottom stem is labeled with a "1". (We could equally well label the stems as "1" and "0", respectively, instead.) After this step, the tree in example 1.4.2 consists of the three symbols a_0, a_1, a_2, a_3 with their respective probabilities.

Step two is then repeated, using the two lowest probability symbols from the new set of symbols. This process continues until all of the original symbols have been combined into a single compound symbol, $a_0 a_1 a_2 a_3$, having probability 1. The final tree is shown in Example 1.4.2.

Code words are assigned by reading the labels of the tree stems from right to left back to the original symbol. For instance, in Example 1.4.2, the symbol a_3 is assigned code word 111. Code-word assignments for the other symbols are done in the same way. Note that the final result is the same as the code word assignments in Example 1.3.1.

EXAMPLE 1.4.2

Construct a Huffman code for the 4-ary source alphabet of Example 1.3.2.

Solution: See the tree construction in the following Figure:

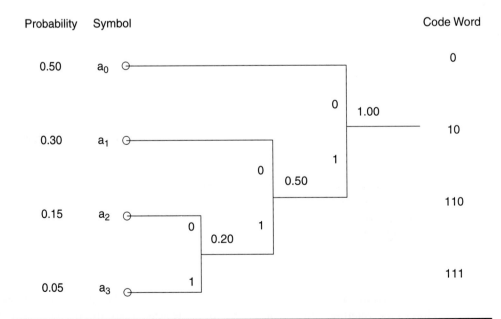

The basic encoding algorithm illustrated in Example 1.4.2 can be extended and made more efficient by encoding source symbols in pairs, triplets, etc. This was illustrated in Example 1.3.2 earlier. The code given in that example is a Huffman code generated by the same procedure as just shown, except that the source symbols were grouped into 16 ordered pairs (with associated double-symbol probabilities) prior to constructing the table. Actual code construction for Example 1.3.2 is left to the reader as a homework exercise.

When constructing the Huffman coding tree, it may sometimes happen that more than two symbols have the same probability. When more than two symbols are tied for lowest probability, any two of these symbols may be used to form the next compound symbol. All that is required is that the two symbols being combined have the two lowest probabilities available at each step in the construction of the tree. When probability ties occur, there is not a unique Huffman code. Rather, two or more codes may be possible. There is absolutely no difference in the average bits per code word among these candidate codes although it may happen that one of the possible codes may have a smaller maximum code word length than the others.

Decoding of Huffman codes is done by parsing the received bit stream as discussed at the beginning of this section. An alternate way of looking at the decoding operation is to view it as tracing back through the tree, beginning at the tree root on the right, according to the sequence of received bits. The parsing of a code word is complete when the traceback arrives at a tree leaf, i.e., at a source symbol at the left side of the tree.

1.4.3 Hardware Implementation Approaches

Hardware implementation of Huffman encoders and decoders can be carried out using digital finite-state machines. The design is quite straightforward for small source alphabets, but can become quite involved for large source alphabets. New architectures and designs for Huffman encoders and decoders are frequently reported in the current technical literature. For the code in Example 1.4.2, the encoder can be constructed using a read-only-memory (ROM) lookup table having, for each source symbol, an entry giving the length of the code word and the code word. For symbol a_0 in the example above, the lookup table would contain and entry of 1,0XX for the length and code word, respectively. The "X" indicates a "don't care" entry and the X bits are only included because the ROM must have a word width equal to the longest code word. The source symbols address the appropriate ROM entry. The code word stored in the ROM is parallel loaded into a shift register and the word length is used to tell a simple sequential circuit how many bits in the shift register are to be shifted out for the code word. Several of the homework problems at the end of the chapter call for you to try your hand at designing some simple Huffman encoder circuits.

Decoders for Huffman codes are frequently somewhat more challenging to efficiently design than are the encoders. The decoder must implement the parsing algorithm discussed earlier or, equivalently, must implement the algorithm for tracing back through the decoding tree. One approach that might be used for simple codes is as follows. As the bits are received serially, they are shifted, one at a time, into a shift register. The shift register is cleared at the start of each code word and its contents are applied to a ROM lookup table. The lookup table contains a "pattern match" bit which indicates whether or not the current contents of the shift register match a legal code word. The table also contains the decoded representation of the source symbol. For example, if source symbol a_0

is represented as "00" by the information source, the table would contain an entry "1, 00" at ROM address 000 to indicate a code word matching this source symbol. The address and lookup table entries for the code in Example 1.4.2 could be as follows:

address	match	decoded symbol
000	1	00
100	0	XX
010	1	01
110	0	XX
001	0	XX
101	0	XX
011	1	10
111	1	11

Each time a symbol match is found, the input shift register is cleared and the decoding cycle begins again. This simple scheme will work for arbitrarily large codes, but it clearly becomes increasingly inefficient as the maximum code word length increases.

Rather than using a ROM lookup table, it is possible to use a content-addressable memory (CAM) as a decoder lookup table. The algorithm given above remains unchanged but the CAM need only store legal code words and need not provide storage space for "illegal" code words as the ROM implementation above does. Likewise, programmable logic arrays (PLAs) may also be used. Either of these approaches can result in more efficient (and oftentimes faster) circuitry for decoding the Huffman code. Other approaches exist as well, and some references to the literature describing these approaches are given at the end of the chapter.

1.4.4 Robustness of Huffman Coding Efficiency

As you have undoubtedly noticed, the design of a Huffman code requires us to know the probabilities of the individual symbols. In practice, we typically must determine the probabilities through an estimation process of some kind and, therefore, what we actually have to work with is a set of *estimated* probabilities, $\hat{p}_i = p_i + e_i$, where p_i is the true probability and e_i is the error in our estimate of the probability. What do these errors do to the efficiency of the Huffman code? Can small errors lead to large increases in the average number of bits per symbol?

There is, in general, no universal answer to these questions. However, we can make some general observations. First, note that the sum of all the e_i must be zero since both the $\hat{p}_i$ and the p_i must sum to unity. Suppose our estimated probabilities lead us to design a code having code-word lengths $\hat{\ell}_i$, while the optimum Huffman code based on the true probabilities would have symbol lengths ℓ_i. If the source alphabet has M symbols, the average number of bits per code word for the two cases would be

$$\hat{L} = \frac{1}{M} \sum_i \hat{p}_i \hat{\ell}_i$$

$$L = \frac{1}{M} \sum_i p_i \ell_i$$

which, after a small amount of algebra, yields

$$\Delta L = \hat{L} - L = \frac{1}{M}\sum_i p_i(\hat{\ell}_i - \ell_i) + \frac{1}{M}\sum_i e_i \hat{\ell}_i \qquad 1.4.1$$

Now, the code word lengths must always be integers, and if we have done a reasonable job of estimating symbol probabilities, it is very likely that the differences $\hat{\ell}_i - \ell_i$ will be zero among the most probable source symbols or, at the least, that, between symbols of nearly equal probability, the differences $\hat{\ell}_i - \ell_i$ will be ± 1 for symbols with adjacent probabilities. In this event, the first term on the right-hand side of Equation 1.4.1 will tend to be very small relative to the second term. Consequently,

$$\Delta L \approx \frac{1}{M}\sum_i e_i \hat{\ell}_i. \qquad 1.4.2$$

If the mean-squared error in our probability estimates is σ^2, then by the method of Lagrange multipliers it can be shown that in the most extreme cases

$$(\Delta L)^2 \approx \left[\frac{1}{M}\sum_i \hat{\ell}_i^2 - \left(\frac{1}{M}\sum_i \hat{\ell}_i\right)^2\right]\sigma^2, \qquad 1.4.3$$

which is the product of the variance in the $\hat{\ell}_i$ and the variance of the e_i. Equation 1.4.3 provides a (normally) pessimistic estimate of the loss of optimality in the code due to errors in our estimates of the symbol probabilities. At the lower end of the extreme, it is quite possible $\Delta L = 0$. As a result, Huffman codes are found to be fairly robust; small errors in the symbol probabilities do not seriously degrade the average number of bits per symbol. However, reasonable care must be taken in estimating the probabilities, since large errors can (and do) lead to serious loss of optimality. For instance, a Huffman code designed for English text might have a serious loss in optimality if it is used for Turkish text.

1.5 DICTIONARY CODES AND LEMPEL–ZIV CODING

1.5.1 The Rationale Behind Dynamic Dictionary Coding

Huffman codes require us to have a fairly reasonable idea of how the source symbol probabilities are distributed. There are a number of applications where this is possible but there are also many applications where such a characterization is impractical. One common example can be found in compression of data files in a computer. Many different kinds of data are stored by computer systems, and even if the data is represented in ASCII format, the ASCII symbol probabilities can vary widely from one file to the next.

Dictionary codes are compression codes that dynamically construct their own coding and decoding tables "on the fly" by looking at the data stream itself. Because they have this capability, it is not necessary for us to know the symbol probabilities beforehand. These codes take advantage of the fact that, quite often, certain strings of symbols are frequently repeated and these strings can be assigned code words that represent the entire string of symbols. The disadvantage of dictionary codes, if one chooses to call it a disadvantage, is that these codes are usually efficient only for long files or messages. Short files or messages can actually result in the transmission (or storage) of *more* bits, on the average, than were contained in the original message.

Nonetheless, dictionary codes are important and useful in many applications. In this section, we shall examine a particular class of dictionary codes known as the Lempel–Ziv (LZ) codes. LZ codes are known to asymptotically approach the source entropy for long messages. They also have the advantage that the receiver does not require prior knowledge of the coding table constructed by the transmitter. Sufficient information is contained in the transmitted code sequences to allow the receiver to construct its own decoding table "on the fly." Most of the information required to do this is transmitted early in the coded message. That is why these codes initially "expand" rather than compress the data: extra information is being supplied to the receiver so that it can construct its decoding table. As time progresses, less and less information need be sent to aid the receiver in constructing its decoding table and the code can get down to the business of compressing the transmitted data.

LZ codes also suffer no significant "decoding delay" at the receiver in the sense that, whatever code symbol is currently being received, the receiver already contains the decoding information in its local dictionary to decode what is being received as it comes in. This is important because it eliminates the need for large buffers to store the received code words until such time as the decoding dictionary is complete enough to decode them. Every received code word results in decoding right away. It's a sort of "just in time" principle for data compression.

There are a number of LZ coding algorithms but they tend to be of a rather similar nature and an example of one LZ algorithm can serve to illustrate the basic ideas. In this section, we will take an in-depth look at the Lempel–Ziv–Welch (LZW) algorithm. This algorithm is better known as the "compress" command in UNIX-based computers and is also the compression algorithm used by most personal computers.

It should be noted at this point that the algorithm as presented below is actually a mild modification of the actual LZW algorithm in that the basic structure of the dictionary and the decoding algorithm have been simplified for explanation purposes and the actual LZW algorithm first presented by Welch is slightly more computer efficient. However, the difference in what follows is very, very slight, no significant features of LZ coding are omitted, and there are no performance differences between the "real" LZW and what follows in terms of the compression that is achieved. The author feels that ease-of-understanding of the fundamental concept more than makes up for his slight distortion of Welch's algorithm.

1.5.2 A Linked-List LZ Algorithm

We begin by defining the structure of the dictionary. Each entry in the dictionary is given an address m. Each entry consists of an ordered pair $\langle n, a_i \rangle$, where n is a *pointer* to another location in the dictionary and a_i is a symbol drawn from the source alphabet A. In the parlance of computer science, the ordered pairs in the dictionary make up what is known as a *linked list*. The pointer variables n also serve as the transmitted code words. The representation of n is a fixed-length binary word of b bits such that the dictionary contains a total number of entries less than or equal to 2^b.

Since the total number of dictionary entries is going to exceed the number of symbols, M, in the source alphabet, each transmitted code word is actually going to contain more bits than it would take to represent the alphabet A. However, the secret of LZ coding is that most of the code words actually represent *strings* of source symbols and in a long message it is more economical to encode these strings than it is to encode the individual symbols. In a way, this strategy is like that of Example 1.3.2, where we saw that it was better to encode our source two symbols at a time than it was to do so one symbol at a time as was done in Example 1.3.1. The LZ algorithm uses this principle with a vengeance and with the added twist that *the strings can be variable length*. No matter how long the string may be, its coded representation n always has the same number of bits. (In another way, this strategy is sort of a mirror-image of a Huffman code where the encoded symbols are always of the same length while the code words are variable length).

The algorithm is initialized by constructing the first $M+1$ entries in the dictionary as follows.

address	dictionary entry	
0	0,	null
1	0,	a_0
2	0,	a_1
...	...	...
m	0,	a_{m-1}
...	...	...
M	0,	a_{M-1}

The 0-address entry in the dictionary is a *null* symbol. It is used to let the decoder know where the end of the string is. In a way, this entry is a kind of punctuation mark. The pointers n in these first $M+1$ entries are zero. They "point" to the null entry at address 0.

The initialization also initializes pointer variable $n=0$ and address pointer $m=M+1$. The address pointer m points to the next "blank" location in the dictionary. Following initialization, the encoder iteratively executes the following steps.

1. Fetch next source symbol a;
2. If the ordered pair $\langle n, a \rangle$ is already in the dictionary then
   $n =$ dictionary address of entry $\langle n, a \rangle$;
   else
   transmit n
   create new dictionary entry $\langle n, a \rangle$ at dictionary address m
   $m = m + 1$
   $n =$ dictionary address of entry $\langle 0, a \rangle$;
3. Return to step 1.

If $\langle n, a \rangle$ is already in the dictionary in step 2, the encoder is processing a string of symbols that has occurred at least once previously. Setting the next value of n to this address constructs a linked list that allows the string of symbols to be traced.

If $\langle n, a \rangle$ is not already in the dictionary in step 2, the encoder is encountering a new string that has not been previously processed. It transmits code symbol n, which lets the receiver know the dictionary address of the *last* source symbol in the previous string. Whenever the encoder transmits a code symbol, it also creates a new dictionary entry $\langle n, a \rangle$. In this entry, n is a pointer to the last source symbol in the previous string of source symbols and a is the root symbol which begins a new string. Code word n is then reset to the address of $\langle 0, a \rangle$. The "0" pointer in this entry points to the null character which indicates the beginning of the new string. The encoder's dictionary-building and code symbol transmission process is illustrated in the following example.

EXAMPLE 1.5.1

A binary information source emits the sequence of symbols 110 001 011 001 011 100 011 11 etc. Construct the encoding dictionary and determine the sequence of transmitted code symbols.

Solution: Initialize the dictionary as shown above. Since the source is binary, the initial dictionary will contain only the null entry and the entries $\langle 0, 0 \rangle$ and $\langle 0, 1 \rangle$ at dictionary addresses 0, 1, and 2, respectively. The initial values for n and m are $n = 0$ and $m = 3$. The encoder's operation is then described in the following table.

source symbol	present n	present m	transmit	next n	dictionary entry
1	0	3		2	
1	2	3	2	2	2, 1
0	2	4	2	1	2, 0
0	1	5	1	1	1, 0
0	1	6		5	
1	5	6	5	2	5, 1
0	2	7		4	
1	4	7	4	2	4, 1
1	2	8		3	
0	3	8	3	1	3, 0
0	1	9		5	
1	5	9		6	
0	6	9	6	1	6, 0
1	1	10	1	2	1, 1
1	2	11		3	
1	3	11	3	2	3, 1
0	2	12		4	
0	4	12	4	1	4, 0
0	1	13		5	
1	5	13		6	
1	6	13	6	2	6, 1
1	2	14		3	
1	3	14		11	

The encoder's dictionary to this point is as follows:

dictionary address	dictionary entry
0	0, *null*
1	0, 0
2	0, 1
3	2, 1
4	2, 0
5	1, 0
6	5, 1
7	4, 1
8	3, 0
9	6, 0
10	1, 1
11	3, 1
12	4, 0
13	6, 1
14	no entry yet

1.5.3 The Decoding Process

The decoder at the receiver must also construct a dictionary identical to the one at the transmitter and it must decode the received code symbols. Notice from the previous example how the encoder constructs code symbols for strings and that the encoder does not transmit as many code words as it has source symbols. Operation of the decoder is governed by the following observations:

1. Reception of any code word means that a new dictionary entry must be constructed;
2. Pointer n for this new dictionary entry is the same as the received code word n;
3. Source symbol a for this entry is not yet known, since it is the *root* symbol of the next string (which has not yet been transmitted by the encoder).

If the address of this next dictionary entry is m, we see that the decoder can only construct a partial entry $\langle n, ? \rangle$ since it must await the next received code word to find the root symbol a for this entry. It can, however, fill in the missing symbol a in its *previous* dictionary entry at address $m - 1$. It can also decode the source symbol string associated with received code word n. To see how this is done, let us *decode* the transmissions sent by the encoder of Example 1.5.1.

EXAMPLE 1.5.2

Decode the received code words transmitted in Example 1.5.1. (*Note:* As you follow along with this example, you may find it useful to write down the dictionary construction as it occurs; that will help you better see what's going on.)

Solution: The decoder begins by constructing the same first three entries as the encoder. It can do this because the source alphabet is known *a priori* by the decoder. The decoder's initialized value for the next dictionary entry is $m = 3$.

The first received code word is $n = 2$. The decoder creates a dictionary entry of $\langle 2, ? \rangle$ at address $m = 3$ and then increments m. Since this is the first received code word, there is no previous partial dictionary entry to "fill in," so the decoder decodes the received symbol. Code word $n = 2$ points to address 2 of the table, which contains the entry $\langle 0, 1 \rangle$. The last symbol of the coded string is therefore "1" and the pointer "0" indicates that this is also the first symbol of the string. Therefore, the entire "string" consists only of the symbol "1".

The next received code word is $n = 2$. The decoder creates a partial dictionary entry at location $m = 4$ of $\langle 2, ? \rangle$. It must now complete the previous entry at location $m = 3$. It does this by looking at the dictionary entry at address $n = 2$. This entry is $\langle 0, 1 \rangle$ and the "0" pointer tells it that the symbol "1" is the root symbol of the string. Therefore, the entry at $m = 3$ is $\langle 2, 1 \rangle$. The decoder then increments m to $m = 5$.

The next code word is $n = 1$ and the decoder constructs a new partial dictionary entry at $m = 5$ of $\langle 1, ? \rangle$. Address $n = 1$ of the dictionary contains $\langle 0, 0 \rangle$. This is a root symbol so the dictionary entry at address 4 is updated to $\langle 2, 0 \rangle$. The string designated by the code word is decoded as "0," since the source symbol at address $n = 1$ is a root symbol. Pointer m increments to 6.

The next code word, $n = 5$, is used to create the partial dictionary entry 6: $\langle 5, ? \rangle$. The decoder completes the entry at dictionary address $m - 1 = 5$ by accessing the dictionary entry at address $n = 5$. This entry is 5: $\langle 1, ? \rangle$. Since this entry has a non-zero pointer, the symbol (which does not happen to be known yet!) is *not* the root symbol of the string. The decoder therefore links to the address specified by the pointer. This is address 1 which contains the entry $\langle 0, 0 \rangle$. The zero pointer for this entry indicates that symbol "0" is the root of the string so the entry at dictionary location 5 is updated to 5: $\langle 1, 0 \rangle$. The decoder then decodes the received string. It begins by looking at the entry in location $n = 5$. This entry indicates that the last character in the string is a "0" and that there is a previous character at dictionary address 1. This address contains a "0" symbol with a pointer of zero. Therefore, the decoded string is "00".

The next received code word is $n = 4$ which produces partial dictionary entry 7: $\langle 4, ? \rangle$. The entry at address 6 is completed by looking up the entry 4: $\langle 2, 0 \rangle$. This is *not* a root symbol so the decoder links to dictionary entry 2: $\langle 0, 1 \rangle$. This *is* a root symbol, so address 6 gets 6: $\langle 5, 1 \rangle$. The decoded symbol string is specified by 4: $\langle 2, 0 \rangle$ and 2: $\langle 0, 1 \rangle$ as "10". (Note that the root symbol comes first in the string!).

This process is repeated with $n = 3$, producing 8: $\langle 3, ? \rangle$, 7: $\langle 4, 1 \rangle$, and decoded output string "11". Code word $n = 6$ produces 9: $\langle 6, ? \rangle$ and 8: $\langle 3, 0 \rangle$. The decoded string has three links this time, 6: $\langle 5, 1 \rangle$, 5: $\langle 1, 0 \rangle$, and 1: $\langle 0, 0 \rangle$, which define the string "001". You are invited to continue this decoding process to complete the dictionary and decoded outputs for the remaining received code words. You will find that the encoder's dictionary is reproduced by this process and that the decoded output symbols are the same as those originally emitted by the information source.

1.5.4 Large-Block Requirement of LZ Compression

A casual inspection of the previous example will show that each code word n must consist of at least five bits (the dictionary has more than 16 entries) and therefore more bits are being transmitted than were in the source message up to that point. Where is the compression? The answer is: We haven't processed enough data yet for the compression to "take hold." The early stages of the LZ algorithm are largely occupied with supplying the receiver with sufficient information to construct its decoder.

How do we know that eventually this situation will turn around and that data compression will occur? Even more to the point, where is the evidence that over the course of the transmission the average number of bits per symbol will approach the entropy of the source?

The LZ algorithm works by parsing the source's t-symbol sequence $s_0 s_1 \cdots s_{t-1}$ into smaller strings of symbols. These strings are commonly called *phrases*. Furthermore, every such phrase is *unique*. No two phrases in the dictionary are alike. The total number of these phrases is equal to the number of dictionary entries. Let $c(t)$ be the total number of phrases resulting from a t-symbol source sequence. The number of bits required to represent the pointer/code word n is therefore $\log_2(c(t))$ and the total number of bits stored in the dictionary is therefore

$$b = c(t)[\log_2(c(t)) + 1],$$

where the "+1" in this expression accounts for the fact that each dictionary entry must include a one-bit source symbol.

The number of phrases depends on the particular source sequence. Lempel and Ziv have proven that $c(t)$ is upper-bounded by

$$c(t) \leq \frac{t}{(1 - \epsilon_t)\log_2(t)},$$

where

$$\epsilon_t = \min\left[1, \frac{\log_2[\log_2(t)] + 4}{\log_2(t)}\right]$$

for $t \geq 4$. Note that

$$\lim_{t \to \infty} \epsilon_t \to 0.$$

Therefore, the asymptotic upper bound on $c(t)$ for large t gives us

$$b = c(t)[\log_2(c(t)) + 1] \leq \frac{t}{\log_2(t)}[\log_2(t) - \log_2[\log_2(t)]] < t.$$

This tells us that compression will eventually occur if t is large enough.

How about the efficiency of the compression? To answer this, recall Equation 1.2.12 from Section 1.2, which stated that the entropy of the sequence $s_0 s_1 \cdots s_{t-1}$ is upper-bounded by

$$H(s_0 s_1 \cdots s_{t-1}) \leq tH(A),$$

where A is the alphabet of our information source. The quantity

$$\lim_{t \to \infty} \frac{H(s_0 s_1 \cdots s_{t-1})}{t}$$

is called the *entropy rate* of the source and is a measure of the average information per transmitted symbol in the sequence $s_0 s_1 \cdots s_{t-1}$. (This assumes the information source probabilities are not functions of time.)

We need one more definition. Suppose $f(x)$ is some function of a variable x for x defined over some specific range of values. The smallest number never exceeded by $f(x)$ over this range is called the *supremum* and is denoted by $\sup f(x)$. It has been shown, *i.e.*, a theorem has been proved, that LZ codes have the property that if the information source is a stationary ergodic process, then

$$\limsup_{t \to \infty} \frac{1}{t} c(t) \log_2[c(t)] \leq \lim_{t \to \infty} \frac{1}{t} H(s_0 s_1 \cdots s_{t-1})$$

with probability 1. This limit tells us that, asymptotically, the average code-word length per source symbol of the LZ code is no greater than the entropy rate of the information source. This is another way of saying that the LZ code is asymptotically as efficient as any algorithm can get. The "investment" it makes in expanding the number of bits sent at the beginning of the message sequence is paid back later if only the message sequence is long enough. Compression studies carried out by Welch on typical computer files have demonstrated that ASCII text files can typically be compressed by a factor of 2 to 1 for files containing 10,000 or more characters using the LZW algorithm.

1.6 ARITHMETIC CODING

1.6.1 Code-Word Length and the Asymptotic Equipartition Property

The efficiency of Huffman coding generally relies on having a source alphabet A with a fairly large number of elements. A is often an alphabet of *compound symbols* strung together from some original binary source. For example, if A is a binary source, we need one bit to represent each symbol, regardless of whatever $H(A)$ might be. In order to compress this data representation, we must form an alphabet of compound symbols

$$(s_0, s_1, \cdots, s_{n-1}), s_i \in A$$

prior to encoding. Let us designate this alphabet of compound n-bit symbols as alphabet A^n.

To obtain an efficient source code, we must calculate (or measure) the probability distribution P_{A^n} for this alphabet. (If we later decide to change our block size, n, we must re do this calculation.) If the original source A is memoryless, this change is not difficult. However, if A is not a memoryless source, this can be a difficult and even impractical undertaking. (The exception to this is if A is a special type of source, known as a Markov process, which we will study in Section 2.4.) Most "real-world" sources, unfortunately, fall into this "hard to do" class. We will have more to say about this in Section 1.7; for now, the important idea is that very large values of n present us with some tough practical realities in obtaining the necessary probabilities needed for Huffman coding.

This quite naturally leads us to ask if there might be some other kind of coding, suitable for intermediate-sized alphabets, which can improve on what we could practically accomplish with Huffman codes. The main issue is that Huffman codes require an integer number of coded bits to represent an integer number of source symbols. What if it were somehow possible to use a fraction of a bit (on the average) to encode a source symbol? If we could do so, we should be able to improve on the Huffman code by getting rid of (or at least reducing) the "quantizing" effect from which the Huffman code suffers with small source alphabets.

Happily for us, *arithmetic coding* is a technique which, essentially, does exactly this. The arithmetic coding technique is designed to obtain the *effect* of fractional-bit encoding of the source symbols (although, of course, the final code uses an *integer* number of bits). Arithmetic coding is essentially an extension of early coding work by Shannon, Fano, and Elias. In its modern form, it was developed largely by Rissanen, Pasco, and Langdon. Arithmetic codes are used extensively for data compression in such applications as image compression, compressed video television, and so on.

To appreciate the value of using "fractions" of a bit, we will take a brief look at an important statistical property of symbol blocks called the *asymptotic equipartition property* or AEP. Let

$$c = (s_0 s_1 \cdots s_{n-1}), s_i \in A \qquad 1.6.1$$

be a compound symbol. The probability of this symbol is given by

$$p_c = \Pr[s_0, s_1, \cdots, s_{n-1}].$$

For the sake of simplicity, let us assume A is a discrete memoryless source with a stationary probability distribution. Then

$$p_c = p_0 p_1 \cdots p_{n-1}, \qquad 1.6.2$$

where each $p_i = \Pr(s_i)$. From this, we have

$$-\frac{1}{n} \log_2(p_c) = -\frac{1}{n} \sum_{i=0}^{n-1} \log_2(p_i). \qquad 1.6.3$$

The right-hand side of Equation 1.6.3 is an unbiased estimate of the expected value of the function $\log_2(1/p_c)$. In the limit where n is very large, this expression is equal to

$$\lim_{n \to \infty} \frac{1}{n} \sum_{i=0}^{n-1} \log_2(1/p_i) \to \sum_{i=0}^{|A|-1} p_i \log_2(1/p_i) = H(A).$$

Therefore,

$$\lim_{n \to \infty} \frac{1}{n} \log_2(1/p_c) \to H(A). \qquad 1.6.4$$

Equation 1.6.4 is a statement of the asymptotic equipartition principle. It says that, provided n is sufficiently large, the logarithm of the probability of the typical compound symbol is approximately given by

$$\log_2(1/p_c) \approx nH(A). \qquad 1.6.5.$$

Although we have derived this result assuming A is memoryless, a similar result is obtained for sources with memory if the entropy $H(A)$ is replaced by the *entropy rate* (which we discuss in Chapter 2). This more general result holds for any ergodic source and is known as the Shannon-McMillan-Breiman theorem. For a memoryless source, the entropy rate and the entropy are equal.

For finite n, the log-probabilities of the symbols in the compound alphabet will be distributed around either side of Equation 1.6.5. It can be shown that a relatively small set of compound source symbols, within some small range about $nH(A)$, makes up the great majority of all blocks emitted by the source. This subset of compound symbols is called the *typical set*. As a consequence, a source encoder that assigns code words to elements of the typical set using $nH(A)$ bits per code word will achieve data compression with very high efficiency. We will make this idea more formal when we discuss Shannon's source-coding theorem in Chapter 9.

Now, generally, $nH(A)$ is not an integer. Huffman codes must "round up" the number of code bits used to represent elements of the typical set to obtain an integer number of bits per code word. Arithmetic codes do not do this.

1.6.2 The Arithmetic Coding Method

How is it possible to have a code word which does not contain an integer number of bits? Suppose each $a_i \in A$ is encoded into a real number β_i lying in the interval

$$0 \leq \beta_i < 1.$$

Clearly this idea might require more bits to represent β_i, since a larger number of bits could be necessary to represent a real number. Now suppose that we encode the sequence $s_0 s_1 \cdots$ by adding together scaled versions of the β_i according to

$$b = \beta_0 + \Delta_1 \beta_1 + \Delta_2 \beta_2 + \cdots,$$

where each β_i is the code number corresponding to s_i and where the Δ_i are monotonically decreasing scale factors also lying in the interval between 0 and 1. If we pick the Δ_i in such a way that it is possible to later decompose b back into the original sequence of β_is, this code can be decoded. Now, imagine we represent each β_i in binary fixed-point format. For now, we'll also assume we can use extremely long fixed-point numbers to do this. As we add the successive terms of b, the bits representing the different scaled β_i are added together. Thus, it is the sum of these representation which represents the entire coded block. In a very real way, we can regard the bit positions used to represent b as being "shared" among several different β_i. The net effect of this sharing is that each β_i effectively uses a non-integer number of bits (on the average).

We now turn to the details of how to carry this out. Arithmetic coding begins by assigning each element of A a subinterval of the real numbers between 0 and 1. The length of the assigned interval is set equal to the symbol probability.

EXAMPLE 1.6.1

Let A be the information source given in Example 1.2.1. Assign each $a_i \in A$ a fraction of the real number interval $0 \leq \beta_i < 1$.

Solution:

$p_0 = 0.5 \Rightarrow$ assign a_0 to the interval $I_0 = [0, 0.5)$,

$p_1 = 0.3 \Rightarrow$ assign a_1 to the interval $I_1 = [0.5, 0.8)$,

$p_2 = 0.15 \Rightarrow$ assign a_2 to the interval $I_2 = [0.8, 0.95)$,

$p_3 = 0.05 \Rightarrow$ assign a_3 to the interval $I_3 = [0.95, 1)$.

The arithmetic encoding process consists of constructing a code interval (rather than a code number b) which *uniquely* describes a block of successive source symbols. When this interval, I_b, is completed, it will have upper and lower bounds within which any b will satisfy the coding condition discussed above, *i.e.*,

$$I_b = [L, H), L \le b < H. \qquad 1.7.6$$

Any convenient b within this range is a suitable code word representing the entire block of symbols.

The calculation of these intervals is based on the following simple algorithm. Assume each $a_i \in A$ has been assigned an interval $I_i = [S_{\ell_i}, S_{h_i})$. Carry out the following steps:

$j = 0$
$L_j = 0$
$H_j = 1$
REPEAT
    $\Delta = H_j - L_j$
    read next a_i
    $L_{j+1} = L_j + \Delta \cdot S_{\ell_i}$
    $H_{j+1} = L_j + \Delta \cdot S_{h_i}$
    $j = j + 1$
UNTIL all a_i have been encoded.

Let us illustrate this procedure by means of an example.

EXAMPLE 1.6.2

For the source of example 1.6.1, encode the sequence

$$a_1\ a_0\ a_0\ a_3\ a_2.$$

Solution: Using the intervals found in Example 1.6.1, and applying the algorithm, we obtain the results shown in the following table:

j	a_i	L_j	H_j	Δ	L_{j+1}	H_{j+1}
0	a_1	0	1	1	0.5	0.8
1	a_0	0.5	0.8	0.3	0.5	0.65
2	a_0	0.5	0.65	0.15	0.5	0.575
3	a_3	0.5	0.575	0.075	0.57125	0.575
4	a_2	0.57125	0.575	0.00375	0.57425	0.5748125

Any b within the final interval will suffice for a code word. One convenient choice might be

$$b = 0.57470703125,$$

which can be represented using an 11–bit fixed-point binary number format.

The encoding algorithm can clearly be carried out for however long a string of source symbols we wish, so long as our computer has sufficient numerical accuracy to do the calculations. In practice, of course, we must be concerned with numerical accuracy in any real computer or in a hardware encoder. However, there are a number of judicious numerical "tricks" that can be employed to get the *effect* of having an extremely long computer word length without actually having to have a very big computer. Many of these tricks are discussed in the Rubin paper given in the references at the end of the chapter.

One such trick is to begin shifting out the result before the entire source sequence is encoded. For example, consider the results at step 4 of the previous example. The "0.57" part of the result has "stabilized" by this step. What I mean by this is that even if we add the Δ given in this step to L and H, the 0.57 part of the result remains unchanged. The next Δ will be even smaller. Therefore, there is no need to keep this part of the result. If our input sequence contained another symbol and we were "running out of room" in the computer's hardware, we could output the 0.57 part and "scale" what was left. (If the computer worked in base 10, this would amount to multiplying Δ, the remaining part of L, and the remaining part of H by 100.) By using this "rescaling" trick, we can extend the computation's effective accuracy.

If we use this trick, we must be certain that the bits we're shifting out have actually stabilized (in the sense that no possible future carry operations could change the bits we have "output" from the algorithm). The possibility of future carries rippling into the higher-order digits is called the "carryover problem." Solutions for this are discussed in detail in the arithmetic coding references given at the end of the chapter.

1.6.3 Decoding Arithmetic Codes

The decoding operation is essentially the inverse of the encoding operation. Given the code value b, the procedure is as follows:

$L = 0$
$H = 1$
$\Delta = H - L$
REPEAT
    Find i such that
$$\frac{b - L}{\Delta} \epsilon I_i$$
    OUTPUT symbol a_i
    $H = L + \Delta \cdot S_{h_i}$
    $L = L + \Delta \cdot S_{\ell_i}$
    $\Delta = H - L$
UNTIL last symbol is decoded.

If the source sequence has a fixed number of source symbols in a block, the decoder merely has to keep a count of the number of decoded symbols to determine when to stop. More often, though, it is desirable to use the same compression algorithm for source sequences of variable length. In this case, a special STOP symbol is added to source alphabet A. This symbol is treated exactly like all the other source symbols and is assigned its own interval I. When the decoder detects that it has decoded the STOP symbol, it exits the decoding loop.

EXAMPLE 1.6.3

For the source and encoder of example 1.6.2, decode $b = 0.57470703125$.

Solution: The step-by-step results of the algorithm are given in the following table:

L	H	Δ	I_i	next H	next L	next Δ	a_i
0	1	1	I_1	0.8	0.5	0.3	a_1
0.5	0.8	0.3	I_0	0.65	0.5	0.15	a_0
0.5	0.65	0.15	I_0	0.575	0.5	0.075	a_0
0.5	0.575	0.075	I_3	0.575	0.57125	0.00375	a_3
0.57125	0.575	0.00375	I_2	0.5748125	0.57425	0.0005625	a_2

In tracing through the steps of example 1.6.3, you must pay attention to the precision with which you calculate the quantity $(b - L)/\Delta$ and the other variables. Roundoff error in this calculation can lead to an erroneous answer.

1.6.4 Other Issues in Arithmetic Coding

In the practical application of arithmetic codes, there are several other issues that the code implementation usually needs to address. One set of issues arises from the fact that computers and hardware encoders always have finite precision in their arithmetic. We must therefore pay attention to the issues of numerical overflow and underflow. Let us suppose that H and L are represented as n-bit binary numbers. Let us further suppose we represent the subinterval calculations $\Delta \cdot S_h$ and $\Delta \cdot S_\ell$ using f bits of precision. Finally, let us suppose the machine's numerical precision is limited to p bits. It has been shown that avoidance of overflow or underflow errors requires

$$f \leq n - 2$$

and

$$f + n \leq p.$$

This limits the size of the source alphabet A to

$$|A| \leq 2^f,$$

since S_h, S_ℓ are limited to f bits of precision. For example, suppose $p = 16$ bits and $n = 9$ bits. Then $f \leq 7$ and our alphabet A is therefore limited to 128 symbols.

Another issue that usually is important in practice is the need for being able to transmit and decode the information "on the fly." In the algorithms presented earlier, we must read in the entire block of source symbols before being able to compute the code word. Likewise, we must receive the entire code word b before we can begin decoding. In many applications, this delay is intolerable.

Earlier, we discussed extending the "apparent" length of the computer's representation of b by outputting part of b before the entire calculation of b was complete. The need for incremental encoding and decoding is, in a sense, a similar requirement. The literature on arithmetic coding cited at the end of this chapter also discusses methods of dealing with the issue of incremental encoding and decoding. The paper by Witten et al. contains a computer program, written in the C programming language, that implements a practical arithmetic coding algorithm.

1.7 SOURCE MODELS AND ADAPTIVE SOURCE CODING

In our discussion of Huffman coding and arithmetic coding, we have assumed the source symbols had known probabilities and that these probabilities did not change with time. This is referred to as a fixed-source model. As mentioned earlier, there are a number of practical applications where the source's symbol probabilities are either non-stationary or else obtaining a sufficiently accurate source probability model is impractical. For example, image compression algorithms are frequently called on to compress both text and picture data. These two data sources have very different symbol probabilities and these probabilities can be very difficult or even impossible to obtain beforehand. A fixed-source model can (and often does) lead to a serious loss of compression performance compared to what is "really" possible. Dynamic dictionary codes, such as Lempel–Ziv, are one approach to dealing with the issue of unknown source probabilities. Unfortunately, LZ codes are not always well-suited to some applications.

To cope with these kinds of issues, *adaptive* source models and adaptive coding methods have been developed. Adaptive Huffman coding was proposed by Gallager in 1978. More recently, Lu and Gough proposed a fast-adaptive Huffman coding algorithm (1993). The basic idea in adaptive Huffman coding is to maintain an occurrence count of the source symbols and dynamically re-design the Huffman encoder. Symbols with the highest occurrence counts are assigned short code words while symbols with low occurrence counts are given longer code words. As was the case in Lempel–Ziv coding, the decoder must also keep track of the decoded symbol counts and dynamically alter its decoding tree as it goes.

Lu and Gough accomplished this by using two encoding trees, called the front tree and the back tree. All source symbols are initially encoded using the back tree. As specific symbols occur, they are moved to the front tree. The most commonly occurring symbols migrate to the front tree, which is reconfigured each time a symbol is added to it and each time a symbol contained in it is sent. After a certain number of total symbols (called the "time constant") have occurred, the occurrence count of the least-recently-used symbol is decremented. If the count for this symbol reaches zero, it is returned to the back tree. The back tree is reconfigured every time a symbol is removed from it and every time a symbol is returned to it. Usually, the back tree assumes all symbols are equally probable since the occurrence counts for symbols in the back tree are always zero. A special symbol is used as a prefix whenever a symbol from the back tree is encoded and sent.

In Huffman coding, the source model and the encoding and decoding trees are intimately tied together. Putting this another way, if the structure of the source model is changed, the structure of the encoding and decoding algorithms must also change. In the case of arithmetic coding, it is possible to separate the source model from the encoding and decoding algorithms. This is because the source model is basically given by the probability intervals I_i while the actual encoding and decoding steps are arithmetic. Adaptive arithmetic coding also usually involves keeping an occurrence count of source symbols. This count is used to alter the symbol intervals but does not change the actual calculation part of the algorithm. The paper by Witten et al. provides a C routine for implementing adaptive source modeling for their arithmetic coding algorithm.

SUMMARY

In this chapter, we have introduced the idea of the discrete memoryless information source and discussed how it is modeled. The average information per symbol emitted by the source is measured by the entropy, Equation 1.2.5. Example 1.2.2 showed us the entropy of a source is maximized when every symbol in the source alphabet is equally probable.

We have also seen how the notion of the simple discrete source can be built upon to produce source models in which the symbols occur in blocks. The entropy of these multiple-symbol models was illustrated for the 2–symbol case by Equation 1.2.6.

The symbols in a block of symbols may be statistically independent, or they may be statistically dependent. This disjunction leads us to define the conditional entropy associated with symbol pairs a_i, b_j in terms of their joint and conditional probabilities:

$$H(B|A) = \sum_{i=0}^{M_A-1} \sum_{j=0}^{M_B-1} p_{i,j} \log_2(1/p_{j|i}). \qquad 1.8.1$$

From this equation, we can describe the joint entropy by Equation 1.2.8. We have seen that a number of interesting relationships arise from this chain rule. Many of these relationships are expressed as inequalities, such as in Equations 1.2.9, 1.2.11, and 1.2.12. We have seen that joint entropies obey a general chain rule given by Equation 1.2.10.

To achieve efficient communication or data storage, it is desirable to encode the symbols emitted by a source. This is called source coding. In Section 1.3, an overview of what a source code does was provided, and we learned that, by encoding blocks of source symbols, the coded representation of the source data can be made to have an average number of bits per symbol that approaches the entropy of the source. We have also seen that if the source coding is to be information lossless, the coding function must map the source alphabet A to the code alphabet B as a one-to-one and onto function.

The idea of source coding also led us to define the mutual information, $I(A; B)$, between two symbol alphabets. $I(A; B)$ is a measure of how much the uncertainty about alphabet A is reduced by knowledge of symbols drawn from alphabet B. For mutual information, $I(A; B) = I(B; A)$, so mutual information is defined by Equation 1.3.4. Mutual information is related to entropy and conditional entropy by Equation 1.3.3. The idea of mutual information will be pivotal in understanding how communication channels limit the rate at which information can be transmitted or stored.

Finally we discussed three practical and widely used lossless compression codes. Huffman codes are prefix codes which are optimal in the sense that the code words generated by a Huffman code have an average number of bits per symbol which asymptotically approaches the entropy of the source arbitrarily closely. Huffman coding requires knowledge of the source's symbol probability but is reasonably robust to small errors in the estimation of these probabilities.

Lempel–Ziv codes are an interesting and useful example of dictionary-based coding. The LZ codes do not require prior knowledge of the source's symbol probability. Instead, they dynamically construct their own code dictionary, and as the message sequence becomes large, they too asymptotically approach optimum performance bounded by the fundamental entropy rate of the source.

Arithmetic codes are an alternative to Huffman coding. They are attractive in a number of applications where the size of the source alphabet is too small to permit good compression using Huffman coding and where obtaining a sufficient compound symbol alphabet is impractical. Arithmetic codes obtain the effect of using fractions of a bit to represent a coded symbol, thus avoiding the "quantization" problems of standard Huffman coding.

REFERENCES

C. E. Shannon, "A mathematical theory of communication," *Bell System Technical Journal,* vol. 27, pp. 379–423, July, 1948.

C. E. Shannon, "A mathematical theory of communication," *Bell System Technical Journal,* vol. 27, pp. 623–656, Oct., 1948.

C. E. Shannon, "Prediction and entropy of printed English," *Bell System Technical Journal,* vol. 30, pp. 50–64, Jan., 1951.

D. A. Huffman, "A method for the construction of minimum-redundancy codes," *Proceedings of the IRE,* vol. 40, pp. 1098–1101, Sept., 1952.

B. McMillan, "The basic theorems of information theory," *Ann. Math. Stat.,* vol. 24, pp. 196–219, June, 1953.

V. Bhaskaran and K. Konstantinides, *Image and Video Compression Standards,* Boston, MA: Kluwer Academic Publishers, 1995.

W. Kou, *Digital Image Compression Algorithms and Standards,* Boston, MA: Kluwer Academic Publishers, 1995.

J. Ziv and A. Lempel, "A universal algorithm for sequential data compression," *IEEE Transactions on Information Theory,* vol. IT-23, pp. 337–343, 1977.

J. Ziv and A. Lempel, "Compression of individual sequences by variable rate coding," *IEEE Transactions on Information Theory,* vol. IT-24, pp. 530–536, 1978.

T. A. Welch, "A technique for high-performance data compression," *Computer,* vol. 17, pp. 8–19, 1984.

K. Parhi, "High-speed VLSI architectures for Huffman and Viterbi decoders," *IEEE Transactions on Circuits and Systems-II: Analog and Digital Signal Processing,* vol. 29, no. 6, pp. 385–391, June, 1992.

R. Hashemian, "Memory efficient and high-speed search Huffman coding," *IEEE Transactions on Communications,* vol. 43, no. 10, pp. 2576–2581, Oct., 1995.

I. Witten, R. Neal, and J. Cleary, "Arithmetic coding for data compression," *Communications of the ACM,* vol. 30, no. 6, pp. 520–540, June, 1987.

J. Rissanen and G. Langdon, "Arithmetic coding," *IBM Journal of Research and Development,* vol. 23, no. 2, pp. 149–162, Mar., 1979.

G. Langdon, "An introduction to arithmetic coding," *IBM Journal of Research and Development,* vol. 28, no. 2, pp. 135–149, Mar., 1984.

F. Rubin, "Arithmetic stream coding using fixed-precision registers," *IEEE Transactions on Information Theory,* vol. IT-25, no. 6, pp. 672–675, Nov., 1979.

W.-W. Lu and M. Gough, "A fast-adaptive Huffman coding algorithm," *IEEE Transactions on Communications,* vol. 41, no. 4, pp. 535–538, Apr., 1993.

R. Gallager, "Variations on a theme by Huffman," *IEEE Transactions on Information Theory,* vol. IT-24, no. 6, pp. 668–674, Nov., 1978.

EXERCISES

1.2.1: What is the cardinality for each of the following sets?
   a) $A = \{0, 1, 2, 3\}$
   b) $A = \{\text{cat, dog, mouse}\}$
   c) $A = \{\text{states in the United States of America}\}$
   d) $A = \{x \text{ such that } 2 < x < 40 \text{ and } x \text{ is a prime number}\}$.

1.2.2: The sets A and B are defined
$$A = \{0, 1, 2, 3, 4\} \quad B = \{0, 2, 4, 6, 8\}.$$
Find the cardinality of
   a) $A \cup B$ b) $A \cap B$.

1.2.3: Let A and B be two sets. The set difference $A - B$ is defined
$$A - B \equiv \{x \in A | x \notin B\}.$$
For find $A = \{a, b, c, d, 0, 1, 2, 3\}, \quad B = \{+, /, a, 2, \#\}$ find
   a) $A - B$ b) $B - A$.

1.2.4: A binary source has symbol probabilities $p_0 = 0.9, p_1 = 0.1$. Find the source entropy.

1.2.5: A binary source has symbol probabilities $p_0 = p, p_1 = 1 - p$. Plot the entropy as a function of p.

1.2.6: A source has an alphabet with $|A| = 8$ and $P_A = \{.25, .20, .15, .12, .10, .08, .05, .05\}$. Find the information efficiency of this source.

1.2.7: Given two information sources with $|A| = 4, |B| = 3$. The joint probabilities of symbols from these sources are given in the following table:

	b_0	b_1	b_2
a_0	0.10	0.08	0.13
a_1	0.05	0.03	0.09
a_2	0.05	0.12	0.14
a_3	0.11	0.04	0.06

$$\Pr(a_i, b_j)$$

Find
   a) $H(A)$ b) $H(B)$ c) $H(A, B)$.

1.2.8: Find the conditional entropies $H(A|B), H(B|A)$ for the sources in Exercise 1.2.7.

1.2.9: Prove: $H(A, B) = H(B, A)$.

1.2.10: Prove: $H(A, B) = H(B) + H(A|B)$.

1.2.11: A discrete memoryless source has a symbol alphabet with $|A|=10$. Find the upper and lower bounds of its entropy.

1.2.12: The entropy of printed English can be upper bounded by assuming each letter emitted by the source is statistically independent. Find the upper bound on entropy for printed English from the measured probabilities given in the following table:

Letter	Probability	Letter	Probability	Letter	Probability
A	0.0642	J	0.0008	S	0.0514
B	0.0127	K	0.0049	T	0.0796
C	0.0218	L	0.0321	U	0.0228
D	0.0317	M	0.0198	V	0.0083
E	0.1031	N	0.0574	W	0.0175
F	0.0208	O	0.0632	X	0.0013
G	0.0152	P	0.0152	Y	0.0164
H	0.0467	Q	0.0008	Z	0.0005
I	0.0575	R	0.0484	space	0.1859

1.2.13: Explain why your result in Exercise 1.2.12 is an upper bound.

1.3.1: For the 4-ary source and encoder of Example 1.3.1, let $A\{a_0 \; a_1 \; a_2 \; a_3\}$. If the source emits the symbol sequence $a_0 \; a_0 \; a_1 \; a_0 \; a_1 \; a_0 \; a_0 \; a_2 \; a_3$, what encoded sequence is output by the encoder C?

1.3.2: Design an algorithm for a decoder for the system of Example 1.3.1. Using your decoder algorithm, decode the sequence 0 1 0 1 1 0 0 0 1 1 0 1 1 1 1 1 1 0.

1.4.1: Construct the Huffman coding tree for Example 1.3.2.

1.4.2: A discrete memoryless source has an alphabet $\{a, b, c, d\}$ with symbol probabilities 0.2, 0.4, 0.2, 0.2, respectively.

   a) Find the entropy of this source
   b) Construct a Huffman code for this source
   c) Calculate the efficiency of the code.

1.4.3: Construct a Huffman code for the source in Exercise 1.4.2, which encodes two source symbols at a time. What is the efficiency of this "expanded" code.

1.4.4: Write a computer program to perform Huffman encoding of the English text alphabet given in Exercise 1.2.12, and have your program calculate $\overline{L}$.

1.4.5: Specify a complete ROM lookup table for a Huffman encoder implementation of Example 1.4.2.

1.4.6: (Lab exercise) Design a Huffman encoder circuit for the code in Example 1.3.2. Build and test your design to verify its completeness and correctness.

1.4.7: Specify a ROM lookup table and draw a block diagram for an encoder for Example 1.3.2.

1.5.1: A discrete memoryless source with $A = \{a, b, c\}$ emits the following string.

$$b\;c\;c\;a\;c\;b\;c\;c\;c\;c\;c\;c\;c\;c\;c\;c\;a\;c\;c\;a$$

Using the Lempel-Ziv algorithm, encode this sequence and find the code dictionary and the transmitted sequence.

1.5.2: A source with $A = \{a, b, c\}$ is encoded using the Lempel–Ziv algorithm. The transmitted code word sequence is

$$2, 3, 3, 1, 3, 4, 5, 10, 11, 6, 10.$$

Construct the dictionary and decode this sequence.

CHAPTER 2

Channels and Channel Capacity

2.1 THE DISCRETE MEMORYLESS CHANNEL MODEL

2.1.1 The Transition Probability Matrix

In information theory, the processes of signal modulation, transmission, reception, and demodulation are represented by a *channel model*. This produces the high-level system model depicted previously in Figure 1.1. Mathematically, we can view the channel as a probabilistic function that transforms a sequence of (usually coded) input symbols, c, into a sequence of channel output symbols, y. Because of noise and other impairments in the communication system, this transformation is typically *not* a one-to-one mapping from the set of input symbols, C, to the set of output symbols, Y. Instead, any particular $c \in C$ may have some probability, $p_{y|c}$, of being transformed to the output symbol $y \in Y$. Each conditional probability $p_{y|c}$ is called a *forward transition probability*.

In a great many communication or storage systems, the signal processing associated with modulation, transmission, reception, and demodulation are designed so that the sequence of output symbols $\bar{y} \equiv (y_0, y_1, \cdots, y_t)$ are statistically independent if the symbols in the input sequence $\bar{c} \equiv (c_0, c_1, \cdots, c_t)$ are statistically independent. (In this notation, the subscript is a time index.) If the output set Y consists of discrete output symbols, and if the property of statistical independence of the output sequence holds, the channel is called a *discrete memoryless channel*, or DMC. The DMC plays a central and important role in coding and information theory.

Let p_c be the probability that symbol c is transmitted. For a DMC, the probability that the received symbol is y is given in terms of the transition probabilities by

$$q_y = \sum_{c \in C} p_{y|c} p_c, \qquad 2.1.1$$

where q_y is the probability of symbol y and the summation is taken over all elements of C. The probability distribution of the output set Y, denoted by Q_Y, may be easily calculated in matrix form as

$$Q_Y \equiv \begin{bmatrix} q_0 \\ q_1 \\ \vdots \\ q_{|Y|-1} \end{bmatrix} = \begin{bmatrix} p_{y_0|c_0} & p_{y_0|c_1} & \cdots & p_{y_0|c_{|C|-1}} \\ p_{y_1|c_0} & & & \vdots \\ \vdots & & & \\ p_{y_{|Y|-1}|c_0} & \cdots & & p_{|Y|-1|c_{|C|-1}} \end{bmatrix} \begin{bmatrix} p_0 \\ \vdots \\ p_{|C|-1} \end{bmatrix}, \quad 2.1.2$$

or, more compactly,

$$Q_Y = P_{Y|C} P_C, \quad 2.1.3$$

where P_C is the probability distribution of the input alphabet and $|A|$ represents the cardinality, or number of elements, in an alphabet A.

EXAMPLE 2.1.1

Let C be a binary alphabet $C = \{0, 1\}$ of equally probable symbols, and let Y be a three-element set $Y = \{y_0, y_1, y_2\}$. Let the channel have transition probability matrix

$$P_{Y|C} = \begin{bmatrix} 0.8 & 0.05 \\ 0.15 & 0.15 \\ 0.05 & 0.8 \end{bmatrix}.$$

Find Q_Y.

Solution: Since the symbols in C are equally probable, $p_0 = p_1 = 0.5$. From Equation 2.1.3, we have

$$Q_Y = \begin{bmatrix} 0.425 \\ 0.15 \\ 0.425 \end{bmatrix}.$$

Numerical values for the transition probability matrix are determined by analysis of the noise and transmission impairment properties of the channel and the modulation/demodulation method used. How these calculations are performed is one of the subjects covered in that branch of communication theory known as *modulation theory*. This is a subject which makes use of signal processing theory and probability theory and is beyond the scope of this text. The interested student can learn more about this subject in a course in digital communication theory or can consult a number of excellent textbooks, which your instructor can probably recommend.

We need not concern ourselves here with how these numbers are determined but we *do* need to be concerned about some of the properties of $P_{Y|C}$. Referring to Example 2.1.1, notice that the *columns* of $P_{Y|C}$ sum to unity. This is because, no matter what symbol c is sent, *some* output symbol y must result. This is a consequence of the fact that

$$\sum_{y \in Y} P_{y|c} P_c = \sum_{y \in Y} P_{y,c} = P_c,$$

where $p_{y,c}$ is the joint probability of y and c. This relationship establishes the property that each column of the transition probability matrix must sum to one.

In Example 2.1.1, we have $|Y| \neq |C|$. In many communication systems $|Y|=|C|$, and such systems are typically said to employ *hard-decision decoding*. This phrase does not (necessarily!) mean that the received symbols are difficult to decode. Instead, it refers to communication systems where the demodulator makes a firm, *i.e.*, "hard," decision of what symbol c was probably transmitted. If this decision is wrong, the demodulator is said to have made a detection error. In the case where $|Y| > |C|$, the demodulator is said to make *soft decisions*. The final decoding decision is left to the receiver's decoder block in Figure 1.1.1. Both types of systems occur in practice, so our theory must be flexible enough to deal with both situations.

2.1.2 Output Entropy and Mutual Information

The entropy

$$H(Y) = \sum_{y \in Y} q_y \log_2(1/q_y)$$

is a measure of the information per symbol in the channel output set Y. Let us examine some of the properties of this entropy.

EXAMPLE 2.1.2

Calculate the entropy of Y for the system in Example 2.1.1. Compare this with the entropy of source C.

Solution:

$$H(Y) = -[.425 \log_2(.425) \cdot 2 + .15 \log_2(.15)] = 1.4598$$

and

$$H(C) = -2 \cdot 0.5 \log_2(0.5) = 1.$$

We see that $H(Y) > H(C)$! How can this be? Let us think back to our discussion in section 1.3, when we examined the notion of a source encoder with output alphabet $|B| > |A|$. We saw that it *was* possible for the output entropy to be greater than the input entropy but that the "additional" information carried in the output was not related to the information from the source. Rather, it was due to the fact that the mapping from A to B was one-to-many, and the "extra" information in the output was due to the randomness of the encoding process itself.

In a similar fashion, the result in Example 2.1.2 comes from the presence of random noise in the channel during transmission and not from the source C. Unlike our case in Chapter one, where we deliberately introduced a larger output alphabet for purposes of encryption, the system in Examples 2.1.1 and 2.1.2 suffers a one-to-many mapping because of undesirable noise effects in the channel over which we have no control. The "extra" information carried in Y is truly "useless" to us and, in fact, is harmful because it produces uncertainty about what symbols were being transmitted.

Can we solve this problem by using only systems which employ hard-decision decoding, that is, systems where $|Y| = |C|$? The answer to this question is "no," as will now be demonstrated.

EXAMPLE 2.1.3

Let C be the binary source of Example 2.1.1, and let the channel have binary outputs $Y = \{0, 1\}$ and transition probability matrix

$$P_{Y|C} = \begin{bmatrix} 0.98 & 0.05 \\ 0.02 & 0.95 \end{bmatrix}.$$

Find $H(Y)$ and compare it with the source entropy.

Solution: By direct calculation, $Q_Y = [0.515\ 0.485]^T$ and $H(Y) = 0.99935 < 1 = H(C)$.

This example shows that Y carries less information than was transmitted by the source. Where did it go? It was *lost* during the transmission process. The channel is information lossy.

So far, we've looked at two examples where the output entropy was either greater than or less than the input entropy. What we have *not* considered yet is what effect all this has on the ability to tell from observing Y what original information was transmitted. After all, the purpose of the receiver is to recover the original transmitted information. What does our observation of Y tell us about the transmitted information sequence?

In Section 1.3, we defined the mutual information as a measure of how much our uncertainty of one variable is reduced by knowledge of another. For our communication system, we are interested in how much observations of the received symbols y tell us about transmitted symbols c. In other words, we want to know the mutual information

$$I(C;Y) = \sum_{c \in C} \sum_{y \in Y} p_{c,y} \log_2\left(\frac{p_{c,y}}{q_y p_c}\right). \qquad 2.1.4$$

If $I(C;Y) = 0$, then Y tells us nothing at all about C. For instance, suppose Y and C are statistically independent. (Perhaps some unscrupulous individual has cut the telephone wire and none of the transmitted signal is getting through.) Then $p_{c,y} = p_c q_y$, and Equation 2.1.4 tells us the mutual information is zero. On the other hand, we know from Equation 1.3.3 that

$$I(C;Y) = H(C) - H(C|Y), \qquad 2.1.5$$

which tells us that the mutual information is upper bounded by

$$I(C;Y) \leq H(C), \qquad 2.1.6$$

with equality if and only if the conditional entropy $H(C|Y) = 0$. The conditional entropy is a measure of how much information loss occurs in the channel, and if it is equal to zero, then it seems reasonable to suspect $I(C;Y) = H(C)$ means that Y contains sufficient information to tell us what the transmitted sequence was. It turns out that something very close to this is true, although an exactly correct statement of this idea must wait awhile longer while we build up some additional analysis tools and concepts.

EXAMPLE 2.1.4

Calculate the mutual information for the system of Example 2.1.1.

Solution: Since the joint probability $p_{c,y} = p_{y|c} p_c$, Equation 2.1.4 can be put into a more convenient form for calculation:

$$I(C;Y) = \sum_{c \in C} p_c \sum_{y \in Y} p_{y|c} \log_2 \left(\frac{p_{y|c}}{q_y} \right) \equiv \sum_{c \in C} p_c I(Y;c), \qquad 2.1.7$$

where the "partial mutual information" $I(Y; c)$ is

$$I(Y;c) \equiv \sum_{y \in Y} p_{y|c} \log_2 \left(\frac{p_{y|c}}{q_y} \right). \qquad 2.1.8$$

From the given transition probabilities and the q_y found in Example 2.1.1, Equations 2.1.7 and 2.1.8 give

$$I(C;Y) = 2 \cdot (.5) \cdot \left[.8 \log_2 \left(\frac{.8}{.425} \right) + .15 \log_2 \left(\frac{.15}{.15} \right) + .05 \log_2 \left(\frac{.05}{.425} \right) \right] = .57566.$$

Note in this example how output symbol y_1 makes no contribution to $I(C; Y)$. This is because the probability of getting y_1 because a "0" was sent is exactly the same as the probability of getting y_1 when a "1" is sent. Since "0" and "1" have the same probability of occurrence, the fact that we observe output symbol y_1 tells us nothing at all about what c might have been sent. The mutual information for this system is well below the entropy of the source and so this channel has a high level of information loss.

EXAMPLE 2.1.5

Calculate the mutual information for the system of Example 2.1.3.

Solution: Using Equations 2.1.7 and 2.1.8 and the results from Example 2.1.3, we have

$$I(C;Y) = .5 \left[.98 \log_2 \left(\frac{.98}{.515} \right) + .02 \log_2 \left(\frac{.02}{.485} \right) + .05 \log_2 \left(\frac{.05}{.515} \right) + .95 \log_2 \left(\frac{.95}{.485} \right) \right]$$

$$= .78543.$$

This channel is quite lossy also. Notice how even though $H(Y)$ was almost equal to $H(C)$ in Example 2.1.3, the mutual information is considerably less than $H(C)$. We can not tell how much information loss we are dealing with simply by comparing the input and output entropies.

2.2 CHANNEL CAPACITY AND THE BINARY SYMMETRIC CHANNEL

2.2.1 Maximization of Mutual Information and Channel Capacity

In the previous section, we learned how to calculate the amount of information that gets through from the source to the receiver. What we have not yet considered is how to determine what the maximum information rate that can be supported by the channel is.

Each time the transmitter sends a symbol, it is said to *use* the channel. The *channel capacity* is the maximum average amount of information that can be sent per channel use. Why is this not the same as the mutual information? The answer is that the mutual information is a function of the probability distribution of *C*. By changing P_C, we get different results from Equation 2.1.4. For a fixed transition probability matrix, a change in P_C also results in a different output symbol distribution Q_Y. This tells us that mutual information is a complicated function of the source's probability distribution. The maximum mutual information achieved for a given transition probability matrix is the channel capacity

$$C_c = \max_{P_C} I(C;Y). \qquad 2.2.1$$

Note that the channel capacity is found by maximizing $I(C; Y)$ with respect to P_C. C_c, having units of bits per channel use.

A quick glance back at Equation 2.1.4 or 2.1.7 might produce a feeling that C_c is difficult to find. One must maximize the mutual information with respect to the discrete set of probabilities of the source symbols *c*, and for a given transition probability matrix, changing P_C also changes the set of probabilities q_y. It is, in fact, true that an analytical closed-form solution to this problem is difficult to achieve for an arbitrary channel. Fortunately, an efficient numerical technique for finding C_c was derived in 1972 by Blahut and, independently, Arimoto. This algorithm is presented below and is based on the fact that channel capacity can be both upper and lower bounded by some simple functions of P_C. The algorithm recursively updates P_C based on these bounds.

Arimoto–Blahut Algorithm

Let $|C| = M, |Y| = N$, and let $F = [f_0\ f_1\ \cdots\ f_{M-1}]$. Let ϵ be some small positive number.

Let *j* and *k* be indices having ranges $j \in [0,\cdots, M-1], k \in [0,\cdots, N-1]$.

Initialize P_C with element values $p_j = 1/M$, and initialize $Q_Y = P_{Y|C} \cdot P_C$.

REPEAT UNTIL stopping point is reached:

$$f_j = \exp\left\{\sum_k \left[p_{k|j} \ln\left(\frac{p_{k|j}}{q_k}\right)\right]\right\} \text{ for } j \in [0,\cdots, M-1]$$

$x = F \cdot P_C$
$I_L = \log_2(x)$
$I_U = \log_2(\max_j(f_j))$
IF $(I_U - I_L) < \epsilon$ THEN
  $C_c = I_L$
  STOP
ELSE
  $p_j = f_j p_j / x$ for $j = 0, \cdots, M-1$
  $Q_Y = P_{Y|C} \cdot P_C$
END IF
END REPEAT

Upon termination, this algorithm provides an estimate of the channel capacity accurate to within the stopping factor and the input probability distribution that achieves this capacity.

Section 2.2 Channel Capacity and the Binary Symmetric Channel

The channel capacity proves to be a sensitive function of the transition probability matrix $P_{Y|C}$ but, in many cases, a fairly weak function of P_C. This is illustrated in the following examples.

EXAMPLE 2.2.1

Using the Arimoto–Blahut algorithm, find the channel capacity, the input and output probability distributions that achieve the channel capacity, and the mutual information given a uniform P_C for channels with the following transition probability matrices:

a) $P_{Y|C} = \begin{bmatrix} .98 & .05 \\ .02 & .95 \end{bmatrix}$ b) $P_{Y|C} = \begin{bmatrix} .8 & .05 \\ .2 & .95 \end{bmatrix}$ c) $P_{Y|C} = \begin{bmatrix} .8 & .1 \\ .2 & .9 \end{bmatrix}$

d) $P_{Y|C} = \begin{bmatrix} .6 & .01 \\ .4 & .99 \end{bmatrix}$ e) $P_{Y|C} = \begin{bmatrix} .8 & .3 \\ .2 & .7 \end{bmatrix}$ f) $P_{Y|C} = \begin{bmatrix} .8 & .05 \\ .15 & .15 \\ .05 & .8 \end{bmatrix}$.

Solutions: Using the algorithm given above with $\epsilon = 10^{-7}$, the following computer solutions are obtained:

a) $C_c = .78585$, $P_C = [.51289 \ .48711]^T$, $Q_Y = [.52698 \ .47302]^T$, $I(C;Y) = .78543$

b) $C_c = .48130$, $P_C = [.46761 \ .53239]^T$, $Q_Y = [.4007 \ .5993]^T$, $I(C;Y) = .47955$

c) $C_c = .39775$, $P_C = [.4824 \ .5176]^T$, $Q_Y = [.4377 \ .5623]^T$, $I(C;Y) = .39731$

d) $C_c = .36877$, $P_C = [.4238 \ .5762]^T$, $Q_Y = [.26 \ .74]^T$, $I(C;Y) = .36145$

e) $C_c = .191238$, $P_C = [.510 \ .490]^T$, $Q_Y = [.555 \ .445]^T$, $I(C;Y) = .191165$

f) $C_c = I(C;Y) = .57566$, $P_C[.5 \ .5]^T$, $Q_Y = [.425 \ .15 \ .425]^T$.

Example 2.2.1 illustrates a wide range of channel capacities in cases (a) through (e). Comparison of C_c with the accompanying value of $I(C; Y)$ demonstrates, however, that the maximum mutual information is actually quite close to the uniform-P_C case. The cases considered in this example therefore show relatively small sensitivity to P_C, since the percent change in the P_C from the uniform distribution to that which maximizes mutual information is greater than the percent change in mutual information that it produces. On the other hand, values of C_c vary significantly from one case to another. This indicates the strong effect $P_{Y|C}$ has on determining C_c.

Case (f) is interesting, since, for this transition probability matrix, the uniform input distribution produces the maximum mutual information. This case is an example of a *symmetric channel*. Note that the columns of its transition probability matrix are permutations of each other. Likewise, the top and bottom rows of $P_{Y|C}$ are permutations of each other. The center row, which is *not* a permutation of the other rows, corresponds to output symbol y_1, which, as we saw in example 2.1.4, makes no contribution to the mutual information.

2.2.2 Symmetric Channels

Symmetric channels play an important role in communication systems and many such systems attempt, by design, to achieve a symmetric channel function. (This is not always possible, such as in multilevel rectangular quadrature amplitude modulation or in multilevel pulse amplitude modulation; however, these systems are generally designed to be as close to having a symmetric transition probability matrix as possible.) The reason for the importance of the symmetric channel is that when such a channel is possible, it frequently has greater channel capacity than an otherwise equivalent nonsymmetric channel would have. The effect of perturbing the channel of Example 2.2.1(f) to a nonsymmetric case is illustrated in the following example.

EXAMPLE 2.2.2

Repeat Example 2.2.1 for a channel with transition probability matrix

$$P_{Y|C} = \begin{bmatrix} .79 & .05 \\ .16 & .15 \\ .05 & .8 \end{bmatrix}.$$

Solution: Using the channel capacity algorithm, we get

$$C_c = .571215, \quad P_C = [.50095 \quad .49905]^T, \quad Q_Y = [.4207 \quad .1550 \quad .4243]^T, \quad I(C;Y) = .5712.$$

The channel capacity for this example is roughly 1% less than the symmetric case, although the actual changes made to the transition probability matrix were very slight.

EXAMPLE 2.2.3

Quadrature phase-shift keying (QPSK) is a modulation method that produces a symmetric channel. For QPSK, $|C|=|Y|=4$. Using the Arimoto—Blahut algorithm, determine the channel capacity and the input symbol probability distribution that achieves this capacity if the transition probability matrix is

$$P_{Y|C} = \begin{bmatrix} .95 & .024 & .024 & .002 \\ .024 & .95 & .002 & .024 \\ .024 & .002 & .95 & .024 \\ .002 & .024 & .024 & .95 \end{bmatrix}.$$

Solution: $C_c = 1.653488, \quad P_C = [.25 \quad .25 \quad .25 \quad .25]^T.$

Because symmetric channels play a very important role in practical communication systems, it is worthwhile to examine some of their special properties. We begin with the channel in Example 2.2.3. Notice that the capacity for this channel is achieved when P_C is uniformly distributed. This is always the case for a symmetric channel as we shall now see.

First, notice that every column of $P_{Y|C}$ is a permutation of the first column. Also notice that every row of $P_{Y|C}$ is a permutation of the first row. When $P_{Y|C}$ is a square matrix, this permutation property of the columns *and* the rows is a sufficient condition for a uniformly distributed input alphabet to achieve the maximum mutual information. Indeed, the permutation condition is what is gives rise to the term "symmetric channel."

Section 2.2 Channel Capacity and the Binary Symmetric Channel

To see this, observe the following properties of the channel in Example 2.2.3. First, if P_C is uniformly distributed, then it is obvious that Q_Y is *also* uniformly distributed, since the product of any row of $P_{Y|C}$ with P_C is the same as the product of the first row with P_C. Hence, all elements of Q_Y must be equal.

Now let us look at the mutual information expression given in Equation 2.1.7. Since all of the input probabilities are equal,

$$I(C;Y) = \frac{1}{|C|} \sum_{y \in Y} I(Y;c),$$

where $I(Y;c)$ is defined by Equation 2.1.8. Notice, however, that since all of the q_y probabilities are equal, the values of $I(Y;c)$ are all equal for every $c \in C$. Therefore,

$$I(C;Y) = I(Y;c_0) = \cdots = I(Y;c_{|C|-1}). \qquad 2.2.2$$

It can be shown that when Equation 2.2.2 is satisfied, $C_c = I(C;Y)$. We will not prove this here, but this fact is a consequence of a theorem known as the Kuhn–Tucker theorem. (It is assumed that the source alphabet does not contain any symbols with zero probability.) When the transition probability matrix is not square, Equation 2.2.2 provides a means to check if a uniform input probability distribution maximizes the channel capacity. If Equation 2.2.2 holds for the uniform input distribution, then the channel is symmetric.

A symmetric channel of considerable importance, both theoretically and practically, is the binary symmetric channel (BSC) for which

$$P_{K|C} = \begin{bmatrix} 1-p & p \\ p & 1-p \end{bmatrix}. \qquad 2.2.3$$

The parameter p is known as the crossover probability, and is equal to the probability that the demodulator/detector will make a hard-decision decoding error. The BSC is the model for essentially all binary-pulse transmission systems of practical importance.

Since the BSC is a symmetric channel, it satisfies Equation 2.2.2 with a uniform input probability distribution. Its capacity is therefore given by Equation 2.1.8 as

$$\begin{aligned} C_c &= (1-p)\log_2(2(1-p)) + p\log_2(2p) \\ &= 1 + (1-p)\log_2(1-p) + p\log_2(p). \end{aligned} \qquad 2.2.4$$

Because the BSC occurs frequently in theory and in practice, Equation 2.2.4 is often abbreviated as

$$C_c = 1 - H(p), \qquad 2.2.5$$

where the notation $H(p)$ arises from the fact that the terms involving p in Equation 2.2.4 are minus one times the entropy of a binary alphabet having elements with probabilities p and $1-p$.

Examination of Equation 2.2.4 reveals some interesting properties of the capacity of the BSC. We observe that the capacity is bounded by the range $0 \le C_c \le 1$. The upper bound is achieved only if $p=0$ or $p=1$. The $p=0$ case is not surprising, since this corresponds to a channel which does not make errors (known as a "noiseless channel"). The $p=1$ case is initially startling because this corresponds to a channel which *always* makes errors. However, if we *know* that the channel output is always wrong, we can easily set things right by decoding the *opposite* of what the channel output symbol is. (Some investment advisors have this property as well; they are to be treasured.)

The zero capacity case occurs when $p = 0.5$. In this case, the channel output symbol is as likely to be correct as it is to be incorrect. Under this condition, the information loss in the channel is total. The capacity equation for the BSC is a concave-upward function possessing a single minimum at $p = 0.5$.

We note from Equation 2.2.4 that, except for the $p = 0$ and $p = 1$ cases, the capacity of the BSC is always less than the source entropy. If we try to send information through the channel using the maximum amount of information per symbol, some of this information will be lost, and decoding errors at the receiver will result. However, we will later see that, if we add sufficient redundancy to the transmitted data stream, it is possible to reduce the probability of lost information to an arbitrarily low level.

2.3 BLOCK CODING AND SHANNON'S SECOND THEOREM

2.3.1 Equivocation

In the previous section, we saw that there is a maximum amount of information per channel use that can be supported by a channel. Any attempt to exceed this channel capacity will result in information being lost during transmission. From Equation 2.1.5, we have

$$I(C;Y) = H(C) - H(C|Y),$$

and so

$$C_c = \max_{P_C}(H(C) - H(C|Y)) \qquad 2.3.1$$

The conditional entropy $H(C|Y)$ is defined by Equation (1.3.2) and corresponds to our uncertainty about what the input to the channel was, given our observation of the channel output. It is a measure of the information lost during the transmission. This conditional entropy is often called the *equivocation* for this reason. From its definition in Section 1.3.2, it is clear that equivocation is a non-negative function. What is less clear, but nonetheless true, is

$$H(C|Y) \leq H(C). \qquad 2.3.2$$

EXAMPLE 2.3.1

Derive Equation 2.3.2.

Solution: Using Equation 2.1.4, we can write Equation 2.1.5 as

$$H(C) - H(C|Y) = \sum_{c \in C} \sum_{y \in Y} p_{c,y} \log_2\left(\frac{p_{c,y}}{p_c \, q_y}\right).$$

Since $\ln(1/x) \geq 1 - x$ for $x > 0$, we can re-write this as

$$H(C) - H(C|Y) \geq \sum_{c \in C} \sum_{y \in Y} \frac{p_{c,y}}{\ln(2)} \left(1 - \frac{p_c q_y}{p_{c,y}}\right).$$

Since

$$\sum_{c \in C} \sum_{y \in Y} p_{c,y} = 1, \quad \sum_{c \in C} \sum_{y \in Y} p_c q_y = \sum_{c \in C} p_c \sum_{y \in Y} q_y = 1,$$

we have

$$H(C) - H(C|Y) \geq 0,$$

which establishes Equation 2.3.2. *QED*

The equivocation is zero if and only if the transition probabilities $p_{y|c}$ are either zero or one for all pairs ($y \in Y, c \in C$). This can be seen from Equation 1.3.2. However, this condition applies only to a *noiseless* channel[1] so we see that noisy channels entail information loss. In Section 2.2, every example we looked at had $C_c < H(C)$. We now see that this was due not merely to an unfortunate set of special cases but, rather, to a fundamental property of noisy channels.

2.3.2 Entropy Rate and the Channel-Coding Theorem

This sad fact was understood intuitively (and without any high-powered mathematics) by engineers from the earliest days of the science and practice of communications. What was *not* understood until the work of Shannon in 1948 was that this is not the last word on the subject. In chapter one, we looked at the entropy of a *block* of n symbols $(s_0, s_1, \cdots, s_{n-1})$. If each of these symbols is drawn from the source alphabet, i.e., $s_i \in C$, then from Equation 1.2.12 we have the entropy of the block is

$$H(C_0, C_1, \cdots, C_{n-1}) \leq nH(C), \qquad 1.2.12$$

with equality if and only if C is a memoryless source. In transmitting a block of n symbols, we use the channel n times. Recall that channel capacity has units of bits per channel *use*, and refers to an average amount of information per channel use. Since $H(C_0, C_1, \cdots, C_{n-1})$ is the average information contained in the n-symbol block, it follows that the average information per channel use would be

$$R = \frac{H(C_0, C_1, \cdots, C_{n-1})}{n} \leq H(C). \qquad 2.3.3$$

---

[1] As in interesting footnote, it turns out that the idea of a "noiseless" channel includes come cases where the channel actually *is* noisy but the noise is bounded in magnitude in a special way. Such channels are more accurately called "errorless channels," but we will stick to the typical lingo. For the terminally curious, a discussion of the special conditions required for a "noiseless noisy channel" may be found in R. Wells, "Applications of set-membership techniques to symbol-by-symbol decoding for binary data transmission systems," *IEEE Trans. Inform. Th.*, vol. 42, no. 4, pp. 1285–1290, July, 1996.

As it happens, R in Equation 2.3.3 is an unbiased estimate of the average bits per channel use, rather than exactly the expected number of bits per channel use. Formally, the true average bits per channel use is achieved in the limit

$$R = \lim_{n \to \infty} \frac{H(C_0, \cdots, C_{n-1})}{n} \leq H(C). \qquad 2.3.4$$

R is therefore called the *entropy rate*.

Now, $R \leq H(C)$, with equality if and only if all the symbols s_t are statistically independent. Suppose they are not; suppose, in our transmission of the block, we deliberately introduce *redundant symbols*. Then $R < H(C)$. Taking this further, suppose we introduce a sufficient number of redundant symbols in the block so that $R \leq C_c$. Does this mean that transmission without information loss, *i.e.*, with zero equivocation, is possible?

Remarkably enough, the answer to this question is "yes"! What is the implication of doing so? The implication is: If we signal without information loss, then it is possible to send information through the noisy channel with an arbitrarily low probability of error (even though individual symbols may be decoded incorrectly!). The process of adding redundancy to a block of transmitted symbols is called *channel coding*. Does there exist a channel code that will accomplish this purpose? The answer to this question is given by Shannon's second theorem:

Theorem: Suppose $R < C_c$, where C_c is the capacity of a memoryless channel. Then for any $\epsilon > 0$, there exists a block length n and a code of block length n and rate R whose probability of block decoding error p_e satisfies $p_e \leq \epsilon$ when the code is used on this channel.

Shannon's second theorem (also called Shannon's main theorem) tells us that it is possible to transmit information over a noisy channel with arbitrarily small probability of error. This is such a remarkable statement that it deserves close examination.

The theorem says that if the entropy rate R in a block of n symbols is smaller than the channel capacity, then we can make the probability of error arbitrarily small. What error are we talking about? Recall Example 1.2.3 where our source alphabet consisted of 2 "information" bits plus a redundant parity bit. The entropy for this block of $n = 3$ bits was 2.

Now suppose we send a block of n bits in which $k < n$ of these bits are statistically independent "information" bits and $R = n - k$ are redundant "parity" bits computed from the k information bits according to some "coding rule." The entropy of the block will then be k bits and the average information in bits per channel use will be $R = k/n$. If this entropy rate is less than the channel capacity, Shannon's theorem says we can make the probability of error in recovering our original k information bits arbitrarily small. The channel will make errors within our block of n bits, but the redundancy built into the block will be sufficient to *correct* these errors and recover the k bits of *information* we transmitted.

Shannon's theorem does *not* say that we can do this for *just any* block length n we might care to choose. The theorem says *there exists* a block length n for which *there is* a code of rate R. The required size of the block length n depends on the upper bound ϵ we pick for our error probability. We shall defer the proof of Shannon's theorem until chapter nine of the text, since we'll need a few more tools and concepts before we're ready to tackle the proof of this remarkable theorem. However, when we eventually *do* see this proof, we shall also see that the proof of Shannon's theorem implies very strongly that the block length n is going to be very large if R is to approach C_c to within an arbitrarily small distance with an arbitrarily small probability of error.

The complexity and expense of an error-correcting channel code are believed to grow rapidly as R approaches the channel capacity and the probability of a block decoding error is made arbitrarily small. It is believed by many that beyond a particular rate called the *cutoff rate*, R_o, it is prohibitively expensive to use the channel. In the case of the binary symmetric channel, this rate is given by

$$R_o = -\log_2\left(0.5 + \sqrt{p(1-p)}\right). \qquad 2.3.5$$

The belief that R_o is some kind of "sound barrier" for practical error correcting codes stems from the fact that for certain kinds of decoding methods the complexity of the decoder grows extremely rapidly as R exceeds R_o. However, a general proof that this must be so, or even of what the precise significance of the cutoff rate is, has never been discovered. In fact, it is known that there are channels for which the very concept of the cutoff rate has no meaning. (This is one reason why in this text we do not go deeper into how R_o is defined for general channels.) Furthermore, there have been some recent developments in coding theory, in particular a new class of error correcting codes discovered in 1993 and known as *turbo codes*, which seem to have finally challenged the notion that R_o represents some sort of practical limit. Nonetheless, for a large number of "older" codes of great practical interest, R_o *does* provide a measure of the upper limit of the rate achievable by these older codes, and it is for that reason that we introduce it here.

EXAMPLE 2.3.2

Calculate the capacity and the cutoff rate for the BSC with

a) $p = 0.1$; b) $p = 0.01$; c) $p = 0.001$ d) $p = 0.0001$.

Solution: $Cc = 1 + p\log_2(p) + (1-p)\log_2(1-p)$ and $R_o = -\log_2\left(0.5 + \sqrt{p(1-p)}\right)$ for the binary symmetric channel. So,

a) $C_c = 0.531$ $R_o = 0.32193$ b) $C_c = 0.91921$ $R_o = 0.73817$

c) $C_c = 0.98859$ $R_o = 0.91157$ d) $C_c = 0.99853$ $R_o = 0.97143$.

Note how, in this example, the cutoff rate approaches the capacity as the crossover probability p grows smaller and, on the other hand, the capacity and the cutoff rate diverge more and more widely as p is increased.

2.4 MARKOV PROCESSES AND SOURCES WITH MEMORY

2.4.1 Markov Processes

Thus far, we have been occupied with memoryless sources and channels. We must now turn our attention to sources with memory. By this, we mean information sources, where the successive symbols in a transmitted sequence are correlated with each other, i.e., the sources in a sense "remember" what symbols they have previously emitted, and the probability of their next symbol depends on this history.

Sources with memory arise in a number of ways. First, natural languages such as English have this property. For example, the letter "q" in English is almost always followed by the letter "u." Similarly, the letter "t" is followed by the letter "h" approximately 37% of the time in English text.

Many real-world signals, such as a speech waveform, are also heavily time correlated. If a signal is bandlimited, that is, if most of the energy of the signal is limited to a specific range of frequencies in its Fourier spectrum, then this signal will be time-correlated. Any time-correlated signal is a source with memory. Finally, we sometimes wish to *deliberately* introduce correlation (redundancy) in a source for purposes of block coding, as discussed in the previous section.

Let $A = \{a\}$ be the alphabet of a discrete source having $|A|$ symbols, and suppose this source emits a time sequence of symbols $(s_0, s_1, \cdots, s_t, \cdots)$ with each $s_t \in A$. If the conditional probability $p(s_t|s_{t-1}, s_{t-2}, \cdots, s_0)$ depends only on j previous symbols so that

$$\Pr(s_t|s_{t-1},\cdots,s_0) = p(s_t|s_{t-1}, s_{t-2},\cdots,s_{t-j}), \qquad 2.4.1$$

then A is called a jth-order Markov process.

The string of j symbols $S_t = (s_{t-1}, \cdots, s_{t-j})$ is called the *state* of the Markov process at time t. A jth-order Markov process therefore has $N = |A|^j$ possible states. Let us number these possible states from 0 to $N-1$ and let $\pi_n(t)$ represent the probability of being in state n at time t. The probability distribution of the system at time t can then be represented by the vector

$$\Pi_t \equiv \begin{bmatrix} \pi_0(t) \\ \pi_1(t) \\ \vdots \\ \pi_{N-2}(t) \\ \pi_{N-1}(t) \end{bmatrix}. \qquad 2.4.2$$

For each state at time t, there are $|A|$ possible *next* states at time $t+1$, depending on which symbol a is emitted next by the source. If we let $p_{i|k}$ be the conditional probability of going to state i given that the present state is state k, the state probability distribution at time $t+1$ is governed by the transition probability matrix

$$P_{A|\Pi} \equiv \begin{bmatrix} p_{0|0} & p_{0|1} & \cdots & p_{0|N-1} \\ p_{1|0} & p_{1|1} & \cdots & p_{1|n-1} \\ \vdots & & \ddots & \\ p_{N-1|0} & & \cdots & p_{N-1|N-1} \end{bmatrix}, \qquad 2.4.3$$

and is given by

$$\Pi_{t+1} = P_{A|\Pi} \Pi_t. \qquad 2.4.4$$

EXAMPLE 2.4.1

Let A be a binary 1st-order Markov source with $A = \{0, 1\}$. This source has 2 states labeled "0" and "1". Let the transition probabilities be

$$p_{0|0} = 0.3, \quad p_{1|0} = 0.7, \quad p_{0|1} = 0.4, \quad p_{1|1} = 0.6.$$

What is the equation for the next probability state and find the state probabilities at time $t = 2$, given that the probabilities at time $t = 0$ are $\pi_0 = 1, \pi_1 = 0$.

Solution: $\Pi_{t+1} = \begin{bmatrix} 0.3 & 0.4 \\ 0.7 & 0.6 \end{bmatrix} \Pi_t$ is the next-state equation for the state probabilities. The state probabilities at $t = 2$ are found from

$$\Pi_1 = \begin{bmatrix} 0.3 & 0.4 \\ 0.7 & 0.6 \end{bmatrix} \begin{bmatrix} 1 \\ 0 \end{bmatrix} = \begin{bmatrix} 0.3 \\ 0.7 \end{bmatrix}, \quad \Pi_2 = \begin{bmatrix} 0.3 & 0.4 \\ 0.7 & 0.6 \end{bmatrix} \begin{bmatrix} 0.3 \\ 0.7 \end{bmatrix} = \begin{bmatrix} 0.37 \\ 0.63 \end{bmatrix}.$$

EXAMPLE 2.4.2

Let A be a second-order binary Markov process with transition probabilities

$$\Pr(a = 0|0,0) = 0.2, \quad \Pr(a = 1|0,0) = 0.8,$$
$$\Pr(a = 0|0,1) = 0.4, \quad \Pr(a = 1|0,1) = 0.6,$$
$$\Pr(a = 0|1,0) = 0.0, \quad \Pr(a = 1|1,0) = 1.0,$$
$$\Pr(a = 0|1,1) = 0.5, \quad \Pr(a = 1|1,1) = 0.5,$$

and assume all states are equally probable at time $t = 0$. What are the state probabilities at $t = 1$?

Solution: Define the states as $S_0 = (0,0), S_1 = (0,1)$, etc. The possible state transitions and their associated transition probabilities can be represented using a state diagram. For this problem, the state diagram is

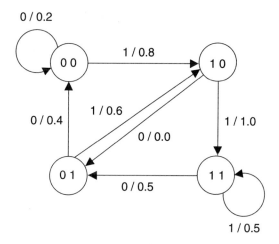

and the next-state probability equation is

$$\Pi_{t+1} = \begin{bmatrix} 0.2 & 0.4 & 0 & 0 \\ 0 & 0 & 0 & 0.5 \\ 0.8 & 0.6 & 0 & 0 \\ 0 & 0 & 1.0 & 0.5 \end{bmatrix} \Pi_t.$$

With all states equally probable at $t = 0$, we have $\Pi_0^T = [0.25 \ \ 0.25 \ \ 0.25 \ \ 0.25]$, so

$$\Pi_1^T = [.15 \ \ .125 \ \ .35 \ \ .375].$$

Did you notice that every column of $P_{A|\Pi}$ adds to 1.0 in each of the previous two examples? This is not accidental. Rather, it is a consequence of the requirement that, at every time t, the probabilities of all the states must sum to unity. Every properly constructed transition probability matrix has this property. Notice, however, that the *rows* of the matrix do *not* have to sum to unity. Can you explain why?

2.4.2 Steady-State Probability and the Entropy Rate

The equation for the state probabilities is a homogeneous difference equation. It can be simply shown by induction that the state probabilities at time t are given by

$$\Pi_t = (P_{A|\Pi})^t \Pi_0.$$

A Markov process is said to be *ergodic* if we can get from any initial state to any other state in some number of steps *and* if, for large t, Π_t approaches a steady-state value that is *independent* of the initial probability distribution Π_0. The steady-state value is reached when $\Pi_{t+1} = \Pi_t$. The Markov processes which model information sources are always ergodic.

EXAMPLE 2.4.3

Find the steady-state probability distribution for the source in Example 2.4.2.

Solution: In the steady state, the state probabilities become

$$\pi_0 = 0.2\pi_0 + 0.4\pi_1,$$

$$\pi_1 = 0.5\pi_3,$$

$$\pi_2 = 0.8\pi_0 + 0.6\pi_1,$$

$$\pi_3 = \pi_2 + 0.5\pi_3.$$

It appears from this that we have four equations and four unknowns, so solving for the four probabilities is no problem. However, if we look closely, we will see that only *three* of the equations above are linearly independent. To solve for the probabilities, we can use any three of the above equations and the constraint equation $\pi_0 + \pi_1 + \pi_2 + \pi_3 = 1$. (This equation is a consequence of the fact that the total probability must sum to unity; it is certain the system is in *some* state!). Dropping the first equation above and using the constraint, we have

Section 2.4 Markov Processes and Sources with Memory

$$\pi_1 - 0.5\pi_3 = 0,$$
$$0.8\pi_0 + 0.6\pi_1 - \pi_2 = 0,$$
$$\pi_2 - 0.5\pi_3 = 0,$$
$$\pi_0 + \pi_1 + \pi_2 + \pi_3 = 1,$$

which has solution

$$\pi_0 = 1/9, \pi_1 = \pi_2 = 2/9, \pi_3 = 4/9.$$

This solution is independent of the initial probability distribution.

This situation illustrated in the previous example, where only $N-1$ of the equations resulting from the transition probability expression are linearly independent and we must use the "sum to unity" equation to obtain the solution, always occurs in the steady-state probability solution of an ergodic Markov process.

Now let's consider the entropy rate of an ergodic Markov process. From Equation 2.3.4, we have

$$R = \lim_{t \to \infty} \frac{1}{t} H(A_0, A_1, \cdots, A_{t-1}).$$

However, as t grows very large, the state probabilities converge to a steady-state value, π_n, for each of the N possible states. Since R is the average information per symbol in this block of symbols, as t becomes large this average will be determined by the probability of occurrence of the symbols in A after the state probabilities have converged to their steady-state values.

Suppose we are in state S_n at time t. The conditional entropy of the next symbol a is given by

$$H(A|S_n) = \sum_{a \in A} \Pr(a|S_n) \log_2(1/\Pr(a|S_n)).$$

Since each possible symbol a leads to a unique next state, S_n can lead to $|A|$ possible next states. The remaining $N-|A|$ states cannot be reached from S_n, and for these states the transition probability $p_{i|n} = 0$ in Equation 2.4.3. Therefore, the conditional entropy expression above can be expressed in terms of the transition probabilities as

$$H(A|S_n) = \sum_{i=0}^{N-1} p_{i|n} \log_2(1/p_{i|n}).$$

For large t, the probability of being in state S_n is given by its steady-state probability π_n. Therefore, the entropy rate of the system becomes

$$R = \sum_{n=0}^{N-1} \pi_n H(A|S_n). \qquad 2.4.5$$

This expression, in turn, is equivalent to

$$R = \sum_{n=0}^{N-1} \pi_n \sum_{i=0}^{N-1} p_{i|n} \log_2(1/p_{i|n}), \qquad 2.4.6$$

where the $p_{i|n}$ the are the entries in the transition probability matrix and the π_i are the steady-state probabilities. The entropy rate of an ergodic Markov process is a function only of its steady state probability distribution and the transition probabilities.

EXAMPLE 2.4.4

Find the entropy rate for the source in Example 2.4.2. Calculate the steady state probability of the source emitting a "0" and the steady-state probability of the source emitting a "1". Calculate the entropy of a memoryless source having these symbol probabilities and compare the result with the entropy rate of the Markov source.

Solution: We have the steady-state probabilities from Example 2.4.3. The entropy rate is

$$R = \frac{1}{9}(.2 \log_2(5) + .8 \log_2(1.25)) + \frac{2}{9}(.4 \log_2(2.5) + .6 \log_2(1.667))$$

$$+ \frac{2}{9}\log_2(1) + \frac{4}{9}(2 \cdot 0.5 \log_2(2)) = .740.$$

The steady-state symbol probabilities are

$$\Pr(0) = \sum_{n=0}^{3} \pi_n \Pr(0|S_n) = \frac{.2}{9} + \frac{0.4(2)}{9} + \frac{0.5(4)}{9} = \frac{1}{3},$$

$$\Pr(1) = 1 - \Pr(0) = \frac{2}{3}.$$

The entropy of a memoryless source having this symbol distribution is

$$H(A) = \sum_{a=0}^{1} \Pr(a)\log_2(1/\Pr(a)) = 0.9183.$$

From this we have $R < H(A)$. This inequality was predicted by Equation 2.3.4.

In Section 2.3, we discussed how introducing redundancy into a block of symbols might be used to reduce the entropy rate to a level below the channel capacity and how this technique might be used for the correction of errors in the information bits to achieve an arbitrarily small information bit error rate. In this section, we have seen that a Markov process also introduces redundancy into the symbol block. Can this redundancy be introduced in such a way as to be useful for error correction? The answer to this question is "yes." This is the principle underlying a class of error correcting codes known as convolutional codes. It is a subject we will explore in some depth later in this text.

2.5 MARKOV CHAINS AND DATA PROCESSING

In the first two sections of this chapter, we examined the process of transmitting information C through the channel to produce channel output Y and discovered that a noisy channel entails some information loss if the entropy rate exceeds the channel capacity. It is only natural to wonder if there might exist some (possibly complicated) form of data processing which can be performed on Y to recover the lost information. Unfortunately, as we will now show, the answer to this question is "no." Once the information has been lost, it's gone.

To show this, suppose we operate on the received signal y by some function f to produce a result $z = f(y)$. We may use any conceivable function. Let the set of possible values of z which can be produced from the domain Y be denoted as Z. Let the joint probability of the ordered triplet $\langle c, y, z \rangle$ be $p_{c,y,z}$. Then

$$p_{c,y,z} = p_{c,y} p_{z|c,y}.$$

Since z is a function of y, the probability of z given c and y is just $p_{z|c,y} = p_{z|y}$, and therefore,

$$p_{c,y,z} = p_{c,y} p_{z|y}. \qquad 2.5.1$$

Three variables that satisfy Equation 2.5.1 are said to form a *Markov chain*. We now ask if the mutual information $I(C;(Y,Z)) \equiv I(C;Y,Z)$ is larger than $I(C;Y)$. Let us define the *conditional mutual information*

$$I(C;Y|Z) \equiv H(C|Z) - H(C|Y,Z). \qquad 2.5.2$$

The conditional mutual information is the reduction in our uncertainty of C due to our knowledge of Y when Z is given to us. The reason for defining this concept now will be clear in a moment.

From our earlier definition of mutual information,

$$I(C;Y,Z) = H(C) - H(C|Y,Z),$$

and substituting for $H(C|Y,Z)$ in terms of the conditional mutual information in Equation 2.5.2, we have

$$I(C;Y,Z) = H(C) - H(C|Z) + I(C;Y|Z).$$

Since

$$H(C) - H(C|Z) = I(C;Z),$$

we have

$$I(C;Y,Z) = I(C;Z) + I(C;Y|Z). \qquad 2.5.3$$

Now, since the joint probability $p_{y,z} = p_{z,y}$, we can also write

$$I(C;Y,Z) = I(C;Z,Y) = I(C;Y) + I(C;Z|Y). \qquad 2.5.4$$

But, from Definition 2.5.2,

$$I(C;Z|Y) = H(C|Y) - H(C|Y,Z).$$

Since $z = f(y)$, if we are given y, z is completely determined and $H(C|Y, Z) = H(C|Y)$, which means $I(C; Z|Y) = 0$. This is a consequence of the fact that c, y, and z form a Markov chain. On the other hand, $I(C; Y|Z)$ will *not* be zero unless the function $f(y)$ is one-to-one and onto. Since we are considering *any* possible function, we must allow any conceivable $f(y)$ so, combining Equations 2.5.3 and 2.5.4 gives us

$$I(C;Z) + I(C;Y|Z) = I(C;Y). \qquad 2.5.5$$

Since $I(C; Y|Z) \geq 0$,

$$I(C;Y) \geq I(C;Z). \quad 2.5.6$$

This is known as the *data-processing inequality*. It states that additional processing of the channel output y can at best result in no further loss of information and may even result in additional information loss. The latter can result if, for example, $f(y)$ is a many-to-one mapping from Y to Z such as $z = y^2$ or $z = |y|$. A very common example of this kind of information loss occurs when a real-valued (that is, an analog) channel output is quantized by an analog to digital converter (ADC). If the ADC has too few quantization levels, the resulting information loss can be quite severe. (In fact, this information loss is often called "quantizing noise".) Another source of information loss can be roundoff or truncation error during digital signal processing in a computer or microprocessor. Since digital processing of signals is cost effective and very commonplace, designers of these systems need to have an awareness of the possible impact of such design decisions as the word length of the digital signal processor or the number of bits of quantization in analog to digital converters have on the information content.

2.6 CONSTRAINED CHANNELS

2.6.1 Modulation Theory and Channel Constraints

So far, we have considered only memoryless channels corrupted by noise, which we modeled as discrete memoryless channels. While such channels appear in numerous applications, it is also very common to encounter channels which place *constraints* on what information sequences are allowable. Let us consider a more detailed model of the channel. Figure 2.6.1 depicts the major elements common to many practical communication channels.

The *coded* information sequence a_t is presented to a modulator. The modulator's task is to transform this sequence of symbols into a continuous-valued *waveform signal* designed to be compatible with the physical transmission medium represented by the bandlimited channel block. Examples of bandlimited channels include wireless channels, made up of antennas and electromagnetic waves propagated through the atmos-

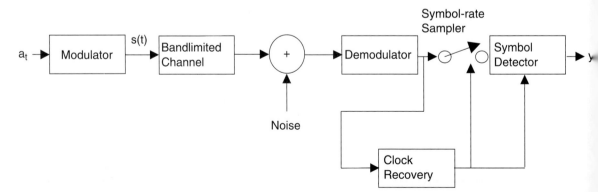

FIGURE 2.6.1 Block diagram of a typical digital communication system.

phere, wire channels, such as commercial telephone lines and cable TV cables, and magnetic recording heads and media, such as found in disk or tape drives.

During transmission, the information-bearing waveform signal is distorted by the channel and corrupted with noise. The received signal is presented first to a demodulator, which attempts to combat this distortion and minimize the effects of noise. The output of the demodulator is sampled, usually at the symbol rate, and a detector attempts to reconstruct the original symbol sequence by producing a discrete output y_t that represents the system's best estimate of the original coded sequence a_t.

In many systems, the rate at which symbols are sent and the precise times when it is best to sample the received waveform are not known with sufficient precision at the receiver and must be *recovered* from the received waveform itself. The clock recovery block performs this task. In many high-speed communication systems, the performance of the clock recovery block is crucial to obtaining acceptable reliability in recovering the transmitted information.

There are many technical details involved in the specification and design of the blocks shown in Figure 2.6.1, and we will not go into these details in this text. The theory and practice of the design of the constituent blocks in the figure is the province of *modulation theory* and many excellent books exist on this subject. (Many of them go under the title of "Digital Communications" or some variation of that title.) However, we *are* concerned in this text with the information theory aspects of this process. What are the issues?

One of them arises from the clock recovery block. As stated above, the task of this subsystem is the recovery of precise symbol-timing information. In order to accomplish this, the clock recovery block works with the information provided to it at the output of the demodulator. The received signal *must* contain adequate information within it for the clock recovery block to do its job. Up until now, we have only considered the information requirements necessary to permit reliable recovery of the original information sequence by the detection and decoding process. When the system must also recover timing information, it is clear that *additional* information devoted to this timing recovery task must be included in the transmitted sequence. Since we have already seen that the maximum information rate is limited by the channel capacity, the clear implication of the timing recovery requirement is that the information needed by the clock recovery block must be included *at the expense* of user information. This may (and often does) require that the sequence of transmitted symbols be *constrained* in such a way as to guarantee the presence of timing recovery information embedded within the transmitted coded sequence.

Another aspect arises from the type and severity of signal distortion imposed by the physical bandlimited channel. We can think of the physical channel as performing a kind of data processing on the information-bearing waveform presented to it by the modulator. In the previous section, we saw that any kind of data processing might result in information loss. A given physical communication channel may therefore place its own constraints on the allowable symbol sequences which it can process *without* information loss.

Modulation theory teaches that it is possible, and usually desirable, to model the communication channel as a cascade of a noise-free constrained channel and an unconstrained noisy channel. Figure 2.6.2 illustrates this concept. The noiseless channel model is frequently cast in a form called the *equivalent baseband channel,* which

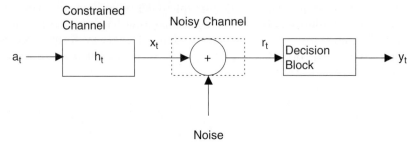

Figure 2.6.2 General model of a physical channel.

includes all effects associated with modulation, channel distortion, demodulation, and the clock recovery and detection processes. (We have been implicitly using such a model all along in this text, except that we have not hitherto considered any constraints on the input symbol sequence.) The process of constructing such a model also provides a *noise model,* which is used as the unconstrained noisy channel.

2.6.2 Linear and Time-Invariant Channels

The most common type of constrained channel model is the *linear time-invariant* or LTI channel. In this model, the input sequence consists of a finite set of symbols $a_t \in A$, where the symbol alphabet A is a subset of either the set of real numbers or the set of complex numbers (depending on whether the constrained channel model is described using real or complex numbers). The constrained channel output is a sequence of symbols $x_t \in X$, where channel output alphabet X is a finite subset of either the set of real numbers or the set of complex numbers (again depending on how the constrained channel model was constructed). The set X and the set A may contain different elements, i.e., A is not necessarily equal to X. The LTI channel is specified by a set of parameters h_t, which, collectively, are usually called the channel's *impulse response*. The number of parameters may be finite, in which case the channel is called a *finite impulse response,* or FIR, channel, or the set of parameters may have a countable infinity of terms, in which case the channel is called an *infinite impulse response,* or IIR, channel.

The channel's output sequence is related to the channel's input sequence by a convolution sum

$$x_t = \sum_{k=-\infty}^{\infty} h_k a_{t-k}, \qquad 2.6.1$$

where, in the case of an FIR channel, $h_k = 0$ for terms in Equation 2.6.1 outside the span of the channel's finite response length. If the channel is an FIR channel with $h_k = 0$ for $k < 0$ and $k > K$, for some integer K, the channel can be modeled using a Markov process. This is frequently done in many practical channel models.

The decision block is presented with a noisy signal

$$r_t = x_t + \eta_t, \qquad 2.6.2$$

where η_t is the noise added by the noisy channel. Often, the set of possible values of r_t has infinitely many elements. The decision block takes these inputs and produces an output symbol y_t drawn from a finite alphabet Y with $|Y| \geq |A|$. If $Y = A$ so that y_t is an estimate of a transmitted symbol, a_t, the decision block is said to have made a hard decision. Otherwise, if the alphabet $Y \neq A$ and $|Y| > |A|$, the decision block makes a soft decision, and final decoding and estimation of the transmitted a_t is made by a decoder as shown in Figure 1.1.1 in Chapter 1.

EXAMPLE 2.6.1

Let A be a memoryless binary source with equiprobable symbols $A = \{-1, +1\}$, and let the band-limited channel have impulse response $\{h_0 = 1, h_1 = 0, h_2 = -1\}$. Calculate the steady-state entropy of the channel's output symbols and the entropy rate of the sequence x_t.

Solution: Define the *state* of the channel at time t as $S_t = \langle a_{t-1}, a_{t-2} \rangle$. Using Equation 2.6.1, the channel can be represented as a Markov process with the following state diagram:

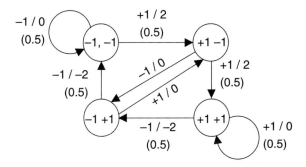

All transition probabilities, shown in parentheses, are equal to 0.5 since the source is memoryless with equiprobable symbols. The arrows in this diagram are labeled as a_t/x_t. From Equation 2.6.1, we find $X = \{-2, 0, +2\}$. Label the states with the subscripts 0 for $(-1\ -1)$, 1 for $(1, -1)$, 2 for $(-1, 1)$ and 3 for $(1, 1)$, respectively. The state probabilities are then described by Equation 2.4.4 as

$$\Pi_{t+1} = \begin{bmatrix} 0.5 & 0 & 0.5 & 0 \\ 0.5 & 0 & 0.5 & 0 \\ 0 & 0.5 & 0 & 0.5 \\ 0 & 0.5 & 0 & 0.5 \end{bmatrix} \Pi_t,$$

from which we find the steady state probabilities are equiprobable $\pi_i = 0.25$ with for each state.
The output symbol probabilities are related to the state probabilities by

$$\Pr(x_t) = \sum_{i=0}^{3} \Pr(x_t | S_i) \pi_i.$$

In the steady state, this gives us

$$\Pr(-2) = 0.25 \cdot 0.5 + 0.25 \cdot 0.5 = 0.25,$$
$$\Pr(0) = 4 \cdot 0.25 \cdot 0.5 = 0.5,$$
$$\Pr(+2) = 2 \cdot 0.25 \cdot 0.5 = 0.25.$$

The steady-state entropy of the channel output is therefore

$$H(X) = \sum_{x \in X} \Pr(x) \log_2(1/\Pr(x)) = .5(1) + 2 \cdot 0.25(2) = 1.5.$$

To find the entropy rate, we apply Equation 2.4.6. This gives us

$$R = \sum_{i=0}^{3} \pi_i \sum_{x \in X} \Pr(x|S_i) \log_2\left(\frac{1}{\Pr(x|S_i)}\right) = 0.25(2 \cdot 0.5) \log_2(2) \cdot 4 = 1,$$

which equals the source entropy. Note that the entropy rate is *not* equal to the steady-state entropy of the channel's output symbols.

While the channel in Example 2.6.1 is lossless, the sequences it produces do *not* carry sufficient information to permit clock recovery for arbitrary input sequences. For example, notice that a long input sequence of "−1", a long sequence of "+1", and a long sequence of alternating symbols "−1, +1" or "+1, −1" all produce a long output sequence of zeroes at the channel output. If these sequences are permitted to occur, timing recovery methods at the receiver may fail with severe effect on the reliability of the communication system.

2.7 AUTOCORRELATION AND POWER SPECTRUM OF SEQUENCES

2.7.1 Statistics of Time Sequences

The constraints placed by a channel on allowable input sequences are often easier to understand using what is known as *frequency domain analysis*. We saw in the last example that a channel may be information lossless, but still have undesirable characteristics that can impact the ability of the receiver to correctly process the received signal. In Example 2.6.1, the undesirable characteristic arose from the fact that certain input sequences produce an all-zeroes channel output. In other channels, there may be severe distortion of other kinds that produce unacceptably high errors in the outputs of the decision block.

In an LTI channel, these effects are often due to how successive symbols x_t are correlated with each other. There are two primary tools useful for analysis of correlation in a sequence. These are the *autocorrelation function* of a sequence and the *power spectrum* of the sequence. In this section, we will define and explore these useful tools.

We begin with the notion of the *statistical expected value* of a function $f(x)$ of a random variable x having probability distribution p_x. For the discrete-time sequences with which we are concerned, the expected value of a function of x is defined by

$$E[f(x)] \equiv \sum_{x \in X} f(x) p_x. \qquad 2.7.1$$

Section 2.7 Autocorrelation and Power Spectrum of Sequences

Some of the more important expectations are as follows:

mean value:
$$\langle x \rangle \equiv E[x] = \sum_{x \in X} x\, p_x, \qquad 2.7.2$$

mean-squared value:
$$\langle x^2 \rangle \equiv E[x^2] = \sum_{x \in X} x^2\, p_x, \qquad 2.7.3$$

variance:
$$\langle (x - \langle x \rangle)^2 \rangle \equiv E[(x - \langle x \rangle)^2] = \sum_{x \in X} (x - \langle x \rangle)^2 p_x$$
$$= E[x^2] - (E[x])^2, \qquad 2.7.4$$

entropy:
$$H(X) \equiv E[\log_2(1/p_x)] = \sum_{x \in X} p_x \log_2(1/p_x). \qquad 2.7.5$$

Did Equation 2.7.5 surprise you? We've actually been using statistical expectation (without calling it that) all along!

We are presently concerned with the behavior of sequences having probabilities that do not change with time or, at least, that achieve a steady-state which does not change with time. Consider two symbols, x_t, x_{t+m}, in a symbol sequence. These two symbols have some joint probability $p_{x_t, x_{t+m}}$. Since we are assuming that probabilities do not change with time, a condition known as stationarity, this joint probability is not a function of t, but only a function of the difference, m, in the time index of the two symbols. The *autocorrelation function* of such a system is a measure of how a symbol at time t is correlated with another symbol at time $t + m$. It is defined as

$$\phi_{xx}(m) \equiv E[x_t \cdot x_{t+m}] = \sum_{x_t \in X} \sum_{x_{t+m} \in X} x_t x_{t+m} p_{x_t, x_{t+m}}. \qquad 2.7.6$$

The double-summation is necessary because we are dealing with a function of two variables.

Given the joint probability, calculation of the autocorrelation is straightforward. Unfortunately, in the most general kind of case, the joint probability may be very difficult to obtain. Fortunately, we are not particularly interested in the most general kinds of cases but, rather, with the specific case of a sequence generated by our communication system. For the great majority of situations we shall face in this text, the sequences that interest us are either uncorrelated in time *or* are correlated by the mechanism of the noiseless constrained channel of the previous section. (The more formidable cases which occasionally do pop up in the real world are for the fellow who is a great expert in the field; we will not begrudge this here, for, after all, even gurus need to eat, and as the attending physician once said to the intern, "Learn all you can about appendicitis and leave the dum-dum fever to the experts.")

For that most important of special cases, the uncorrelated sequence, the joint probability is given by

$$p_{x_t, x_{t+m}} = \begin{cases} p_{x_t} \delta(x_t - x_{t+m}), & m = 0 \\ p_{x_t} p_{x_{t+m}}, & m \neq 0 \end{cases}. \qquad 2.7.7$$

where $\delta(x)$ is the Kronecker delta function. The autocorrelation of an uncorrelated sequence is therefore

$$\phi_{xx}(m) = \begin{cases} \langle x_t \rangle^2, & m \neq 0 \\ E[x_t x_{t+0}] = \langle x_t^2 \rangle, & m = 0 \end{cases}. \qquad 2.7.8$$

A minor inspection of the various definitions we have presented should tell us that it is always the case that $\phi_{xx}(0) = \langle x_t^2 \rangle$, and since we're dealing with stationary processes, ϕ is not a function of t.

Now consider the case where x_t is the output of a linear, time-invariant noiseless channel driven by an uncorrelated input sequence with zero-mean. x_t is given by Equation 2.6.1, so

$$\phi_{xx}(m) = E\left[\sum_{k=-\infty}^{\infty} h_k a_{t-k} \sum_{\ell=-\infty}^{\infty} h_\ell a_{t+m-\ell}\right].$$

Rearranging things a bit and recognizing that the h terms are constants, so that we can pull the expectation inside the summations, we have

$$\phi_{xx}(m) = \sum_{k=-\infty}^{\infty} \sum_{\ell=-\infty}^{\infty} h_k h_\ell E[a_{t-k} a_{t+m-\ell}].$$

Since the a terms are uncorrelated and have a mean of zero,

$$E[a_{t-k} a_{t+m-\ell}] = \begin{cases} 0, & \ell \neq m + k \\ \langle a^2 \rangle, & \ell = m + k \end{cases}.$$

Therefore,

$$\phi_{xx}(m) = \langle a^2 \rangle \sum_{k=-\infty}^{\infty} h_k h_{k+m}. \qquad 2.7.9$$

The autocorrelation for this special case is therefore a function only of the mean-squared value of the input signal a_t and the coefficients in the LTI channel model. Note that the autocorrelation is an even function, i.e., $\phi_{xx}(-m) = \phi_{xx}(m)$. The proof of this is quite straightforward and is left as a homework exercise.

2.7.2 The Power Spectrum

Useful as the autocorrelation function is, it is frequently even more useful to look at it in the frequency domain (which we are about to define). When we do so, we are looking at what is known as the *power spectrum* of x_t. The power spectrum of a sequence is defined as

$$\Phi_{xx}(\theta) \equiv \sum_{m=-\infty}^{\infty} \phi_{xx}(m) e^{-im\theta}, \qquad 2.7.10$$

where $i = \sqrt{-1}$ and θ is called the *digital frequency* and takes on values over the range from $-\pi$ to π. Equation 2.7.10 is derived from the theory of Fourier analysis, and much of the terminology used (digital frequency, spectrum, frequency domain, and so forth) arises from this theory. For example, suppose we have a discrete sequence composed of samples of a cosine function that is periodic after M samples

$$f(m) = \cos\left(\frac{2\pi}{M} m\right) = \cos(\theta m), \quad \theta \equiv \frac{2\pi}{M}.$$

Section 2.7 Autocorrelation and Power Spectrum of Sequences

By analogy with the continuous-time function $\cos(2\pi t/T)$, we see that θ can be interpreted as 2π divided by the period of the signal, and therefore, plays the role of frequency. The transformation of a discrete-time sequence f_m

$$F(\theta) = \sum_{m=-\infty}^{\infty} f_m e^{-i\theta m} \qquad 2.7.11$$

is called the *discrete Fourier transform*.

Since the autocorrelation function is an even function of m, Equation 2.7.10 can also be expressed in an alternate form. Since

$$e^{-i\theta m} + e^{-i\theta(-m)} = 2\cos(\theta m),$$

we have

$$\Phi_{xx}(\theta) = \phi_{xx}(0) + 2\sum_{m=1}^{\infty} \phi_{xx}(m)\cos(\theta m). \qquad 2.7.12$$

EXAMPLE 2.7.1

Suppose a sequence x_t has autocorrelation function

$$\phi_{xx}(m) = \alpha^{|m|}\cos(2\pi m/M),$$

where $0 < \alpha < 1$. Calculate the power spectrum of x_t.

Solution: Using equation 2.7.12, we have

$$\Phi_{xx}(\theta) = 1 + 2\sum_{m=1}^{\infty} \alpha^m \cos(2\pi m/M)\cos(\theta m).$$

Using the trigonometric identity $\cos(A)\cos(B) = \frac{1}{2}[\cos(A-B) + \cos(A+B)]$, we have

$$\Phi_{xx}(\theta) = 1 + \sum_{m=1}^{\infty} \alpha^m \cos\left(\left(\frac{2\pi}{M}-\theta\right)m\right) + \sum_{m=1}^{\infty} \alpha^m \cos\left(\left(\frac{2\pi}{M}+\theta\right)m\right).$$

These summations look challenging, but they are standard sums, which may be looked up in a table of infinite sums. Defining

$$\omega_1 \equiv \frac{2\pi}{M} - \theta, \quad \omega_2 \equiv \frac{2\pi}{M} + \theta,$$

and evaluating the expression above, we get

$$\Phi_{xx}(\theta) = 1 + \frac{\alpha(\cos\omega_1 - \alpha)}{1 - 2\alpha\cos\omega_1 + \alpha^2} + \frac{\alpha(\cos\omega_2 - \alpha)}{1 - 2\alpha\cos\omega_2 + \alpha^2}.$$

From this expression, we have $\Phi_{xx}(-\theta) = \Phi_{xx}(\theta)$, so it is sufficient to examine this expression over the range $0 \leq \theta \leq \pi$. A graph of this function for $\alpha = 0.6$ and $M = 5$ is as follows:

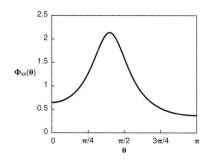

For $M = 5$, the autocorrelation function in this example has a sinusoidal term at digital frequency $\theta_0 = 2\pi/5 = 1.2566$. Note that this corresponds to the peak in the plot of $\Phi_{xx}(\theta)$. The power spectrum of a sequence provides us with information about periodicities in the autocorrelation of the sequence. It tells us how the "energy" per unit time of a sequence is distributed in the "frequency domain." The concept of the "energy" or "power" of a signal is something which arises from signal processing theory; those of us already familiar with this notion find the previous statement meaningful and comfortable; for those of us who are unfamiliar with Fourier analysis and signal processing theory, think of the power spectrum as something which gives us an indication of whether the sequence x_t repeats itself over certain periods more often than over other periods. The relative values of $\Phi_{xx}(\theta)$ indicate the relative levels of correlation arising from strings of symbols which repeat with frequencies θ.

EXAMPLE 2.7.2

Find the power spectrum of a binary memoryless source with alphabet $A = \{-1, +1\}$.

Solution: Since the source is memoryless, a_t is uncorrelated, and

$$\phi_{aa}(m) = \begin{cases} 1, & m = 0 \\ 0, & m \neq 0 \end{cases},$$

and $\Phi_{aa}(\theta) = 1$. The power spectrum is uniform over all values of θ. This is often called a "white" spectrum, a reference to the fact that white light contains all the different frequencies (colors) of light.

EXAMPLE 2.7.3

Find the power spectrum for the system in Example 2.6.1.

Solution: Using the h_k for this example and the results of Example 2.7.2, Equation 2.7.9 gives us

$$\phi_{xx}(m) = \begin{cases} 2, & m = 0 \\ -1, & |m| = 2 \\ 0, & \text{otherwise} \end{cases}.$$

From Equation 2.7.12,
$$\Phi_{xx}(\theta) = 2 - 2\cos(2\theta) = 4\sin^2(\theta).$$

A plot of this response is as follows:

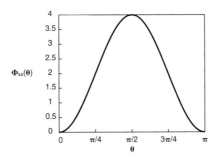

Note that $\Phi_{xx}(0) = \Phi_{xx}(\pi) = 0$. This means that constant sequences, such as $-1,-1,-1,\ldots$ or $+1,+1,+1,\ldots$, and alternating sequences, such as $+1,-1,+1,-1,\ldots$, emitted from the memoryless source are *not* transmitted by the channel. We saw this behavior earlier in Example 2.6.1.

The *transfer function* of a constrained LTI channel with parameters h_t is defined as

$$T(\theta) \equiv \sum_{t=-\infty}^{\infty} h_t e^{-i\theta t}. \qquad 2.7.13$$

There is an interesting relationship between the transfer function of a channel and the power spectrum of its output. Earlier, we saw that

$$\phi_{xx}(m) = \sum_{k=-\infty}^{\infty}\sum_{\ell=-\infty}^{\infty} h_k h_\ell E[a_{t-k}a_{t+m-\ell}] = \sum_k \sum_\ell h_k h_\ell \phi_{aa}(m+k-\ell). \qquad 2.7.14$$

Applying the definition of power spectrum,

$$\Phi_{xx}(\theta) = \sum_{m=-\infty}^{\infty} \sum_k \sum_\ell h_k h_\ell \phi_{aa}(m+k-\ell) e^{-i\theta m}.$$

Making the substitution $n = m + k - \ell$ and rearranging terms gives us

$$\Phi_{xx}(\theta) = \sum_n \phi_{aa}(n) e^{-i\theta n} \sum_k h_k e^{+i\theta k} \sum_\ell h_\ell e^{-i\theta \ell} = \Phi_{aa}(\theta)|T(\theta)|^2. \qquad 2.7.15$$

Equation 2.7.15 tells us that the power spectrum of the output of an LTI channel driven by a source with power spectrum $\Phi_{aa}(\theta)$ is simply the product of the input power spectrum with the squared-magnitude of the channel's transfer function.

This relationship allows us to break the calculation of power spectra into two parts. We may analyze the power spectrum of the coded input sequence, the transfer function of the channel, and then multiply the results together, as in Equation 2.7.15. This can be useful, for instance, in analyzing the suitability of a particular coding method for use in a given channel. For example, if the power spectrum of a coded input sequence contained a very high peak at frequency θ_0, but $|T(\theta_0)|^2 = 0$, we might (correctly!) expect that the system would have very poor performance, since the channel would not pass information associated with frequency θ_0, and a great deal of the input information would be associated with this frequency because of the high peak in its power spectrum at this frequency.

This assertion as presented in the previous paragraph is, of course, non-rigorous. However, it can be made rigorous (and has been!) through a somewhat involved mathematical proof given by Shannon. It would be premature to go into those details at this point, because at this stage of our knowledge, we (the new students of the subject) lack some key mathematical tools needed to carry out the proof. However, the principle is easy to state: The power spectrum of coded input should match the channel's transfer function. In particular, it is known (Karabed and Siegel) that if the channel has nulls in its response, *i.e.,* frequencies where $|T(\theta_0)|^2 = 0$, the coded input should have a power spectrum that also has nulls at these frequencies if the best overall system performance is to be obtained.

We need to say a word of reconciliation about this principle. In Example 2.6.1 there appeared the statement, "The channel is information lossless." That seems to flatly contradict the statement in the previous paragraph. However, it does *not,* because the statement about the channel being lossless only applied to the *noiseless constrained channel* and *not* the complete channel, which included a noisy part. (Information theory would be a trivial and uninteresting subject if we only had to deal with noiseless communication channels.)

2.8 DATA TRANSLATION CODES

2.8.1 Constraints on Data Sequences

The constraints placed on the transmitted coded signal by the channel can often be expressed in terms of forbidden sequences of input symbols. For example, the channel described in the previous two sections can not pass very long runs of "−1" or " +1" inputs, so a reasonable constraint on the input sequence is to forbid input symbol sequences containing more than some maximum number of identical symbols in a row. Likewise, the channel can not pass a long string of alternating symbols, such as +1, −1, +1, −1 ..., and so it might be wise to ensure that no such string of symbols is permitted beyond some maximum length.

Data translation codes are codes that map the original sequence of symbols from the information source into a new sequence of symbols such that certain coded sequences are forbidden and do not occur, no matter what symbol sequence is emitted by the information source. In this text, we will concern ourselves only with binary information sources and binary data translation codes. Because the purpose of such a code is to match the transmitted bit sequence to the constraints placed on sequences by the channel, such codes are often called modulation codes, line codes, run-length-limited codes, or recording codes (in the case of magnetic or optical recording).

The general idea is shown in Figure 2.8.1. An information source C with a binary alphabet $\{0, 1\}$ emits a sequence of symbols. These are encoded by the data translation encoder (DTC) to produce a sequence of output symbols $\beta_t \in \{0, 1\}$. One or more symbols β_t are produced for each symbol emitted by the information source in a manner similar to the encoding technique we looked at earlier for Huffman codes. The β_t are mapped by a digital modulator one at a time to a set of symbols $a_t \in \{-1, +1\}$ before being presented to the channel. This is done to avoid transmitting a large mean value $\langle a_t \rangle$ through the channel. (The mean value of a signal does not change, and therefore contributes zero information to the transmitted sequence; however, it does require *power* to transmit the mean value, and since no information is transmitted by the mean, this power would be wasted.) The mapping $\beta_t \to a_t$ is simply $0 \to -1, 1 \to +1$.

EXAMPLE 2.8.1

Sometimes the purpose of the DTC is to make implementation of the symbol detector shown in Figure 2.6.1 simple. Consider the system in Example 2.6.1. Examination of the state diagram in this example shows that the channel output is nonzero only if the current input bit differs from the rightmost state bit. Since the state bits are merely past input bits, we can simplify the symbol by coding the sequence of information bits using what is known as a Non-Return to Zero-Inverted or NRZI code. Define the *encoder state* by the vector $S_t = [s_0 \ s_1]^T$, where $s_0 = \beta_{t-1}$ and $s_1 = \beta_{t-2} = s_0(t-1)$, where the output of the encoder is

$$\beta_t = c_t \oplus s_1,$$

c_t is the bit emitted by the information source at time t, and $\oplus$ is the exclusive-or operator defined by

$$0 \oplus 0 = 1 \oplus 1 = 0,$$
$$0 \oplus 1 = 1 \oplus 0 = 1.$$

The encoder's output a_t is now "+1" if c_t differs from the value of β_{t-2} and "−1" otherwise. Assume we initialize the encoder so that $S_{t=-1} = [0 \ 0]^T$ prior to the start of the data sequence from C. Then the channel output x_t will be either +2 or −2 if $c_t = 1$ and 0 otherwise. The symbol detector can decode the received signal as $c_t = 0$ if its input is $|r_t| < 1$ and $c_t = 1$ if $|r_t| > 1$.

Example 2.8.1 illustrates that even a simple DTC such as NRZI requires the encoder to remember its past inputs. The operation of the encoder can be described by a state diagram in much the same way that we have used state diagrams to describe Markov processes.

Our NRZI encoder example puts out one bit for each input bit and is information lossless. Therefore, if we think about an encoder as a funny kind of "channel," its channel capacity would be 1 bit per "channel" use. Its code rate, which is defined as the average number of information bits inputted per code bits outputted, would be 1. In

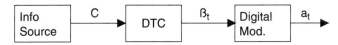

Figure 2.8.1 Encoding for a data translation code in a digital transmitter.

point of fact, any encoder can be viewed as if it were a channel and we can use our tools of mutual information, channel capacity, and so on to describe it. Encoders, in fact, are often referred to as noiseless channels in some of the coding literature.

A DTC used to constrain the sequence of coded bits to fit a binary channel will rarely have a capacity of 1 bit per channel use. This is because its entire purpose is to prevent certain symbol sequences from occurring. Suppose the information source emits a sequence of k bits, and assume this source is memoryless. The total number of possible k bit sequences it could emit is clearly 2^k, and the encoder must be ready to supply a coded output sequence for each of them. However, since some sequences are forbidden, the coded sequence must consist of some $n > k$ bits. If, on the average, k source bits results in n code bits, the average *code rate* of the DTC is $R = k/n$. Further, since the DTC uses the physical channel n times to transmit k information bits, the channel capacity *of the DTC* must be less than one bit per channel use, assuming binary transmission. One of our tasks, then, must be to find the capacity of a data translation code.

2.8.2 State Space and Trellis Descriptions of Codes

We begin by looking at how to mathematically describe the operation of the code. In our NRZI example, we defined the *state* of the code in terms of the previously transmitted bits β_t. We will continue with this strategy since, if the code is to prevent certain bit sequences, it must retain the knowledge of what sequence it has sent up to this moment in time. Assume that the code allows K possible bit sequences. The encoder must then have K possible states (one for each legal string of outputs). An example will help illustrate this point.

EXAMPLE 2.8.2

Define the required states and describe the state transitions for an encoder which forbids any sequence containing a run of more than three repetitions of the same symbol.

Solution: Since the constraint being enforced by this code is to forbid runs of more than three of the same symbol, let each state represent the most recent run of symbols of the same kind. There are six such strings possible, and each one will be assigned a state. The six possibilities are

$$S_0 = 111,$$
$$S_1 = 11,$$
$$S_2 = 1,$$
$$S_3 = 0,$$
$$S_4 = 00,$$
$$S_5 = 000.$$

Suppose we are in state S_2 and the information source emits a "0". This breaks the previous string of repetitions, and therefore, we can transmit the "0", since it does not violate any constraint. Since it begins a new string, the state of the encoder changes to S_3. On the other hand, suppose the information source had emitted a "1". This will take us from a previous string "1" to a string "1 1", which is also a legal string. Therefore, the encoder goes ahead and transmits the source's "1" bit and changes to state S_1.

Now suppose we are in state S_1 and the information source emits another "1". This produces a string of "1 1 1", which is legal, and so we transmit the source's "1". The new state of the encoder is S_0, which represents the maximum run of consecutive "1" bits. The next bit we transmit *must* be a "0". If the information source next emits a "0", all is well. We transmit that "0" and go to state S_3. However, if the source emits a "1", we must *insert* a "0" codebit prior to transmitting the source's information bit. Our insertion of the "0" takes us to state S_3, the source's "1" bit is then transmitted, and we go to state S_2.

As far as the output of the encoder is concerned, we can describe all possible sequences of bits that could be transmitted by the encoder using a state diagram. An interesting and often useful variation on this theme is to split the state diagram into two columns. The first column depicts the "present state," and the second column depicts the "next state." This is equivalent to adding a time axis to the state diagram. Such a diagram is called a *trellis diagram*. The trellis for this encoder is shown below.

A trellis diagram shows the allowable state transitions from the present state to the next state. The encoder's outputs associated with each transition are listed next to the trellis nodes. These are associated with specific branches of the trellis by reading the outputs from left to right and pairing them with the order in which branches are encountered when read clockwise from the present state node. We shall have numerous occasions to employ trellis diagrams in this text.

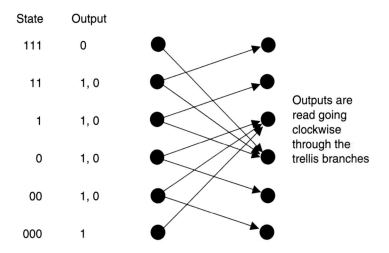

Another useful way of describing the operation of the encoder is by using a state transition matrix (sometimes called a connection matrix). The state transition matrix B is a square matrix having K columns if the encoder has K states. Each element $b_{i,j}$ in the matrix gives the number of trellis branches that run from state S_i to state S_j. The state transition matrix for the encoder in this example can be read from the trellis diagram with the result

$$B = \begin{bmatrix} 0 & 0 & 0 & 1 & 0 & 0 \\ 1 & 0 & 0 & 1 & 0 & 0 \\ 0 & 1 & 0 & 1 & 0 & 0 \\ 0 & 0 & 1 & 0 & 1 & 0 \\ 0 & 0 & 1 & 0 & 0 & 1 \\ 0 & 0 & 1 & 0 & 0 & 0 \end{bmatrix}.$$

The state transition matrix description illustrated in the previous example is a very useful tool in the analysis of the encoder. One of its uses is to determine the number of possible sequences that can reach some state S_j at time t given some initial state S_i at time 0. Let $N(t)$ be a K-element row vector $N(t) = [n_0(t) n_1(t) \cdots n_{K-1}(t)]$, where $n_j(t)$ is the number of paths through the trellis leading to state S_j at time t. At $t=0$, set $n_i(0) = 1$ and all other elements equal to zero. At time t, the number of paths leading to each state is given by

$$N(t) = N(t-1)B. \qquad 2.8.1$$

This is a homogeneous difference equation. By induction, it is easily seen that

$$N(t) = N(0)B^t. \qquad 2.8.2$$

The total number of possible coded sequences of length t is therefore given by

$$n_t = \sum_{k=0}^{K-1} n_k(t). \qquad 2.8.3$$

Equations 2.8.2 and 2.8.3 provide us with an upper bound on the number of possible code words of length t that can be generated by the DTC. The reason this is an upper bound rather than an equality has to do with the details of implementing the encoder. Consider Example 2.8.2. In the description of this encoder's operation, most of the bits from the information source were directly transmitted as code bits. However, in the case where we were in state S_0 and the information source emitted a "1", the encoder was required to generate two bits, "0 1", to satisfy the code constraint. This action eliminates the possibility for this particular source sequence of passing through the set of states $S_0 \to S_3 \to S_4$, even though the trellis and the state matrix both permit this transition. Note also that, had the information source emitted a "0" instead of a "1", this state sequence *would* have been possible.

EXAMPLE 2.8.3

For the code in Example 2.8.2, find the maximum number of possible sequences of length 3 and the initial states which achieve this maximum.

Solution: By direct calculation

$$B^3 = \begin{bmatrix} 0 & 1 & 1 & 1 & 0 & 1 \\ 0 & 1 & 2 & 1 & 1 & 1 \\ 0 & 1 & 2 & 2 & 1 & 1 \\ 1 & 1 & 2 & 2 & 1 & 0 \\ 1 & 1 & 1 & 2 & 1 & 0 \\ 1 & 0 & 1 & 1 & 1 & 0 \end{bmatrix}.$$

The initial state that achieves the largest possible number of sequences will be the one corresponding to the column of $N(0)$ that achieves the largest row-sum in B^3. By inspection of the matrix above, we see that the largest row-sums occur in the rows corresponding to S_2 and S_3 and $n_t = 7$ in both cases.

2.8.3 Capacity of a Data Translation Code

It is clear from the previous example that, for long sequences the direct calculation method using Equations 2.8.2 and 2.8.3 quickly becomes somewhat impractical. However, we are very interested in being able to come up with answers such as in Example 2.8.3 because, as we will see, these answers determine the capacity of the code. A more numerically efficient procedure is required, and that is what we will go after next. We begin with some important results from the theory of linear algebra.

Since the state transition matrix B is a square matrix, it has K eigenvalues and K linearly independent eigenvectors given by the solution to the eigenvalue equation

$$BV_k = \lambda_k V_k,$$

where V_k is the eigenvector and λ_k its associated eigenvalue. Let U be the square matrix whose columns are the K linearly independent eigenvectors. Since the rank of this matrix is K, its inverse U^{-1} exists. Since the eigenvectors are determined only to an arbitrary multiplicative constant, we may assume that the eigenvectors are orthonormal, i.e.,

$$V_k^T V_j = \begin{cases} 1, & k = j \\ 0, & k \neq j \end{cases}.$$

If all of the eigenvalues of B are distinct, then by a well known result from linear algebra, B can be expressed as

$$B = U^{-1} \begin{bmatrix} \lambda_0 & 0 & \cdots & 0 \\ 0 & \lambda_1 & & 0 \\ \vdots & & \ddots & 0 \\ 0 & \cdots & 0 & \lambda_{K-1} \end{bmatrix} U = U^{-1} \Lambda U.$$

Therefore,

$$B^2 = U^{-1} \Lambda U U^{-1} \Lambda U = U^{-1} \Lambda^2 U,$$

and by induction

$$B^t = U^{-1} \Lambda^t U. \qquad 2.8.4$$

From this, Equation 2.8.2 becomes

$$N(t) = N(0) U^{-1} \Lambda^t U. \qquad 2.8.5$$

Since the elements of U are constants that are independent of t, Equation 2.8.3 may be reexpressed as

$$n_t = \sum_{k=0}^{K-1} n_k \lambda_k^t. \qquad 2.8.6$$

Now let λ_s be the largest real and non-negative eigenvalue of B. As t becomes large, this term will come to dominate all others in Equation 2.8.6, since λ_s^t becomes large more rapidly than any other term. Further, we know that its coefficient n_s in Equation 2.8.6 is non-zero since n_s is independent of t, and a zero value of n_s implies B

does not depend on this eigenvalue (which would be a contradiction). Hence, for large t,

$$\lim_{t \to \infty} n_t \to n_s \lambda_s^t. \qquad 2.8.9$$

Hence, the number of possible code words having t code bits is proportional to the largest real, non-negative eigenvalue of B in the limit where t becomes large. The value of λ_s may be readily obtained from a number of available mathematics packages that run on the personal computer.

The determination of n_s is less straightforward, but we shall now show that we do not need to know its value. We are concerned with the channel capacity of the code, not the precise number of code words of some given length. Since we know the encoder is lossless, its channel capacity is found by recognizing that in t channel uses, one of n_t possible information sequences are transmitted, and therefore, $\log_2(n_t)$ information bits are transmitted. The channel capacity of the code is therefore

$$C_c = \lim_{t \to \infty} \frac{\log_2(n_t)}{t} = \lim_{t \to \infty} \frac{\log_2(n_s)}{t} + \frac{\log_2(\lambda_s^t)}{t} = \log_2(\lambda_s). \qquad 2.8.10$$

Equation 2.8.10 is the result we are after. The channel capacity of the DTC is simply the logarithm of the largest real, positive eigenvalue of the state transition matrix B. This is a delightfully simple result and is easily obtainable from the state transition matrix.

EXAMPLE 2.8.4

Find the channel capacity of the code in Example 2.8.2.

Solution: Using MATHCAD™, the largest real and positive eigenvalue of B is $\lambda_s = 1.839$. Therefore, $C_c = \log_2(1.839) = 0.879$ bits per channel use. The MATHCAD™ program for this calculation is

$$B := \begin{bmatrix} 0 & 0 & 0 & 1 & 0 & 0 \\ 1 & 0 & 0 & 1 & 0 & 0 \\ 0 & 1 & 0 & 1 & 0 & 0 \\ 0 & 0 & 1 & 0 & 1 & 0 \\ 0 & 0 & 1 & 0 & 0 & 1 \\ 0 & 0 & 1 & 0 & 0 & 0 \end{bmatrix} \quad V := \text{eigenvals}(B) \quad V = \begin{bmatrix} 1.839 \\ -1i \\ 1i \\ -1 \\ -0.42 - 0.606i \\ -0.42 + 0.606i \end{bmatrix}$$

$$\frac{\log(V_0)}{\log(2)} = 0.879$$

We have derived Equation 2.8.10 assuming the eigenvalues of B are distinct. The case where some of the eigenvalues are repeated is only slightly more involved since in this case B must be expressed in what is known as the Jordan (or "block") form. However, the end result is unchanged by this, and Equation 2.8.10 is the general solution regardless of whether or not some eigenvalues repeat.

2.9 (d,k) SEQUENCES

2.9.1 Run-length-limited Codes and Maxentropic Sequences

In many cases, constraints imposed by the channel can be satisfied using run-length-limited (RLL) codes. It is often the case that RLL sequences are most easily explained and understood in terms of (d, k) sequences. A binary (d, k) sequence is a sequence of symbols drawn from $\{0, 1\}$ that simultaneously satisfy two conditions:

1. A d-constraint that two "1" bits are separated by a run of d or more "0" bits;
2. A k-constraint that no more than a maximum of k consecutive "0" bits may occur.

In general, a (d, k) sequence is not applied directly to the channel. Instead, the (d, k) sequence is processed by an NRZI encoder just prior to digital modulation and transmission. If the current bit in the (d, k) sequence is δ_t, then the output of the NRZI encoder becomes

$$\beta_t = \delta_t \oplus \beta_{t-1}.$$

The digital modulator then translates $\beta_t = 0 \to -1; \beta_t = 1 \to +1$ for transmission through the channel.

The state transition diagram for a (d, k) sequence is shown in Figure 2.9.1. The state names are assigned sequentially from 1 to $k+1$, and the coded output bit from the (d, k) sequence is shown next to the transition arrow. The state transition matrix with elements $b_{i,j}$, denoting a transition from state i to state j, has a particularly simple form for a (d, k) sequence. The entries are

$$b_{i,1} = 1, \quad i \geq d + 1,$$
$$b_{i,j} = 1, \quad j = i + 1,$$
$$b_{i,j} = 0, \quad otherwise.$$

For example, a $(1,3)$ sequence has a state transition matrix

$$B = \begin{bmatrix} 0 & 1 & 0 & 0 \\ 1 & 0 & 1 & 0 \\ 1 & 0 & 0 & 1 \\ 1 & 0 & 0 & 0 \end{bmatrix}.$$

Since the (d, k) sequence is merely a special case of the previous section, the capacity of the sequence is given by

$$C_c = \log_2(\lambda),$$

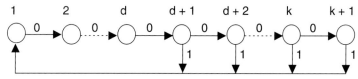

Figure 2.9.1 Canonical state diagram for a (d, k) sequence.

TABLE 2.9.1

k	d = 0	d = 1	d = 2	d = 3	d = 4	d = 5	d = 6
2	.8791	.4057					
3	.9468	.5515	.2878				
4	.9752	.6174	.4057	.2232			
5	.9881	.6509	.4650	.3218	.1823		
6	.9942	.6690	.4979	.3746	.2669	.1542	
7	.9971	.6793	.5174	.4057	.3142	.2281	.1335
8	.9986	.6853	.5293	.4251	.3432	.2709	.1993
9	.9993	.6888	.5369	.4376	.3620	.2979	.2382
10	.9996	.6909	.5418	.4460	.3746	.3158	.2633
11	.9998	.6922	.5450	.4516	.3833	.3282	.2804

where λ is the largest positive real eigenvalue of B. Owing to the simple form of B for a (d, k) sequence, this eigenvalue is the largest real root of the characteristic equation

$$\lambda^{k+2} - \lambda^{k+1} - \lambda^{k-d+1} + 1 = 0. \qquad 2.9.1$$

Capacities for several combinations of (d, k) parameters are shown in Table 2.9.1.

Examining the values given in Table 2.9.1, we see that capacity at a given k is maximized for $d = 0$. Capacity for given d is maximized as $k \to \infty$. However, for $d > 0$, an asymptote is reached and the $k \to \infty$ values are within a few percent of the values in the last row of the table. It is easy to see why this behavior results, since increasing d decreases the possible number of paths through the trellis of the sequence, while increasing k increases the number of possible paths through the trellis of the sequence.

The purpose of the k constraint is to guarantee the existence of a maximum number of symbols that can be transmitted before a change in the level of a_t will occur. This provides information that the receiver can use to recover the symbol clock. The purpose of the d constraint is usually to control the total amount of intersymbol interference between adjacent channel symbols, since the nonzero d constraint guarantees a minimum spacing between changes in the level of the channel symbol.

EXAMPLE 2.9.1

Implement the code of Example 2.8.2 as a (d, k) code.

Solution: The code constraint calls for a maximum symbol run of three and a minimum symbol run of one. This can be implemented as a $(0, 2)$ sequence followed by NRZI encoding. The d constraint of zero produces the minimum runlength sequence

$$1111\cdots \xrightarrow{NRZI} 0101\cdots \quad \text{or} \quad 1010\cdots.$$

The k constraint produces a maximum runlength sequence

$$100100100\cdots \xrightarrow{NRZI} 000111000\cdots \quad \text{or} \quad 111000111\cdots.$$

Using Table 2.9.1 and recalling that NRZI coding is lossless, we see that the capacity of this code is 0.8791, which is identical to our previous result. Notice that a given $(0, 2)$ sequence has two possible outcomes following NRZI encoding. This is because the NRZI encoder has an initial state of either "0" or "1" and its output sequence depends on the initial state of the NRZI encoder.

In some applications, such as digital magnetic recording, it is customary to employ a figure of merit known as the *density ratio*

$$DR \equiv (1 + d)R, \qquad 2.9.2$$

where $R \leq C_c$ is the code rate. The code rate, as you recall, is the ratio of the average number of information bits in a block of length t code bits, divided by t. The density ratio has units of information bits per change in the level of the channel sequence. In digital magnetic recording, these level changes are known as flux reversals, so DR has units of information bits per flux reversal.

2.9.2 Power Spectrum of Maxentropic Sequences

As we will see in the next chapter, practical RLL code capacities typically are less than the ideal upper bounds given previously. A sequence that achieves the ideal capacity above is called a *maxentropic* sequence. The runlengths $T_j \in \{d+1, \cdots, k+1\}$ of a (d, k) sequence are random variables, and in a maxentropic sequence, these random variables are statistically independent and identically distributed (this is abbreviated as "i.i.d."). It has been shown that these T_j follow a probability distribution given by

$$\Pr(T_j) = \lambda^{-T_j},$$

where λ is the largest positive eigenvalue of B. The *average runlength* for a maxentropic (d, k) sequence is therefore given by

$$\overline{T} = \sum_{j=d+1}^{k+1} j\lambda^{-j} = \frac{d\lambda^{-d} - (k+1)\lambda^{-(k+1)}}{\lambda - 1} + \frac{\lambda(\lambda^{-d} - \lambda^{-(k+1)})}{(\lambda - 1)^2}. \qquad 2.9.3$$

In Section 2.7, it was argued that a good data translation code attempts to match the power spectrum of its coded sequences to the transfer function of the constraining channel. It is therefore of interest to us to find the power spectrum of the maxentropic (d, k) sequence. The calculation of power spectra for arbitrary codes is generally quite difficult. However, for (d, k) sequences and run-length-limited codes based on (d, k) sequences, a general expression for calculating the power spectra has been derived by Gallapoulos, Heegard, and Siegel. For the special case of maxentropic (d, k) sequences followed by NRZI encoding, the power spectrum is given by

$$\Phi(\theta) = \frac{1}{\overline{T} \sin^2(\theta/2)} \frac{1 - |G(\theta)|^2}{|1 + G(\theta)|^2}, \qquad 2.9.4$$

where

$$G(\theta) = \sum_{\ell=d+1}^{k+1} \lambda^{-\ell} e^{i\theta\ell} \qquad 2.9.5$$

and $\theta \in [0, \pi]$. Because of the symmetry of the power spectrum, $\Phi(-\theta) = \Phi(\theta)$. It should also be noted that

$$\sum_{\ell=d+1}^{k+1} \lambda^{-\ell} = 1,$$

so $G(0) = 1$, and $\Phi(0)$ remains finite.

EXAMPLE 2.9.2

It is customary to plot power spectra in decibels:

$$\Phi_{dB}(\theta) = 10 \log_{10}(\Phi(\theta)).$$

Calculate and plot the power spectrum of the (0, 2) sequence.

Solution: Applying the equations above, the result is

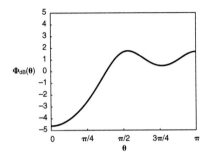

Equations 2.9.4 and 2.9.5 express the power spectrum of the discrete-time sequence. In communication theory, this discrete-time sequence is of fundamental importance. However, it is also often the case that the power spectrum of the transmitted continuous-time analog waveform is also desired. The discrete-time NRZI bit is modulated to become a continuous-time pulse of amplitude β_t and duration T_s, where T_s is the signaling period of the bit sequence. The power spectrum of the analog waveform is related to $\Phi(\theta)$ by

$$\Phi_a(\theta = 2\pi f T_s) = T_s^2 \frac{\sin^2(\theta/2)}{(\theta/2)^2} \Phi(\theta), \qquad 2.9.6$$

with θ evaluated over the range from 0 to 2π. The continuous-waveform spectrum corresponding to Example 2.9.2 is as follows for $T_s = 1$:

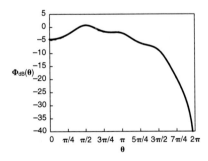

This spectrum is useful in comparing the suitability of the code for use in an analog channel. We will discuss continuous-waveform signaling in more detail later in the text.

Comparing the result from Example 2.9.2 with that of Example 2.7.3, we make the following observations. First, the power spectrum of the (0, 2) code places most of its power at frequencies above $\theta = \pi/2$, while the channel in Example 2.7.3 has its peak response at $\pi/2$ and attenuates the spectrum for frequencies above about $\theta = 0.25$. Second, the primary effect of a (d, k) constraint is to redistribute the power spectrum as a function of frequency. Intuitively, a larger d constraint will tend to distribute power into the lower frequencies by increasing the average runlength between transitions. A smaller k constraint will tend to have the opposite effect by decreasing the average runlength. Let us look at some examples.

EXAMPLE 2.9.3

Plot the power spectrum for a (0,3) sequence.

Solution: Applying Equations 2.9.4 and 2.9.5, we find that the power spectrum in decibels is as follows:

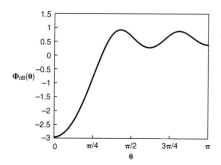

Comparing this with example 2.9.2, we see that the first peak in the response has been shifted lower in frequency from about $\theta = 1.57$ to about $\theta = 1.25$. Also, note that the spectrum near zero frequency is about -3 dB in this example, compared with about -4.6 dB in example 2.9.2. The longer runlengths permitted by the $(0, 3)$ constraint results in more allowable sequences of longer lengths. The consequence of this is more power being distributed into the lower frequency ranges. From Table 2.9.1, the capacity of the $(0, 3)$-constrained sequence is 0.9468, vs. 0.8791 for the $(0, 2)$ sequences.

EXAMPLE 2.9.4

Plot the power spectrum for a (0, 7) sequence.

Solution:

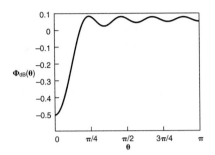

This spectrum is more nearly uniform, beginning to approach that of an unconstrained sequence, with a response near zero frequency of about −0.5 dB. From Table 2.9.1, the capacity of this sequence is 0.9971, which closely approaches the unconstrained sequence capacity of 1.0.

EXAMPLE 2.9.5

Plot the power spectrum of a (1, 7) sequence.

Solution:

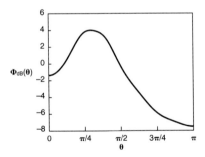

The (1, 7) constraint dramatically shifts the power spectrum into the lower frequencies at the expense of the higher frequencies. The capacity of this sequence is 0.6793. This shift to the lower frequencies occurs because the d constraint produces more longer-runlength sequences while reducing the number of available sequences for a given block length t.

EXAMPLE 2.9.6

Plot the power spectrum for a (2, 7) sequence.

Solution:

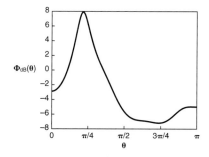

Comparing this result with the previous example, we again see a shift to the lower frequencies and, in addition, a decrease in the range of frequencies for which the power spectrum is above 0 dB. This is a more "narrowband" sequence and is a consequence of the fact that the larger d constraint significantly reduces the number of possible sequences allowed for a given block length t. It is also interesting to compare the high frequency responses for these two examples. Note how the response in Example 2.9.6 reaches a minimum at about $\theta = 2.4$ and then begins to rise slightly while the response in Example 2.9.5 reaches its minimum at $\theta = \pi$. The (2, 7) constraint has fewer allowable sequences for a given block length, and so a somewhat larger percentage of the allowable sequences occur at the minimum runlength, compared with the (1, 7) sequence. The capacity of the (2, 7) sequence is further reduced to 0.5174.

Earlier, in Table 2.9.1, we saw that larger d constraints led to smaller capacities for a given k. We now see that larger d constraints also lead to a "narrowbanding" of the power spectrum. The two effects are related and we can tentatively conclude that capacity and bandwidth are proportional. (A rigorous mathematical treatment verifies this.) Observe that the unconstrained sequence with its constant power spectrum has the greatest capacity and that the higher capacity (0, k) constrained sequences also enjoy wide bandwidth.

EXAMPLE 2.9.7

Plot the power spectrum for a (2, 11) sequence:

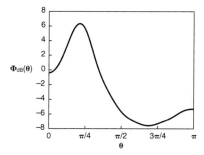

Comparing this case with Example 2.9.6, we see a slight shift in the peak of the spectrum to a higher frequency and a somewhat broader bandwidth, although the bandwidth is still less than that of Example 2.9.5. The capacity of this sequence is 0.5450, which is slightly above that of Example 2.9.6, but still significantly less than the capacity of Example 2.9.5.

SUMMARY

We have introduced many new concepts in this chapter which will have an important bearing on all of the material that follows. We began with the idea of the discrete memoryless channel model and its representation in terms of a transition probability matrix in Equation 2.1.2. The DMC plays a central role in information and coding theory. In particular, it stands as the main model for understanding channel error rates and in analysis of the reliability of transmission in most communication or storage systems.

The DMC led us to the concept of mutual information embodied in Equation 2.1.4, which is a measure of how much of the original transmitted information makes it through the process to the end receiver. Mutual information obeys several important properties of symmetry, relationships to entropy and equivocation, and obeys a chain rule. In Section 2.2 we discussed the maximization of mutual information as a function of the probability distribution of the source symbols. This led us to the definition of channel capacity. Channel capacity for a given DMC can be determined from the Arimoto–Blahut algorithm. We also saw in the case of the symmetric DMC how channel capacity is achieved using a source with a uniform symbol probability distribution and that the capacity in this case can be calculated by simply using Equations 2.1.8 and 2.2.2. We introduced the binary symmetric channel and obtained in Equation 2.2.5 a very simple expression for its capacity.

In Section 2.3 we introduced the notion of block coding of transmitted information and defined the entropy rate of a block of symbols. (See Equation 2.3.3.) The entropy rate is a fundamental relationship governing the reliable transmission of information. This led us to state Shannon's second theorem which is the fundamental theorem governing reliable communication.

We then turned to sources with memory and introduced the Markov process as the main tool for the analysis of this case. For sources of interest to us in communication theory, the Markov process representation of a source achieves a steady-state symbol probability independent of the initial state of the source. This steady-state probability distribution can be calculated directly from the state transition matrix description of the source. In turn, the steady state probability distribution determines the entropy rate of the source through Equation 2.4.6. We also paused briefly in Section 2.5 to develop the data-processing inequality 2.5.6, which states that no method of data processing performed on the channel output can increase the mutual information between the processed data and the information source, although it is possible to decrease or lose information during such processing.

In Section 2.6 we turned our attention to the constrained channel and, in particular, the linear and time-invariant (LTI) channel. We saw it was possible to describe the constrained channel with a state transition diagram representation. We then introduced the important concepts of the autocorrelation and power spectrum of a sequence. These concepts are central to an important and useful description of the communication process.

One key consequence of the constrained channel is the desirability and usefulness of data translation codes (DTCs). These were described in Section 2.8. We saw these codes also have a representation in terms of a state diagram and we introduced the trellis diagram. The trellis diagram is an essential tool in coding theory and will play a prominent role in the theory of convolutional codes later in this text.

Sequences with constraints are best described by means of a special state transition matrix called the connection matrix. We showed how to use the connection matrix to calculate the number of possible different sequences of a given length when the sequences are constrained. We also saw that the entropy rate of a constrained sequence was determined by this matrix through Equation 2.8.10, which is a fundamental result of Section 2.8.

Section 2.9 discussed a special class of constrained sequences called (d, k) sequences. The theory of (d, k) sequences is fundamental to a class of run-length-limited codes known as (d, k) codes. These codes are discussed in Chapter 3. We calculated the capacity of (d, k) constrained sequences and the power spectra of these sequences. Through several examples, we gradually saw a relationship between the characteristics of the power spectrum and the capacity of the sequence. This brings us to the point where we are ready to discuss a number of practical data translation codes in the next chapter.

REFERENCES

C. E. Shannon, "Communication in the presence of noise," *Proceedings of the IRE,* vol. 37, pp. 10–21, Jan., 1949.

C. E. Shannon, "The zero error capacity of a noisy channel," *IRE Transactions on Information Theory,* vol. IT-2, pp. 8–19, Sept., 1956.

C. E. Shannon, "Certain results in coding theory for noisy channels," *Information and Control,* vol. 1, pp. 6–25, Sept., 1957.

D. Blackwell, L. Breiman, and A. Thomasian, "The capacity of a class of channels," *Ann. Math. Stat.,* vol. 30, pp. 1229–1241, Dec., 1959.

D. T. Tang and L. R. Bahl, "Block codes for a class of constrained noiseless channels," *Information and Control,* vol. 17, pp. 436–461, 1970.

K.A.S. Immink, "Runlength-limited sequences," *Proceedings of the IEEE,* vol. 78, no. 11, pp. 1745–1759, Nov., 1990.

S. Arimoto, "An algorithm for computing the capacity of an arbitrary discrete memoryless channel," *IEEE Transactions on Information Theory,* vol. IT-18, pp. 14–20, 1972.

R. Blahut, "Computation of channel capacity and rate distortion functions," *IEEE Transactions on Information Theory,* vol. IT-18, pp. 460–473, 1972.

EXERCISES

2.1.1: A DMC is represented by the transition probability graph.

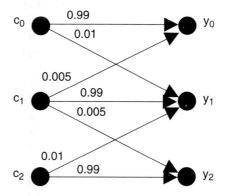

The numbers labeling the arcs in this graph are transition probabilities. Find the transition probability matrix for this channel.

2.1.2: A DMC is described by the transition probability matrix

$$P_{Y|C} = \begin{bmatrix} .9 & .05 & .05 & 0 \\ .05 & .9 & 0 & .05 \\ .05 & 0 & .05 & 0 \\ 0 & .05 & .9 & .95 \end{bmatrix}.$$

a) What is the cardinality of the input alphabet C?
b) What is the cardinality of the output symbol alphabet Y?
c) Draw a graphical representation of this DMC, and label the transition probabilities.

2.1.3: A DMC has the graphical representation.

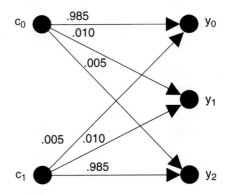

a) Does this channel represent a system using hard-decision or soft-decision decoding? Explain.
b) If it is desired to have $c_0 \mapsto y_0$ and $c_1 \mapsto y_2$, what is the probability of this occurring if the source symbols are equally probable?

2.1.4: For the channel of Exercise 2.1.3, what is the entropy of Y if the input symbols are equally probable?

2.1.5: Calculate the mutual information for the channel of Exercise 2.1.4.

2.1.6: The channel of Exercise 2.1.4 is modified to use hard-decision decoding. The resulting graphical representation of the new channel is

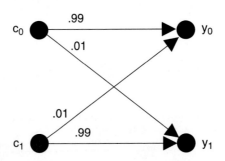

For equally probable input symbols,
   a) Calculate Q_Y
   b) Calculate $H(Y)$
   c) Calculate $I(C;Y)$, and compare this with the result of Exercise 2.1.5.

2.2.1: Write a computer program to implement the Arimoto-Blahut algorithm. Test your program for the channels in Example 2.2.1.

2.2.2: Using the Arimoto–Blahut algorithm, find the channel capacity and the P_C that achieves this capacity for a channel with transition probability matrix

$$P_{Y|C} = \begin{bmatrix} .6 & .3 \\ .3 & .1 \\ .1 & .6 \end{bmatrix}.$$

2.2.3: Repeat Exercise 2.2.2 for a channel with transition probability matrix

$$P_{Y|C} = \begin{bmatrix} .6 & .1 \\ .3 & .3 \\ .1 & .6 \end{bmatrix}.$$

2.2.4: Determine whether the following channels are symmetric:

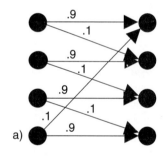
a)

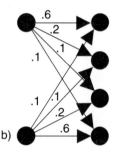
b)

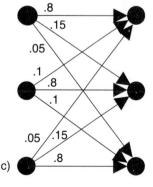
c)

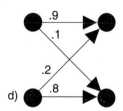
d)

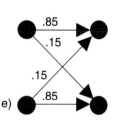
e)

2.2.5: Find the channel capacity and the input probability distribution that achieves this capacity for the channels in Exercise 2.2.4.

2.2.6: Plot the capacity of the BSC as a function of crossover probability p.

2.3.1: In a communication system with an optimum receiver operating over a BSC, the crossover probability of the channel is a function of the signal to noise ratio, γ. Communication theory gives this crossover probability as

$$p = Q(\sqrt{2\gamma}), \qquad \text{2.P.1}$$

where the error rate function Q is defined as

$$Q(x) \equiv \frac{1}{\sqrt{2\pi}} \int_x^\infty e^{-t^2/2}\, dt \approx \frac{1}{\sqrt{2\pi}} \frac{e^{-x^2/2}}{.661x + .339\sqrt{x^2 + 5.51}}. \qquad \text{2.P.2}$$

It is also common practice to express the signal to noise ratio in decibels

$$\gamma_{dB} \equiv 10 \cdot \log_{10}(\gamma). \qquad \text{2.P.3}$$

a) Plot p on a logarithmic scale as a function of γ_{dB} from 0 dB to 15 dB.
b) Plot the channel capacity for the BSC as a function of γ_{dB} from 0 dB to 15 dB.
c) Plot the cutoff rate for the BSC as a function of γ_{dB} from 0 dB to 15 dB.

2.3.2: Binary phase-shift keying (BPSK) is a modulation method that produces a binary symmetric channel. Assuming the transmitter can transmit 1000 binary symbols per second and the receiver is optimum (see Exercise 2.3.1), what is the theoretical maximum number of information bits per second that can be transmitted if the signal to noise ratio at the receiver is 5 dB?

2.3.3: For the communication channel of Exercise 2.3.2, estimate the number of information bits per second that can be transmitted over a practical communication system.

2.3.4: You are given a modem capable of transmitting 56,000 binary symbols per second and a BSC which makes an error every 100 symbols (on the average). What is the maximum error-free information rate you can expect from a practical encoder/decoder using this system?

2.4.1: You are given an information source having a four-symbol alphabet

$$A = \{\text{start}, 0, 1, \text{stop}\}.$$

The source obeys the following rules:

1. a "start" symbol must always be followed by either a "0" or a "1" symbol;
2. the probability A will emit a "0" at any given time is always equal to the probability A will emit a "1" symbol;
3. the probability a "0" or a "1" symbol will be followed by an "end" symbol is 0.1;
4. an "end" symbol is always followed by a "start" symbol.

Draw a state diagram for this source, label its transition probabilities, and find the steady-state probabilities of each state.

2.4.2: Find the entropy rate for the source in Exercise 2.4.1.

2.4.3: For the source of Exercise 2.4.1, calculate $H(A)$ assuming the steady-state symbol probabilities, and compare this with the entropy rate from Exercise 2.4.2.

2.4.4: The design of computer systems often takes advantage of the property of "locality." This property says that the probability that the next data or instruction needed by the computer is largest for data or instructions located in memory near the current data or instruction and that this probability decreases rapidly for data or instructions located farther away. Use information theory to qualitatively explain how computers might be able to take advantage of locality without any significant loss in computer performance.

2.4.5: In Exercise 1.2.12, the entropy of the letters in English text was found to be 4.08 bits per letter if it was assumed that the letters occur independently of each other. Text files stored in a computer system are normally stored using an 8–bit ASCII representation. When the Lempel-Ziv algorithm is used to compress large text files, it is often found to achieve more than a 2:1 compression ratio. Explain how this is possible.

2.5.1: High-level computer languages are often advertised as being "machine independent." Give some example situations where this might be true and some examples where it might be false. Use the data-processing inequality to justify your arguments.

2.6.1: The dicode channel is a LTI channel characterized by impulse response

$$h_0 = 1, \quad h_1 = -1, \quad h_k = 0 \text{ for } k \notin \{0,1\}.$$

Find the transition probability matrix for the dicode channel assuming a discrete memoryless source with alphabet $A = \{-1,+1\}$ and $H(A) = 1$.

2.6.2: Find the entropy rate and the output symbol probabilities for the dicode channel defined in Exercise 2.6.1.

2.6.3: Given a discrete memoryless source with alphabet $A = \{-1,+1\}$ having symbol probabilities $\Pr(-1) = p$ and $\Pr(+1) = 1 - p \equiv q$, and given the dicode channel of Exercise 2.6.1, let state S_0 correspond to $a_{t-1} = -1$ and state S_1 correspond to $a_{t-1} = +1$. Show that, in the steady state, $\pi_0 = p, \pi_1 = q$, and $R = H(A)$.

2.7.1: Show that $\phi_{xx}(-m) = \phi_{xx}(m)$.

2.7.2: Find the power spectrum of a discrete sequence x_t with autocorrelation $\phi_{xx}(m) = \exp(-0.25|m|)$.

2.7.3: Find the power spectrum of a discrete sequence x_t with autocorrelation $\phi_{xx}(m) = (0.1)^{|m|}$.

2.7.4: Given a discrete memoryless source with alphabet $A = \{-1,+1\}$, $H(A) = 1$, and a channel

$$h_k = \begin{cases} 1, & k = 0 \\ -1, & k = 1, \\ 0, & else \end{cases}$$

find the power spectrum of the channel output.

2.8.1: Given an NRZI encoder defined by $\beta_t = c_t \oplus \beta_{t-1}$ with $\beta_{-1} = 0$. Find the encoded output sequence for an input sequence $\{c_t\} = \{1110100001\}$.

2.8.2: The NRZI-encoded sequence from Exercise 2.8.1 is applied to another identical NRZI encoder. Find the output sequence of the second encoder.

2.8.3: The NRZI-encoded output sequence from Exercise 2.8.1 is applied to a decoder

$$\gamma_t = \beta_t \oplus \beta_{t-1}.$$

Find the output sequence of the decoder.

2.8.4: Draw the state diagram and the trellis diagram for the NRZI encoder of Exercise 2.8.1 and determine its connection matrix.

2.8.5: For the NRZI encoder of Exercise 2.8.1, find the number of possible 3–bit output sequences assuming $\beta_{-1} = 0$.

2.8.6: Find the capacity of the NRZI encoder of Exercise 2.8.1.

2.8.7: Find the capacity of the dicode channel of Exercise 2.6.1 assuming a binary input.

2.9.1: Find the state transition matrix for a $(d, k) = (1, 2)$ sequence and verify the channel capacity given in Table 2.9.1.

2.9.2: The capacity of a $(d, k) = (1, 2)$ sequence is 0.4057. Specify an encoder for a rate 1/3, $(d, k) = (1, 2)$ code.

2.9.3: Calculate and plot the power spectrum for a maxentropic $(d, k) = (1, 2)$ sequence.

CHAPTER 3

Run-Length-Limited Codes

3.1 GENERAL CONSIDERATIONS FOR DATA TRANSLATION CODING

This chapter discusses an important class of practical data translation codes (DTCs) known as run-length constrained or run-length-limited (RLL) codes. It might seem that our discussion of constrained channels and (d,k) sequences in chapter two might have exhausted the key information theoretic issues associated with these codes, but that perception is incorrect, and some important issues remain to be dealt with in this chapter.

Let us remind ourselves of the issue being dealt with here. In chapter two, we presented and discussed a general model of the transmission and reception problem. This is depicted in Figure 2.6.1, and a simplified view of this system is given by Figure 2.6.2. The existence of constraints on the transmitted bit sequence imposed by the channel raises the need for translation of the original information sequence into a form compatible with these constraints with the intent to produce an overall resulting system with as high a channel capacity as possible, subject to practical considerations and economic constraints.

The specific details of what constraints are imposed by the channel is properly the subject of that branch of communication theory known as *modulation theory*. This is a topic which involves a great deal of signal processing theory such as may be found in many textbooks on communication systems. The specific details of this theory need not concern us here, and it is sufficient to summarize the main consequences and results. The design of any particular communication system is centered around one of two significantly different symbol detection strategies. The first of these is called *symbol-by-symbol* (SBS) detection, and the second is known as *maximum likelihood sequence estimation* (MLSE).

Historically, the SBS strategy came first, has enjoyed the longest history, and is typically the least expensive approach. The MLSE strategy was first developed in the early 1970s. It involves significantly more complexity than the SBS approach, but offers the promise of higher performance for the system. Both approaches are in use today, and both approaches are still the subject of continued research and development.

The theory and techniques for data translation coding were largely developed for SBS systems, and it is on these systems that we shall concentrate in this chapter. DTC for MLSE systems is still a relatively new enterprise, and many of the codes employed in these systems are carry-overs from codes developed for SBS systems. Research into DTC methods optimized for MLSE systems is presently an active research area.

Within the sphere of SBS systems we can identify two approaches as well. The first approach is based on classical Nyquist theory, where the goal of the modulation and signal processing is to produce an overall discrete memoryless channel. At the modern forefront of these systems is a technique called *decision feedback equalization* (DFE). The DFE is an optimum receiver for SBS detection and is typically used to produce a DMC-like channel response.

The second SBS approach involves what is known as *partial response signaling* (PRS). The technique was invented by Lender in the mid 1960s and is widely found in older telecommunications systems. However, it is fair to say that the bulk of new developments involving partial response signaling also incorporate MLSE detection to form a detection method called partial response maximum-likelihood (PRML) detection. This, too, is an active research area, and techniques for data translation coding for these systems is still in its formative stages. Consequently, this chapter will deal exclusively with codes for SBS systems based on the discrete memoryless channel.

Let us assume a channel of the form given in Figure 2.6.1, and let us further assume the constrained channel portion of Figure 2.6.2 is an LTI channel. This is a channel with memory. How, then, can a DMC channel model be appropriate? The earliest and most straightforward answer to this question was put forth in 1928 by Harry Nyquist. Nyquist showed that signal processing can be used to, in essence, cancel the memory of the LTI channel and achieve SBS detection. Under Nyquist signaling, a DMC channel can be achieved, provided that certain conditions can be met.

For some channels, it is possible to meet these conditions with an unconstrained symbol sequence. However, in many high-performance SBS systems it is possible to achieve these conditions only with constrained sequences. The two primary factors that must be considered are intersymbol interference between waveforms from neighboring transmitted symbols (sometimes known as *crosstalk*) and the recovery of a synchronous clock for symbol detection. Left unchecked, these factors can lead to serious degradation of the error rate performance of the system.

Data translation coding can be used to combat both of these effects. Most popular DTC approaches employ run-length-limited codes and many such codes are (d,k) constrained codes. The d constraint is typically used to combat the intersymbol interference while the k constraint is used to provide clock recovery information. The primary challenge facing the code designer is to come up with a simple and efficient method for mapping the source symbols into the coded sequence for transmission and for mapping the received symbols back into the original source symbols after symbol detection. In this context, a code's efficiency is judged by its code rate and its simplicity is judged in terms of how complex (more accurately, how *not* complex) the encoders and decoders are.

In the *block code* approach, the DTC maps a fixed number of source bits, m, into a fixed number of code bits, n. The efficiency of the code is measured in terms of its code rate $R = m/n$. The entropy rate of such a code is equal to its code rate. The block coding approach, however, is not widely favored for DTCs because good block codes tend to require very large values for n and the encoders and decoders for this are often unacceptably complex.

The more widely used approach is to let variable length blocks of source symbols to be mapped into variable length blocks of code symbols. The usual practice is to employ variable length blocks but to ensure the *ratio* of source block length to code block length is fixed at some particular value of $R = m/n$. When a (d, k) code is used for this purpose, its parameters are often specified using the notation $(d, k; m, n)$.

An *optimum* code for an infinitely long sequence of data would use the set of all possible infinite-length paths defined by code's trellis. (Can you say why this is? Hint: You have seen the answer to this question earlier in the text.) Unfortunately, however, such an optimum encoder and decoder would be impossibly complex. A code is considered to be a good code if it uses *most* of the possible paths, yet is simple to describe and implement. For a given (d, k) constraint, the maximum possible code rate is equal to the capacity defined in chapter two by Equation 2.8.10. This capacity provides an upper bound for the code rate, $m/n < C$. Since m/n must be the ratio of two integers and, usually, the number of code words in a finite block must be a power of two while the number of available paths is not, the constraints placed on codes by the need to have a practical implementation can make finding good codes which come close to the capacity difficult.

3.2 PREFIX CODES AND BLOCK CODES

3.2.1 Fixed-Length Block Codes

The majority of RLL codes employ variable-length coding techniques. The reason for this is that variable-length codes often have less complicated encoders and decoders than an equivalent-rate fixed-length block code would have. For example, suppose we wished to encode m source bits into n code bits using a fixed-length block code. Suppose B is the connection matrix with elements $b_{i,j}$ giving the number of connections from initial state i to final state j. Because we wish to be able to concatenate code words without violating any runlength constraints of the code, we should allow any channel state i to be an initial state and any channel state j to be a final channel state.

From Equations 2.8.2 and 2.8.3, we know that the number of possible paths leading from state i to state j after n codebits are transmitted is equal to the sum of elements in the ith row of B^n. Since any initial state is permitted and since we must have at least 2^m paths for encoding a block of m source bits, every row of B^n must have a row sum of at least 2^m.

EXAMPLE 3.2.1

Find the minimum block lengths required for a $(1, 3)$ RLL code with code rate $m/n = 1/2$.

Solution: From Table 2.9.1, we see that the maximum capacity of a $(1, 3)$ code is 0.5515. A code of rate 1/2 is therefore permitted since this code rate does not exceed the maxentropic capacity. The connection matrix for a $(1, 3)$-constrained code is found from the state diagram of Figure 2.9.1 and is

$$B = \begin{bmatrix} 0 & 1 & 0 & 0 \\ 1 & 0 & 1 & 0 \\ 1 & 0 & 0 & 1 \\ 1 & 0 & 0 & 0 \end{bmatrix}.$$

Since the code rate is 1/2, we know that n must be an even number, and the minimum row sum must equal or exceed $2^{n/2}$. The first even value of n for which this occurs may be found by computing successive values of B^n. Doing so, one finds that the first even n for which this condition is met is $n = 14$

$$B^{14} = \begin{bmatrix} 77 & 52 & 36 & 24 \\ 112 & 77 & 52 & 36 \\ 88 & 60 & 41 & 28 \\ 52 & 36 & 24 & 17 \end{bmatrix}.$$

The smallest row sum is that of the bottom row, which sums to $129 > 2^7 = 128$. Therefore, the minimum block length is $m = 7, n = 14$, and the code would be a (1, 3; 7, 14) code. We will shortly see that an equivalent code implemented as a variable-length prefix code exists that is much simpler than this. Also, it is noteworthy that the exercise we just carried out proves the *possible existence* of the fixed blocklength (1, 3; 7, 14) code but does not tell us what the code actually is, i.e., does not specify the code mapping. All we know is that such a mapping is not forbidden by the runlength constraints. We still do not know for certain that such a code actually exists.

Example 3.2.1 illustrates the basic issue with block RLL codes, namely, very long code words and, consequently, very complicated encoders and decoders. We now turn our attention to variable-length codes.

3.2.2 Variable-Length Block Codes

We encountered variable-length codes back in chapter one when we looked at Huffman codes. In our discussion in Section 1.4, we described the need for such a code to be *self-punctuating*, that is, any code word composed of a string of n_1 bits must not match the first n_1 bits of any longer code word having $n_2 > n_1$ bits. This is called the *prefix condition*. When the prefix condition is satisfied, a code word can be recognized by the decoder when it occurs and the code is said to be *instantaneously decodable*. Such a code is called a prefix code.

If the ratio m/n of information bits to code bits is fixed, the code is called a *fixed-rate* variable-length block code. Such a code must be formulated such that it satisfies the prefix condition and such that any two consecutive code words can be concatenated without violating the code's run-length constraint. Let us look at an example.

EXAMPLE 3.2.2   Franaszek's (2, 7; 1, 2) Code

This code is a fixed-rate (2, 7) code with code rate 1/2. The (2, 7; 1, 2) notation is not to be read as $m = 1$ and $n = 2$, but rather as $m/n = 1/2$. The encoding table for this code is as follows:

Source words	Code words
11	0100
10	1000
000	000100
010	001000
011	100100
0010	00001000
0011	00100100

By inspection of the code words, we see that the prefix condition is satisfied. Also notice that the maximum number of consecutive zeroes at the end of any code word is three while the maximum number of zeroes at the beginning of any code word is four. Therefore, the maximum runlength of zeroes from any code word to any other code word is seven, thus satisfying the k constraint for all possible concatenations of code words. Likewise, the minimum number of consecutive zeroes in any code word is two, thus satisfying the d constraint. From Table 2.9.1, the maximum capacity for a (2, 7) maxentropic sequence was 0.5174, while the rate of this code is 0.500. Franaszek's code is therefore very *efficient*, since the ratio of its code rate to the maximum rate is .5/.5174 or 96.6%.

This code is very easy to implement using shift registers and a lookup table. An encoder implementation is shown in Figure 3.2.1 below. The lookup table can be implemented using a device known as a programmable logic array (PLA) having seven entries. Source bits are shifted in one bit at a time, and code bits are shifted out two bits at a time. The encoder is controlled by a counter that is loaded from the PLA when its count goes to zero and that causes the next code word to be loaded into the code shift register (CSR).

The operation of this encoder can be illustrated by example. Suppose the input sequence consists of the string of source bits 11, 000, 0010, 10, ... , which, from the coding table above, must produce the coded output string 0100, 000100, 00001000, 1000, etc. The counter's terminal count output (TC) is "0" when the contents of the counter are not zero and "1" otherwise. When the TC output becomes a "1", the PLA count (C) and the next code word are loaded into the counter and the code shift register, respectively, on the next clock cycle. When TC is a "0", the output of the PLA is ignored and the contents of the shift registers are shifted to the left. The source word is shifted in one bit per clock cycle, while the code shift register's contents are shifted two bits per clock cycle. Operation of the encoder is illustrated in Table 3.2.2.

Decoding is essentially the inverse of this operation. Code bits are shifted two at a time into a shift register and applied to a PLA lookup table. Since the code satisfies the prefix condition,

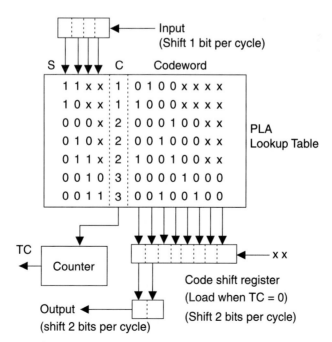

Figure 3.2.1: Encoder for Franaszek's Code

Chapter 3 Run-Length-Limited Codes

TABLE 3.2.2: Example of encoder operation for the (2, 7; 1, 2) Franaszek encoder

Clock	Counter	TC	Source	C	Codeword	CSR	Output
0	0	1	1100	1	0100xxxx	-------- --	
1	1	0	1000	1	1000xxxx	0100xxxx	--
2	0	1	0000	2	000100xx	00xxxxxx	01
3	2	0	0000	2	000100xx	000100xx	00
4	1	0	0001	2	000100xx	0100xxxx	00
5	0	1	0010	3	00001000	00xxxxxx	01
6	3	0	0101	2	001000xx	00001000	00
7	2	0	1010	1	1000xxxx	001000xx	00
8	1	0	010-	2	001000xx	1000xxxx	00
9	0	1	10--	1	1000xxxx	00xxxxxx	10
10	1	0	0----	-	------ --	1000xxxx	00

when a match is found in the lookup table the decoded source word is loaded into an output shift register and the code word input shift register is cleared to begin a new decoding cycle.

The implementation described in this example is not the only possible one for this code. The point, for now, is that encoding and decoding of the Franaszek code is possible and simple.

EXAMPLE 3.2.3

What minimum block length would be required for a fixed-block-length rate 1/2 code with the same runlength constraints as the Franaszek code?

Solution: The (d, k) constraint is $(2, 7)$ so the connection matrix is

$$B = \begin{bmatrix} 0 & 1 & 0 & 0 & 0 & 0 & 0 & 0 \\ 0 & 0 & 1 & 0 & 0 & 0 & 0 & 0 \\ 1 & 0 & 0 & 1 & 0 & 0 & 0 & 0 \\ 1 & 0 & 0 & 0 & 1 & 0 & 0 & 0 \\ 1 & 0 & 0 & 0 & 0 & 1 & 0 & 0 \\ 1 & 0 & 0 & 0 & 0 & 0 & 1 & 0 \\ 1 & 0 & 0 & 0 & 0 & 0 & 0 & 1 \\ 1 & 0 & 0 & 0 & 0 & 0 & 0 & 0 \end{bmatrix}.$$

By repeated calculations of B^n using the computer, we find the minimum block parameters are $m = 31$ and $n = 62$. Such a code is totally impractical.

3.2.3 Prefix Codes and the Kraft Inequality

The Franaszek code in Example 3.2.2 has the property of having both the source words and the code words be variable length. Notice that the *source words* in this code satisfy the prefix condition. Because the source words satisfy the prefix condition, any arbitrary string of binary source symbols can be uniquely punctuated to produce a unique code word. When the source words are variable length, this prefix condition ensures the code is uniquely encodable. Likewise, the code words satisfy the prefix condition and therefore can be punctuated for unique decoding. Because of the *d* constraint in this code, however, certain strings of coded bits are illegal (for example, "1 1") and therefore if such an illegal sequence is seen by the decoder we know that an error has been made in the transmission of the coded sequence. Because certain coded bit sequences are forbidden, the Franaszek code is an example of an *incomplete* code. (This is not bad; it just means some coded bit sequences are forbidden.) Any run-length-limited code is going to be an incomplete code, since the whole idea of such a code is to forbid certain bit sequences.

Huffman codes, which we looked at in chapter one, are also prefix codes. If we glance back at the Huffman code in Example 1.4.2, a little study will show that this code can have *any* semi-infinite string of "0"s and "1"s in its coded output. It is an example of a *complete* prefix code because any semi-infinite string of bits can occur and any semi-infinite string of bits can be uniquely punctuated into code words. Indeed, the *definition* of a complete prefix code is, "a prefix code with the property that *any* semi-infinite string of symbols from the code alphabet can be punctured uniquely into a string of code words."

When and under what conditions do prefix codes exit? For instance, suppose we wanted to know if a prefix code containing M code words constructed as strings of symbols from a symbol alphabet with K symbols existed such that the length of the code words were $\ell_0, \ell_1, \cdots, \ell_{M-1}$. Does such a prefix code exist? The answer to this question is given by a theorem known as Kraft's inequality.

Theorem 3.2.1 (Kraft Inequality): In a symbol alphabet of size K, there exists a prefix code with M code words of length $\ell_0, \ell_1, \cdots, \ell_{M-1}$ if and only if

$$\sum_{j=0}^{M-1} K^{-\ell_j} \leq 1.$$

Note that applying this theorem to the Franaszek code ($K = 2$), we see the Kraft inequality is satisfied by this code. Note also that applying this to the Huffman code of Example 1.4.2, we see the summation above satisfies the relationship with equality. A complete prefix code exists if and only if the relationship in Theorem 3.2.1 is satisfied with equality.

Theorem 3.2.1 can be useful in implementing computer searches to find data translation codes. There is, in general, no known algorithm or design procedure for finding a good prefix code for data translation. The most widely used fixed-rate variable-length block codes are *discovered* rather than invented, and this is usually done using a computer search. The essential properties when looking for a code are:

1. high efficiency (code rate divided by maxentropic capacity);
2. simple encoder and decoder implementations;
3. limited error propagation.

This last point is something we have not yet discussed. Although data translation codes are not typically intended to be structured in such a way that errors during transmission are guaranteed to be detected (let alone corrected!), it *is* very important that, if an error does occur during transmission, the error will not propagate through the decoding process for more than some finite number of bits. The primary weakness of the decoder described earlier for the Franaszek code is its failure to be robust in the presence of a bit error in the received code sequence. However, it will turn out that the Franaszek code itself *is* robust in terms of error propagation if the decoder is sufficiently "smart." We will discuss this in Section 3.4.

3.3 STATE-DEPENDENT FIXED-LENGTH BLOCK CODES

In our discussion of fixed-length block codes in Section 3.2, we placed a constraint on the code that any initial state i was to be allowed to go to any final state j and that code words from successive source words were to be composable, that is, concatenated together, without violating the runlength constraint. Fixed-length block codes designed to obey these constraints are said to be state-independent codes. As we have seen, such codes are frequently impractical.

The same is not true for fixed-block-length codes which are state-*dependent.* In a state-dependent code, the fixed-length code word associated with a fixed-length source word is made to depend on the *previous* code word. Recall that the previous code word defines the *state* of the encoder "channel." The usual approach to state-dependent coding for fixed-length codes is to make the first bit of the code word depend on the previous code word, *i.e.*, on the *state* of the encoder. This is called a *merging rule*. An example will help to illustrate this.

EXAMPLE 3.3.1   The MFM (1,3;1,2) State-Dependent Code

This code was widely used in hard disk drives into the early years of the 1980s. The "MFM" designation stands for "modified frequency modulation" but this code is also known as "delay modulation" and as "Miller code." It's operation is easily explained by means of a state diagram. For MFM, the state diagram is shown in Figure 3.3.1.

The source "word" is one bit in length and the code word is always two bits in length. The encoding rules are very simple. If the source bit is a "1", the code word is a "01". If the source bit is a "0" and the last code bit transmitted was a "0", then we are in state A, and the code word is "10". Otherwise, if the last code bit transmitted was a "1" then we are in state B, and the code word is "00". It is interesting to compare this code against the $(1, 3; 1, 2)$ fixed-length state-independent code of Example 3.2.1, which required $m = 7$ and $n = 14$.

Decoding MFM is very easy. Notice the second code bit is always equal to the source bit. MFM is decoded by discarding the first bit in the code word and keeping the second.

One simple approach to finding state-dependent block codes of a given (d, k) constraint is to first try to construct a state-independent fixed-length code with a constraint of (d, ∞) and then modify this code with a merging rule. Figure 3.3.2 illustrates

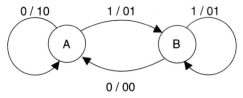

Figure 3.3.1: MFM Encoder

Section 3.3 State-Dependent Fixed-Length Block Codes

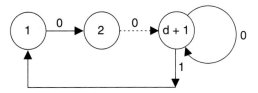

Figure 3.3.2: State diagram of a (d, ∞) sequence.

the state transition diagram for a (d, ∞) code. From this figure, it is easy to show that the transition matrix B has elements

$$b_{i,j} = 1, \quad j = i + 1,$$
$$b_{d+1,1} = b_{d+1,d+1} = 1, \qquad \qquad 3.3.1$$
$$b_{i,j} = 0, \quad \text{otherwise}.$$

To find a state-dependent code with constraints (d, k) for a given code rate m/n, we first find the minimum n for a (d, ∞) code by computing B^n for increasing values of n until the smallest row sum of B^n is greater than 2^m. We then set the first d bits in each code word equal to "0" and construct the required 2^m code words such that the d constraint is satisfied. If code word length exceeds the desired k constraint, we must also take care that the k constraint is satisfied within each code word we are constructing. We then attempt to find a merging rule for the first d bits so that the k constraint is satisfied for all possible sequences of source words. (This may not always be possible; if it isn't, try a larger n.)

EXAMPLE 3.3.2

Design a state-dependent code with rate 0.6 and a (1,6) runlength constraint.

Solution: Since $d = 1$, the state transition (connection) matrix for a (d, ∞) code is

$$B = \begin{bmatrix} 0 & 1 \\ 1 & 1 \end{bmatrix}.$$

By repeated calculation of B^n, we find the smallest n such that all rows of B^n have a row sum greater than 2^m, where m is an integer and $m/n \geq 0.6$ is $n = 5$. This gives us $m = 3$. Since $n < k$, we only need to worry about the d constraint. Keeping the first bit of each code word equal to "0", we construct the eight required code words for the (d, ∞) code:

00000
00001
00010
00100
00101
01000
01001
01010

Table 3.3.1: (1, 6; 3, 5) Code

Source	Codewords	
	State A	State B
000	10000	00000
001	10001	00001
010	10010	00010
011	10100	00100
100	10101	00101
101	01000	01000
110	01001	01001
111	01010	01010

Now define state A as the state when the last code word transmitted ended in "0" and state B as the state when the last transmitted code word ended in "1". Our merging rule is defined as follows. If we are in state B, any of the code words listed above may be concatenated with the previous code word without violating the d constraint (because the first bit in each of the above is equal to "0"). If we are in state A and the *second* bit of the code word above is "0", then set the first bit equal to "1". This satisfies the k constraint for any possible pair of cascaded code words. If the second bit is equal to "1", then leave the first bit equal to "0". This also satisfies the k constraint without violating the d constraint. The final state-dependent code is shown in Table 3.3.1. It is pertinent to note that decoding of this code is *not* state dependent. The final four bits of each code word are unique to their respective source words and the code words differ, if at all, only in the first bit. Therefore, the first bit is ignored (just like in the MFM code earlier), and decoding is done by lookup on the remaining 4 bits. State-independence in the decoding process is important because it guarantees that errors will not be propagated beyond the effected code word.

For relatively simple codes, the method just illustrated in Example 3.3.2 is easy to use. It does not, however, guarantee one will find a code with the minimum possible code length. Algorithmic approaches to finding codes of minimum length have been developed by Franaszek. These algorithms are based on a concept known as *principal states*. We will not go into this method here but, rather, refer the reader to Franaszek's 1969 paper in the reference list. Table 3.3.2 summarizes some results from this procedure taken from Immink's 1990 paper.

The codes tabulated below demonstrate that high efficiency (code rate divided by maxentropic capacity) can be achieved with relatively simple codes in many cases. However, the table also illustrates a number of cases of very complicated codes, such as the (3, 7; 46, 115) code. The table includes a (2, 7; 17, 34) code. We have already seen an example of Franaszek's variable-length (2, 7) code for which the maximum m was only four and the maximum n only 8. The Franaszek code is clearly the simpler of these two, so it is time to return to the subject of variable-length codes.

3.4 VARIABLE-LENGTH FIXED-RATE CODES

Construction of variable-length fixed-rate codes is based on an algorithmic approach developed by Franaszek and based on the notion of principal states. The theory behind

Table 3.3.2: Shortest fixed-length state-dependent block codes

d	k	m	n	efficiency	
0	1	3	5	0.864	
0	2	4	5	0.910	
0	3	9	10	0.951	
1	3	1	2	0.907	(MFM)
1	7	11	33	0.981	
2	5	4	10	0.860	
2	7	17	34	0.962	
2	10	8	16	0.923	
3	7	46	115	0.986	
3	11	8	20	0.886	
4	9	9	27	0.921	
4	14	12	33	0.916	
5	12	9	30	0.890	
5	17	15	45	0.937	

this approach is beyond the scope of this text, but it is possible to provide you with the flavor of the approach. That is our goal in this section.

Let us approach the design of a rate 1/2 (2, 7) variable-length code. In the previous section, we saw that it was possible to approach the design of a (d, k)-constrained state-dependent code by starting with a (d, ∞) code and modifying it to add in the k constraint. Suppose we try this approach again but allow ourselves to consider variable-length code words and source words. The state transition matrix for a $(2, \infty)$ sequence is

$$B = \begin{bmatrix} 0 & 1 & 0 \\ 0 & 0 & 1 \\ 1 & 0 & 1 \end{bmatrix} \Rightarrow B^2 = \begin{bmatrix} 0 & 0 & 1 \\ 1 & 0 & 1 \\ 1 & 1 & 1 \end{bmatrix} \begin{array}{l} \to 0 + 0 + 1 = 1 \\ \to 1 + 0 + 1 = 2, \\ \to 1 + 1 + 1 = 3 \end{array}$$

where the row sums for B^2 are as shown. To realize a fixed-length code of rate with $m = 1$ and $n = 2$, we must have row sums of at least 2 for all rows. We see that B^2 fails to achieve this in the first row (which corresponds to initial state $i = 1$). However, the condition *is* met for the other two initial states.

Since we're willing to consider variable-length code words, why not go for a partial solution with $n = 2$? We will *exclude* state 1 as an initial state for $n = 2$ code words by striking the first row and the first column from B^2 with the result

$$B^2 = \begin{bmatrix} 0 & 0 & 1 \\ 1 & 0 & 1 \\ 1 & 1 & 1 \end{bmatrix} \to \begin{bmatrix} - & - & - \\ - & 0 & 1 \\ - & 1 & 1 \end{bmatrix} \begin{array}{l} \to 0 + 1 = 1. \\ \to 1 + 1 = 2 \end{array}$$

Since the second row now has a row sum of only 1, we can not get two code words out of this reduced matrix. Striking out state 2 by deleting the second row and second column reduces us to the single element $b_{3,3} = 1$, which tells us we can encode one source word with the transition $3 \to 3$, so let us assign the two-bit code word mapping 0: 00. This maps a source "0" to a code word of "00". Our choice of "0" for the source word assignment is arbitrary, but the selection of "00" as the code word is not. "00" is the only two-bit code word that does not violate our d constraint when concatenated with itself.

We now turn to the next source and code word lengths, $m=2$ and $n=4$. We have

$$B^4 = \begin{bmatrix} 1 & 1 & 1 \\ 1 & 1 & 2 \\ 2 & 1 & 3 \end{bmatrix} \begin{matrix} \to 1+1+1=3 \\ \to 1+1+2=4 \\ \to 2+1+3=6 \end{matrix}.$$

Since the minimum row sum is less than 4, there is no fixed-length code that satisfies all four possible 2-bit source words. However, we have already assigned the "0" source word in the previous step so all we need to consider at this step are two bit source words beginning with "1". There are only two of these, and they are "10" and "11". Since we used state $i=3$ in the previous step, we eliminate it from this step by striking out the third row and third column of B^4. The row sums of the remaining terms add up to 2 for each remaining initial state. Therefore, we can make the following code-word assignments:

$$\begin{matrix} 0 & : & 00 \\ 10 & : & 1000 \\ 11 & : & 0100. \end{matrix}$$

In making these assignments, our pairing of the source words "10" and "11" with their respective code words is arbitrary, but the code-words themselves are not. These code words are constructed such that the d constraint is satisfied for any possible concatenation of code words.

We now have a $(2, \infty)$ code with rate 1/2. Our next task is to add in the k constraint. We immediately notice that code word "00" must be excluded, since, if it is concatenated repeatedly with itself, it will eventually violate the constraint of $k=7$. The other two code words can be concatenated in any combination with each other without violating the k constraint, so we keep these code-words in our code. Since these two words by themselves can not represent all possible sequences of source bits, we must increase m to 3 and construct some $n=6$ code words. Two obvious possibilities are

$$\begin{matrix} 010 & : & 001000 \\ 011 & : & 000100. \end{matrix}$$

We must still address the issue of being able to represent all possible source bit sequences. With our code assignments so far, we can not represent an endless sequence of all "0" source bits. To remedy this, we'll make an assignment for the source word "000". In order to satisfy both the d constraint of 2 and the k constraint, let us assign

$$000 \quad : \quad 100100.$$

Gathering together what we have done, our code set so far is now

$$\begin{matrix} 10 & : & 1000 \\ 11 & : & 0100 \\ 010 & : & 001000 \\ 011 & : & 000100 \\ 000 & : & 100100. \end{matrix}$$

Any of these five code words can be concatenated without violating either the d or k constraint. As it stands, though, the maximum run of "0" s permitted by this code is only 6. More importantly, notice that we can not yet represent a source string which begins as, say 11001xxx. Our work is therefore unfinished. We need to be able to introduce a string of two consecutive source "0" bits. We can not directly add "001" to the list above because there is no code word remaining for $n = 6$ that we could add without violating the d constraint within the code word or when concatenated with other code words. We need to go up to $m = 4, n = 8$.

Since the purpose is to get a string of two consecutive "0" source bits, let's add

$$\begin{array}{lcl} 0010 & : & 00001000 \\ 0011 & : & 00100100. \end{array}$$

The "0010" code word assignment was arrived at by looking at the source word as "$0 + 010$" and concatenating the old code $(2, \infty)$ word "00" as a prefix to "001000". The assignment of the code word for "0011", however, bears some explanation. If we append "00" as a prefix to the code word for "011", we would have "00000100" as a code word. However, with five leading zeroes, this word would lead to violation of the k constraint if we happened to have a source word sequence such as "0010, 0011". To prevent this, the "0011" code word is changed to the one given above by making the third bit of the code word a "1".

This completes our (2,7;1,2) variable-length code. If we compare it against the Franaszek code discussed earlier, we see it is the same except for the permutation of some of the assignments between source words and code words. There are a total of 24 possible permutations in the assignment of code words in this code. While equivalent in terms of code rate and run length constraints, these permutations are *not* equivalent in terms of simplicity of decoding and control of error propagation if the received sequence should contain a bit error. One particularly judicious code-word assignment was found by Eggenberger and Hodges. Their assignment is

$$\begin{array}{lcl} 10 & : & 0100 \\ 11 & : & 1000 \\ 011 & : & 001000 \\ 010 & : & 100100 \\ 000 & : & 000100 \\ 0011 & : & 00001000 \\ 0010 & : & 00100100\,. \end{array}$$

The primary benefit of this assignment lies in decoding the received sequence. With this assignment, decoding can be carried out using an eight-bit shift register to hold the received code sequence. The bits are shifted in two at a time, and the corresponding source bits are decoded bit by bit using the Boolean expression

$$s = x_4\bar{x}_1 + x_7\bar{x}_5 x_3 + x_3 x_8 + x_6,$$

where "+" is the logic "OR" operation, concatenation implies logic "AND" and $\bar{x}$ is the complement of code bit x. The subscripts on the x bits denote their location in the shift register with 8 being the oldest bit and 1 being the most recent. In addition to simple decoding, this particular assignment also limits error propagation due to a bit error to at most the next two subsequent decoded bits. As mentioned earlier, control of error propagation is of great importance in variable-length codes.

3.5 LOOK-AHEAD CODES

3.5.1 Code-Word Concatenation

As you may have noticed, much of the work involved in developing an RLL code stems from the requirement to avoid violating (d, k) constraints when short code words are concatenated. Look-ahead (LA) codes provide an alternative approach to the sequence-state approaches we have discussed so far. Look-ahead codes are also known as "future-dependent" codes because the basic idea behind them is to allow the use of alternative code words depending on the upcoming future code words. Several useful LA codes have been developed.

We will illustrate the LA coding method by looking at a specific example. Suppose we want a rate 2/3 RLL code with constraints (1, 7). If we did not concern ourselves with run length violations when code words were concatenated, our list of code words would only need to consist of code words that were internally consistent with the run length constraints. How many code words of length n are there which satisfy a particular (d, k) constraint? This question was asked and answered by Tang and Bahl with the results

$$N(n) = \begin{cases} n + 1, & 1 \leq n \leq d + 1 \\ N(n-1) + N(n-d-1), & d+1 \leq n \leq k \\ d + k + 1 - n + \sum_{i=d}^{k} N(n-i-1), & k < n \leq d + k \\ \sum_{i=d}^{k} N(n-i-1), & n > d + k \end{cases} \quad 3.5.1$$

Equations 3.5.1 provide recursion relationships that allow us to calculate the exact number of sequences of a given length that obey the (d, k) constraint. For the case of a (1,7) constraint, we find $N(3) = 5$. This is enough to encode an $m = 2$ source alphabet. We can easily compile a list of the available codewords by writing down the binary strings omitting those which violate either the d or k constraint. For our short $n = 3$ example, there will be no k violations since $n < k$. The available code words are

$$000, \quad 001, \quad 010, \quad 100, \quad 101.$$

We need to select four of these for our four source words. It is clear by inspection that we're going to face an issue with concatenating some of these words, such as 101, without violating the d constraint. Since 000 can be concatenated with any of the others without a d violation, let's reserve it to use as "glue" later and keep the remaining four. Suppose we make the following code word assignments

Section 3.5 Look-Ahead Codes

00	:	101
01	:	100
10	:	001
11	:	010.

This will constitute our basic code.

We now have to deal with the problem of concatenating code words in a code sequence. This is where the future-dependence or "look-ahead" feature comes in. Our basic code has four cases where concatenating code words would violate the d constraint. These are

00.00	:	101.101
00.01	:	101.100
10.00	:	001.101
10.01	:	001.100.

We can avoid this problem during encoding by "looking ahead" at what source word will be following the present one. For instance, if we are encoding source word "00" now and we see that another source word "00" will follow this one, we can make a substitution of, say, 101.000 to replace the troublesome 101.101 sequence.

We must take care, however, not to repeat ourselves. If we use 101.000 to encode the source sequence 00.00, we must be careful to notice that we can *not* use 101.000 to also encode the source sequence 00.01. Likewise, 100.100 can not be used to encode 00.01 because this is the same as the encoding for 01.01. Some "tinkering" with the code selection for these "violation" sequences is necessary. The following selection, however, works well:

00.00	:	101.000
00.01	:	100.000
10.00	:	001.000
10.01	:	010.000.

Notice that we had to "tweak" the 00.01 and 10.01 cases to avoid duplicate code sequences.

Encoding can be carried out using a state machine. We require one state for each of the four source words plus an additional state when the look-ahead shows us that a code violation will occur for the next source word. Let us make the following state associations for the basic code assignments: (00, A), (01, B), (10, C), and (11, D). We will designate state V as the "violation" state, i.e., the encoder state when code words are to be selected from the alternative code word assignment. Letting the notation ccc/S indicate the selected code word and the next state of the encoder, the code described above is specified by the state assignment table given in Table 3.5.1.

Table 3.5.1 provides the specifications required to construct a sequential circuit for encoding the source sequence. The look-ahead coding method can be viewed as a novel case of state-dependent coding, and in this example, it produces a code of

TABLE 3.5.1 (1, 7; 2, 3) RLL Code

Next Source	A	B	C	D	V
00	101/V	100/A	001/V	010/A	000/A
01	100/V	100/B	010/V	010/B	000/B
10	101/C	100/C	001/C	010/C	000/C
11	101/D	100/D	001/C	010/D	000/D

fixed-length code words. Analysis of this code shows that it has a maximum error propagation length of five source bits, i.e., a single bit error in the coded received sequence will propagate to no more than five decoded source bits.

The code we have been using as our example was first developed by Cohn, Jacoby, and Bates, and we shall refer to it as the CJB code. A number of other LA-RLL codes have also been developed. These include the "3PM" code, which is a (2, 11; 3, 6) code developed by Jacoby, a (2, 7; 3, 6) code by Jacoby and Cohn, which is a modification of the 3PM code, a (1, 7; 2, 3) code by Franaszek, and the so-called "zero modulation" or ZM code developed by Patel.

3.5.2 The k Constraint

Perhaps it did not escape your notice that during our development of the (1, 7; 2, 3) code example, we did not take steps anywhere to enforce the k constraint. Since we had $n < k$, k never entered in to the generation of the "basic" code words. When we turned to the problem of "merging" concatenated code words, our attention was riveted on the d constraint, and we never brought up the matter of the k constraint, except to notice when we were done that the maximum run length was, indeed, $k = 7$. Did we, or, more accurately, Cohn, Jacoby, and Bates, just get lucky? Well, yes and no (sort of).

Did the inventors of this code set out to achieve a (1, 7) RLL constraint or did they set out to achieve the d constraint and just settle for any reasonably limited k which resulted? Cohn, Jacoby, and Bates do not say. But suppose they *did* just settle for $k = 7$ as a "good enough" maximum runlength? Is that so bad? Not if it works. The primary purpose of a k constraint is to ensure that changes occur in the transmitted sequence often enough to guarantee the symbol clock recovery system functions properly. Therefore, we often have a certain amount of flexibility with our k constraint. Most clock recovery systems can function with any k below some specified maximum and so the design criterion for (d, k)-constrained RLL codes is best taken as "k to be in some range $k_{min} \leq k \leq k_{max}$."

Naturally, the code capacity should concern us and we know from chapter two that this capacity increases with increasing k. Therefore, we should try to shoot for longer values of k whenever we can. On the other hand, capacity as a function of k reaches a point of diminishing returns and beyond some point is reasonably insensitive to whether we have k or $k + 1$ or $k - 1$. What makes the difference, finally, is whether our code rate m/n can be achieved with the k we choose.

The (1, 7; 2, 3) code we have just described is a highly successful code in the sense that it has been used in a number of commercially important systems. Its success demonstrates that clock recovery systems with a k constraint of 7 are practical and feasible. Suppose circumstances were to force upon us the need for a k constraint of 6. From Table 2.9.1, we see that a $(1, 6)$ RLL sequence has a capacity of 0.6690, and therefore, a rate 2/3 $(1, 6)$ code should be obtainable. However, at this time, no one has found a practical one. If you examine the steps we took and the reasoning we employed in our $(1, 7; 2, 3)$ example, it should become evident that enforcing a k constraint of 6 is not feasible within the confines of the state-dependent, fixed-length structure we developed for the $(1, 7)$ code.

3.5.3 Informal and Formal Design Methods

Our approach to the construction of RLL codes over the past couple of sections has been a bit informal and ad hoc. Since the author wrote these sections, it should be clear that he feels there is value in this. Of what value are these ad hoc lines of reasoning? The answer, in the mind of the author, is this: By examination of and practice with these lines of reasoning, some big enduring ideas emerge and one can obtain a "feel" for the mathematics underlying this branch of coding theory. What are these ideas? The feasibility of a code is related to the number of possible paths which exist to go from an initial state i to a final state j in n steps. The number of available paths must be large enough to permit unique assignment of code words to 2^m source words. The number of paths is determined by the state transition (or connection) matrix B and its power set.

When the necessary n to obtain the required number of paths from B^n becomes too large, the code may be impractical to construct. In this case, it often happens that only one, or perhaps a few, of the rows of B^n fail to have the required minimum row sum. When this happens, a "divide and conquer" approach may be appropriate. One generates most of the code words needed from that set of initial states which *do* enjoy the required minimum number of paths to reach the needed final states. The remainder of the required code words may then be achievable if we use state-dependent coding (such as MFM or the CJB code above) or if we allow variable-length code words (such as in the Franaszek $(2, 7)$ code). Sometimes this process is simplified if we first design a code without k constraints (a (d, ∞) code) and then try to introduce the k constraints in the construction of a variable-length or a state-dependent code.

It is possible to formalize these notions by introducing the idea of *principal states*. The principal states of a code are the set of initial states that have enough paths to reach any final state j such that 2^m source words may be encoded for sequences that begin in these principal states. The principal states may be identified using the *approximate eigenvector* method.

Let $V = [v_1, v_2, \cdots, v_{k+1}]^T$ be a binary vector, i.e., $v_i \in \{0,1\}$.

The approximate eigenvector of a fixed-length code is that vector V which satisfies the inequality

$$B^n V \geq 2^m V, \qquad 3.5.2$$

for every element in the resulting vector. For some given code rate, m/n, the existence of such an eigenvector is guaranteed by the theory of nonnegative matrices. The set of elements of the approximate eigenvector for which $v_i = 1$ correspond to the principal states i. For state-independent coding, we must have every initial state be a principal state, and this requirement then drives the selection of m and n, as we have previously seen in Example 3.2.1.

Now let us relax the constraint that $v_i \in \{0, 1\}$ and allow v_i to be any positive integer. The existence of the approximate eigenvector is now guaranteed for any selection of m and n. However, if some $v_i > 1$, encoding will require look-ahead for state i.

EXAMPLE 3.5.1

Find the approximate eigenvector for a rate 2/3 (1, 7) RLL code.

Solution: From Section 2.9, the state transition matrix for this run length constraint is

$$B = \begin{bmatrix} 0 & 1 & 0 & 0 & 0 & 0 & 0 & 0 \\ 1 & 0 & 1 & 0 & 0 & 0 & 0 & 0 \\ 1 & 0 & 0 & 1 & 0 & 0 & 0 & 0 \\ 1 & 0 & 0 & 0 & 1 & 0 & 0 & 0 \\ 1 & 0 & 0 & 0 & 0 & 1 & 0 & 0 \\ 1 & 0 & 0 & 0 & 0 & 0 & 1 & 0 \\ 1 & 0 & 0 & 0 & 0 & 0 & 0 & 1 \\ 1 & 0 & 0 & 0 & 0 & 0 & 0 & 0 \end{bmatrix},$$

so

$$B^3 = \begin{bmatrix} 1 & 1 & 0 & 1 & 0 & 0 & 0 & 0 \\ 2 & 1 & 1 & 0 & 1 & 0 & 0 & 0 \\ 2 & 1 & 1 & 0 & 0 & 1 & 0 & 0 \\ 2 & 1 & 1 & 0 & 0 & 0 & 1 & 0 \\ 2 & 1 & 1 & 0 & 0 & 0 & 0 & 1 \\ 2 & 1 & 1 & 0 & 0 & 0 & 0 & 0 \\ 1 & 1 & 1 & 0 & 0 & 0 & 0 & 0 \\ 1 & 0 & 1 & 0 & 0 & 0 & 0 & 0 \end{bmatrix}.$$

After some amount of work, we find that

$$B^3 V = \begin{bmatrix} 1 & 1 & 0 & 1 & 0 & 0 & 0 & 0 \\ 2 & 1 & 1 & 0 & 1 & 0 & 0 & 0 \\ 2 & 1 & 1 & 0 & 0 & 1 & 0 & 0 \\ 2 & 1 & 1 & 0 & 0 & 0 & 1 & 0 \\ 2 & 1 & 1 & 0 & 0 & 0 & 0 & 1 \\ 2 & 1 & 1 & 0 & 0 & 0 & 0 & 0 \\ 1 & 1 & 1 & 0 & 0 & 0 & 0 & 0 \\ 1 & 0 & 1 & 0 & 0 & 0 & 0 & 0 \end{bmatrix} \begin{bmatrix} 2 \\ 3 \\ 3 \\ 3 \\ 2 \\ 2 \\ 2 \\ 1 \end{bmatrix} \geq 2^2 \begin{bmatrix} 2 \\ 3 \\ 3 \\ 3 \\ 2 \\ 2 \\ 2 \\ 1 \end{bmatrix},$$

so $V = [2\ 3\ 3\ 3\ 2\ 2\ 2\ 1]^T$ is the approximate eigenvector.

The approximate eigenvector in the above example has most of its elements greater than 1, indicating that look-ahead coding is necessary. The technical literature, particularly the paper by Lempel and Cohn, describe how the eigenvector's component values can be used as a guide in constructing codes. This description is illustrated by means of example. Unfortunately, a systematic construction procedure for the design of look-ahead codes has not yet been published. It would seem, then, that acquiring familiarity with the type of ad hoc approaches we have described earlier is a good idea for until and unless we develop a better "feel" for how to let the approximate eigenvector guide us, calculating it doesn't bring us much in the way of new information since we already knew from the row sums of B^n that state-dependent, look-ahead, or variable-length coding would be required for a (1, 7; 2, 3) code.

All this is *not* to say that the idea of approximate eigenvectors is without merit. There is an advanced treatment of RLL codes, known as the Sliding Block-Code Algorithm, in which the approximate eigenvector plays an important role. Sliding block-code theory was published in 1983 by Adler, Coppersmith, and Hassner. It involves a number of advanced concepts, such as "state splitting," and provides a systematic procedure for constructing a code at any rate m/n which does not exceed the code capacity. Under this algorithm, the element values of the approximate eigenvector indicate which states must be "split" and how many pieces this vivisection must produce. The sliding block-code algorithm is typically implemented on the computer and set loose to do its code construction. There is a relationship between sliding block codes and look-ahead codes, although what this relationship is in all of its implications is not yet clear and remains one of the outstanding research problems in coding theory.

Since this text is an elementary introduction to the subject, we will not pursue the theory of the sliding block-code algorithm here. It is a topic best put off to a more advanced course. However, some of the flavor of the sliding block-code approach can be seen in the following sections.

3.6 DC-FREE CODES

3.6.1 The Running Digital Sum and the Digital Sum Variation

During the discussion in chapter two on the power spectrum of codes, it was mentioned that it is typically a very good idea to make the power spectrum of the code "match" the power spectrum of the physical channel over which it is to be used. In particular, if the physical channel has frequencies where its transfer function response is zero, it is a very, very good idea to design the code so that its power spectrum also contains nulls at these frequencies.

One important special case is the case where the channel has a null at zero frequency (or "at dc"). An example of such a system is the playback process for a helical-scan tape recorder. In this system, the playback head is connected to the outside world through a rotary transformer. A rotary transformer is incapable of passing zero frequency signals and so produces a null at dc.

Doesn't a finite k constraint automatically produce a null at dc? No. To see this, all we need do is review the power spectrum plots given as examples in chapter two. All of the sequences in these examples have finite k constraints and *none* of them have a null at zero frequency. The reason is simple: The k constraint prevents any code sequence from being *purely* zero frequency but it does *not* prevent sequences which have an average value. Consider the following sequence

100000010100000010100000 ....

After NRZI encoding (assuming an initial channel state of "−1"), this sequence produces the NRZI sequence

$$1111111\text{-}1\text{-}11111111\text{-}1\text{-}1111111\ldots,$$

which clearly has a non-zero average value[1].

What about the RLL codes we've been looking at? The power spectrum of an RLL code can be calculated using the method of Gallapoulos, Heegard, and Siegel. These calculations show that the codes we have been looking at have power spectra that are closely approximated by the simple (d, k)-constrained sequence power spectra in chapter two.

In order to produce a null in the power spectrum of a code, it is necessary to control the code's *running digital sum* (*RDS*). The *RDS* of an sequence of bits $(a_0, a_1, \cdots, a_t), a_i \in \{-1, +1\}$ is simply

$$RDS_t = \sum_{i=0}^{t} a_i. \qquad 3.6.1$$

For the repeating sequence shown above, the *RDS* approaches infinity as the length of the sequence approaches infinity.

If the symbol sequence is composed of binary symbols $\beta_t \in \{0, 1\}$ prior to digital modulation, the *RDS* can be re-expressed in the equivalent form

$$RDS_t = -\sum_{i=0}^{t} (-1)^{\beta_i}. \qquad 3.6.2$$

The maximum variation in the RDS of a sequence is called its *digital sum variation* (*DSV*). It can be shown that the power spectrum of a sequence has a null at dc if and only if its *DSV* is finite.

The state transition diagram of an NRZ-encoded sequence (*not* a (d, k) sequence!) having finite *DSV* and run-length constraints of $d = 0$ and $k = 2$ is shown in Figure 3.6.1. From the figure, we can see the maximum runlength of either symbol is 3 and the symbols can alternate in a pattern of 101010, …. Inspection of the figure shows the maximum *RDS* is plus or minus 3.

We can find the capacity of this code from its state transition matrix

$$B = \begin{bmatrix} 0 & 1 & 0 & 0 \\ 1 & 0 & 1 & 0 \\ 0 & 1 & 0 & 1 \\ 0 & 0 & 1 & 0 \end{bmatrix}.$$

---

[1] Back in his days in the disk drive industry, one of the author's favorite practical jokes was to record a pattern like this on all of the tracks of a sampled-servo disk drive. The side-reading efficiency of a magnetic playback head improves with decreasing frequency, and on a disk drive with embedded servo code, this pattern would produce a very large low frequency interference with the embedded servo code, often enough so that the disk drive could not stay on track. One then goes and gets the servo control designer and asks, "Hey! What's wrong with the servo!?"

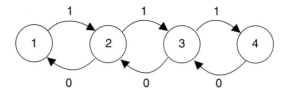

Figure 3.6.1: State diagram of a DSV-constrained sequence.

The maximum eigenvector for this matrix is $\lambda = 1.618$, giving us a capacity of

$$C = \log_2(\lambda) = 0.694.$$

The capacity of a (0, 2) sequence in Table 2.9.1 was 0.8791. Comparing this with the preceding result, we see that limiting the *RDS* to a finite value has dramatically reduced the capacity of the code.

The k constraint of a dc-free code can be increased by allowing a larger, but still finite, *DSV* and adding more states in the diagram of Figure 3.6.1. The general canonical form of an N state, $k = N - 2$ dc-free code is shown in Figure 3.6.2. The state transition matrix for this form has zero elements everywhere, except on the superdiagonals,

$$b_{i,j} = \begin{cases} 1, & j = i \pm 1 \\ 0, & else \end{cases}. \qquad 3.6.3$$

The largest positive eigenvalues of state transition matrices of this form obey a particularly convenient form. It can be shown that the capacity of a code of the form of Figure 3.6.2 is

$$C = \log_2\left[2\cos\left(\frac{\pi}{N+1}\right)\right], \quad N \geq 3. \qquad 3.6.4$$

Since $k = N - 2$, the capacities for runlengths of $k = 2, 3, 4,$ and 5 given by Equation 3.6.4 are 0.694, 0.792, 0.849, and 0.885, respectively. This suggests practical code rates of 2/3, 3/4, 4/5, and 4/5, respectively.

3.6.2 State-Splitting and Matched Spectral Null Codes

Figure 3.6.2 is the starting point for an interesting and useful class of dc-free codes known as Matched Spectral Null (MSN) codes. The theory of MSN codes was developed by Karabed and Siegel. They are sliding block codes, and while the theory behind them is too advanced for this textbook, it is worthwhile to examine one of them in

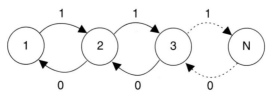

Figure 3.6.2: Canonical state diagram for DSV-constrained sequences.

some detail as an illustration of the concepts of "state-splitting" and other techniques that underlie the theory of sliding block codes.

EXAMPLE 3.6.1

Develop a rate 2/3 (0, 2) dc-free code.

Solution: Let us begin by seeing if we can achieve this objective using a state-independent block code. To encode m information bits, we require $m/n = 2/3$, with n such that the row sums of B^n are greater than or equal to $2^{2n/3}$. By repeated calculation of B^n, we find the smallest possible n to be 18,

$$B = \begin{bmatrix} 0 & 1 & 0 & 0 \\ 1 & 0 & 1 & 0 \\ 0 & 1 & 0 & 1 \\ 0 & 0 & 1 & 0 \end{bmatrix} \Rightarrow B^{18} = \begin{bmatrix} 1597 & 0 & 2584 & 0 \\ 0 & 4181 & 0 & 2584 \\ 2584 & 0 & 4181 & 0 \\ 0 & 2584 & 0 & 1597 \end{bmatrix}.$$

An $m = 12, n = 18$ encoder is clearly impractical, so let us consider a state-dependent block code. Since n must be a multiple of 3, we look at the first few cases of B^n. We find that

$$B^3 = \begin{bmatrix} 0 & 2 & 0 & 1 \\ 2 & 0 & 3 & 0 \\ 0 & 3 & 0 & 2 \\ 1 & 0 & 2 & 0 \end{bmatrix}, \quad B^6 = \begin{bmatrix} 5 & 0 & 8 & 0 \\ 0 & 13 & 0 & 8 \\ 8 & 0 & 13 & 0 \\ 0 & 8 & 0 & 5 \end{bmatrix}.$$

Notice something interesting about the B^6 case. Code words that start in state 1 end either in state 1 or state 3, while code words that start in state 3 can end only in state 1 or state 3. Furthermore, there are 21 paths possible from state 3, which is more than enough to handle the 2^4 code words necessary. Since there are only 13 paths available from state 1, we can not do state-independent coding. However, if we can make the encoding of paths leaving state 1 state-dependent, we should be able to construct a code based around *only* states 1 and 3. States 2 and 4 are superfluous, except as transient states through which a path may pass on its way to its terminal state. Put another way, we need consider only states 1 and 3 as valid initial states in the code.

This suggests that our code requires only three states for a state-dependent implementation: state 1, and two other states that designate the state-dependent encoding of paths leaving state 1. (State 3 has enough exit paths, and therefore, does not need to have any additional state dependence involved in encoding its paths.) If a three-state implementation is possible, we should be able to reduce Figure 3.6.1 to a new state diagram which has only three states. We will now show that this is, indeed, possible.

Since paths leaving state 1 must be made state-dependent, and since such paths can only end up in state 1 or state 3, let's make an equivalent *five*-state diagram by "splitting" state 3. This is shown in Figure 3.6.3. As far as state 1 is concerned, this graph is identical to Figure 3.6.1.

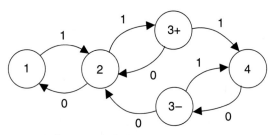

Figure 3.6.3: State-Splitting

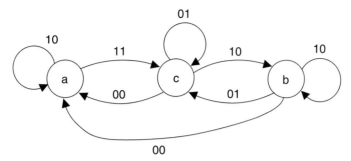

Figure 3.6.4: Sequence state diagram for a rate 2/3 (0, 3) dc-free code.

However, we see state 3+ can only be entered by sequences with pass through state 2, while state 3− can only be entered by paths passing through state 4. We can eliminate states 2 and 4 by considering two-bit sequences. A two-bit sequence beginning in state 1 can terminate only in either state 1 or state 3+. A two-bit sequence beginning in state 3+ can terminate only in states 1, 3+, or 3−. The same is true for a two-bit sequence beginning in state 3−.

Let us re-designate states 1, 3−, and 3+ as states "a", "b", and "c", respectively. We can then represent all two-bit sequences using the 3-state diagram of Figure 3.6.4. We have therefore succeeded in reducing our original four-state diagram to three states.

The state transition matrix and its cube (representing sequences of $n = 6$ bits) for Figure 3.6.4 is given by

$$B_3 = \begin{bmatrix} 1 & 0 & 1 \\ 1 & 1 & 1 \\ 1 & 1 & 1 \end{bmatrix}; \quad B_3^3 = \begin{bmatrix} 5 & 3 & 5 \\ 8 & 5 & 8 \\ 8 & 5 & 8 \end{bmatrix}.$$

We must now assign code words. From B_3^3, we see that there are eight possible six-bit sequences which originate in state "a" and terminate in the other two states. We require 16 sequences which can be emitted originating from state "a." Since we desire as simple an encoder as possible, given the constraints on the code, let us consider using these eight sequences plus an "encoder state" as the code words from state "a." If we do this, three of those eight sequences will terminate in state "b," and five will terminate in state "c." Each of these states has enough paths to allow eight additional sequences that can return to state "a." In the interest of obtaining simple code assignment rules, let us specify the sixteen sequences originating in state "a" as a set $cccccc/ss$, where $cccccc$ is a six-bit sequence and ss is a two-bit encoder state. Let us define the "encoder state" as "00" when the sequence originates in state "a". Using the eight possible sequences originating in "a" and terminating in either "b" or "c," and somewhat arbitrarily pairing these sequences up with the four-bit source words, we have

		Codeword Assignments (state 00)			
Sourceword	Codeword	Terminal Node	Sourceword	Codeword	Terminal Node
0000	101011/10	c	1000	101011/11	c
0001	101101/10	c	1001	101101/11	c
0010	101110/10	b	1010	101110/11	b
0011	110011/10	c	1011	110011/11	c
0100	110101/10	c	1100	110101/11	c
0101	110110/10	b	1101	110110/11	b
0110	111001/10	c	1110	111001/11	c
0111	111010/10	b	1111	111010/11	b

The most significant bit of the source word determines the next encoder state for sequences originating in node "a." The code-word sequences are specified by the remaining three bits of the source word.

Since both "b" and "c" are terminal states for the sequences given above, we can not uniquely associate encoder states "10" or "11" with either state "b" or "c." Observe, however, that we must specify 32 more code words to complete the code (16 code words from state 10 and 16 code words from state 11). Why 32 and not 16? We want the code word that *follows* a code word emitted from state 00 (state "a") to specify the most significant bit of the source word that generated the state 00 code word. In that way, we can implement *state-independent decoding* of the code at the receiver. The strategy will be as follows: When the *decoder* sees a code word belonging to the set above, it will "look ahead" at the following code word to determine how to decode the most-significant bit of the source word associated with the present code word. (The other three bits of the source word are uniquely specified by the above code word.) Therefore, all we have to do is make certain code words generated by a particular source word from encoder state 10 differ from code words generated for the same source word from encoder state 11. The leading "1" in the encoder state instructs the encoder to use all four source word bits *plus* the least-significant encoder state bit to determine the six-bit code word.

Now let us consider code words generated from encoder state 10. Referring to B_3^3, we see we have eight code words available originating from either "b" or "c" which can terminate in "a". The remaining 13 available code words will terminate in either "b" or "c." Let us again use the most-significant bit of the source word to determine if we will return to state 00 and make the code word assignments shown in the following table.

The eight code words assigned to source words 0*sss* in this table are the *only* available source words that can take us from either "b" or "c" back to "a." Examination of Figure 3.6.4 shows that these code words are independent of whether we start from "b" or from "c." This is a consequence of our splitting state 3 into states 3+ and 3− and then associating these split states with "b" and "c." As far as "a" is concerned, "b" and "c" are the "same" and therefore must have the same sequences that terminate in "a."

Code Word Assignments (state 10)					
Source Word	Code Word	Terminal Node	Source Word	Code Word	Terminal Node
0000	001010/00	a	1000	001011/10	c
0001	001100/00	a	1001	001101/10	c
0010	010010/00	a	1010	001110/10	b
0011	010100/00	a	1011	010011/10	c
0100	011000/00	a	1100	001011/11	c
0101	100010/00	a	1101	001101/11	c
0110	100100/00	a	1110	001110/11	b
0111	101000/00	a	1111	010011/11	c

The code words assigned to source words 1*sss* are all unique from the code words assigned to source words 0*sss*. These code words must also be unique from all code words assigned to state 11 in order for the decoder to be able to decode the most significant source bit using the look-ahead technique. The code words were selected from four of the five possible code words which pass through "a," but do not originate in "a" or end in "a." We know there are only five such sequences from examination of the B_3^3 matrix, and we know that no two sequences which pass through "a" can differ *only* because one originates in "b" *vs.* the other originating in "c" (another consequence of our splitting of state 3 into "b" and "c"). Because there are only five available sequences that pass through "a" but end elsewhere but we require a total of eight sequences to complete the table, we again use the trick of assigning identical sequences with different next encoder states. Decoding will employ look-ahead to decode these sequences.

All that remains is to assign code words to the 11 encoder state. We require 16 code words that are distinct from any previous code words. Referring again to the B_3^3 matrix, we have already used all 8 code words that originate in "b" or "c" and terminate in "a." We have also used four of the 13 code words that originate in "b" or "c" and do not terminate in "a." That only leaves us with 9 code words, so we will have to use our encoder state trick again. Also, since there's nothing in particular to be gained from doing otherwise, we might as well use up all 9 available words. The code-word assignments are shown in the following table:

	Code Word Assignments (state 11)				
Source Word	Code Word	Terminal Node	Source Word	Code Word	Terminal Node
0000	010110/10	b	1000	010110/11	b
0001	011001/10	c	1001	011001/11	c
0010	011010/10	b	1010	011010/11	b
0011	100011/10	c	1011	100011/11	c
0100	100101/10	c	1100	100101/11	c
0101	100110/10	b	1101	100110/11	b
0110	101001/10	c	1110	101001/11	c
0111	101010/10	b	1111	010101/10	c

This completes the encoder specification. We have used every possible six-bit sequence at least once. Decoding of this code will be performed by a sliding block decoder such as illustrated in Figure 3.6.5. The decoder uses a 12-bit shift register to hold the current code word and to "look ahead" at the next upcoming code word. Because we have arranged for different encoder states to produce different code words for a given source word, we can do decoding by table lookup using a PLA table having twelve inputs and sixteen lookup terms. The decoding is state-independent and, consequently, error propagation is limited to at most two source words (eight bits).

Perhaps you sensed a certain inevitability while we were coming up with the code in the previous example. Once we had recognized from the state transition matrix that

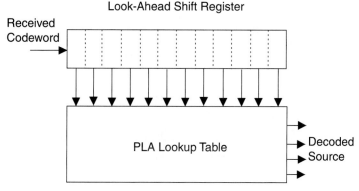

Figure 3.6.5: Sliding Block Decoder

a rate 4/6 code could be based around states 1 and 3 of the state diagram and performed state-splitting on state 3, the rest of the code specification actually held relatively few arbitrary choices (other than pairing source words up with code words). The result is a (0, 2; 4, 6) dc-free code with state-dependent coding and state-independent decoding by means of a sliding block decoder. Even our selection of code words was constrained by the main consideration of where we wished to terminate the sequence given its initial state.

In the example, a primary tool to which we returned again and again was the state transition matrix B_3^3. From this matrix, we learned how many sequences were at our disposal for assigning code words to transitions from the initial state to the final state. We coupled this with the realization that sequences originating outside of "a" could not depend on whether the initial state was "b" or "c" because "b" and "c" were the products of splitting the original state 3 and therefore were the *same* regardless of whether "b" or "c" was the originator. (If this is not obvious to you yet, compare B_3^3 with the original matrix B^6; the row sum of row 3 of B^6 shows only 21 possible sequences originating in state 3; splitting state 3 into a "b" and a "c" can not change the total; the row sums of rows 2 and 3 of B_3^3 add up to 42; therefore, the "extra" 21 sequences must be copies of the original 21 sequences from B^6.) These considerations dictated the structure of our encoder's "next state" scheme and our assignment of code words within it.

The particular code in Example 3.6.1 was developed by Karabed and Siegel and presented in their paper on MSN codes. In this same paper, they also presented a (0, 3; 6, 8) code based on an $N = 5$ state diagram and a (0, 4; 8, 10) code based on an $N = 7$ state diagram. They point out that a rate 4/5 code is possible using only $N = 6$, but they find that such a code is actually more difficult to implement than the one they reported on. These MSN codes have application in digital magnetic recording.

SUMMARY

This chapter has presented an overview of various types of practical data translation codes employing run-length-limit constraints. The primary classes of codes were: 1) state-independent fixed-length block codes; 2) state-independent variable-length prefix codes; 3) state-dependent fixed-length block codes; 4) state-independent fixed-length look-ahead codes, and; 5) state-dependent fixed-length sliding block codes.

Code analysis and design methods are heavily based on the state transition matrix (or "connection matrix") of the code. This matrix provides information on the maximum number of possible code words of length n. It is used to ascertain the code word size and its properties point towards which of the five approaches named above appears most attractive. The code rate must be less than the code's capacity, which is given by Equation (2.8.10).

The purpose of these codes is to respond to constraints imposed by the physical channel. These constraints usually take the form of run-length constraints (d and/or k) or constraints on the power spectrum of the code (such as the need for a dc-free code). The examples presented in this chapter are by no means a complete survey of the many different modulation codes used in practical commercial application but the examples *do* include a number of codes which are of either historical or current commercial importance.

In chapter four, we will turn our attention to another important class of block codes. These codes are called *channel codes* and their purpose is to detect and correct errors made in the transmission of information. Unlike data translation codes, where the purpose is to assist and improve bit detection, these channel codes are used to gain improvements in the reliability of the transmitted information over and above the performance that can be delivered through signal-processing and detection methods.

REFERENCES

P. Franaszek, "Sequence-state encoding for digital transmission," *Bell Syst. Tech. J.,* vol. 47, pp. 143–157, Jan., 1968.

P. Franaszek, "On synchronous variable-length coding for discrete noiseless channels," *Information and Control,* vol. 15, pp. 155–164, 1969.

P. Franaszek, "Sequence-state methods for run-length-limited coding," *IBM J. Res. Develop.,* vol. 14, pp. 376–383, July, 1970.

J. Eggenberger and P. Hodges, "Sequential encoding and decoding of variable word length, fixed rate data codes," U.S. Patent 4,115,768, 1978.

D. Tang and R. Bahl, "Block codes for a class of constrained noiseless channels," *Information and Control,* vol. 17, pp. 436–461, 1970.

K. Immink, "Some statistical properties of maxentropic runlength-limited sequences," *Phillips J. Res.,* vol. 38, pp. 138–149, 1983.

A. Lempel and M. Cohn, "Look-ahead coding for input-restricted channels," *IEEE Trans. Inform. Th.,* vol. IT-28, pp. 933–937, Nov., 1982.

G. Jacoby, "A new look-ahead code for increasing data density," *IEEE Trans. Magn.,* vol. MAG-13, pp. 1202–1204, no. 5, Sept., 1977.

M. Cohn and G. Jacoby, "Run-length reduction of 3PM code with full word look-ahead," *IEEE Trans. Magn.,* vol. MAG-18, pp. 1253–1255, Nov., 1982.

G. Jacoby and R. Kost, "Binary two-thirds rate code with full word look-ahead," *IEEE Trans. Magn.,* vol. MAG-20, pp. 709–714, Sept., 1984.

K. Immink, "Runlength-limited sequences," *Proc. IEEE,* vol. 78, no. 11, pp. 1745–1759, Nov., 1990.

P. Siegel, "Recording codes for digital magnetic storage," *IEEE Trans. Magn.,* vol. MAG-21, no. 5, pp. 1344–1349, Sept., 1985.

B. Marcus and P. Siegel, "On codes with spectral nulls at rational submultiples of the symbol frequency," *IEEE Trans. Inform. Th.,* vol. IT-33, no. 4, pp. 557–568, July, 1987.

R. Karabed and P. Siegel, "Matched spectral-null codes for partial-response channels," *IEEE Trans. Inform. Th.,* vol. 37, no. 3, pp. 818–855, May, 1991.

R. Adler, D. Coppersmith, and M. Hassner, "Algorithms for sliding block codes," *IEEE Trans. Inform. Th.,* vol. IT-29, no. 1, pp. 5–22, Jan., 1983.

EXERCISES

3.2.1: Determine the minimum m and n required for a $(0, 2)$ RLL code with rate $4/5$ using fixed-length state-independent block coding.

3.2.2: Design a state-independent fixed-length block code of rate $1/2$ and $(d, k) = (0, 2)$. What is the efficiency of this code?

3.2.3: Determine the minimum m and n required for a $(2, 7)$ RLL code with rate $1/2$ using fixed-length state-independent block coding.

3.3.1: Draw the state diagram and find the connection matrix for a $(d, k) = (2, \infty)$ code.

3.3.2: Franaszek's $(2, 7; 1, 2)$ code is a rate $1/2$ code. Using state-dependent fixed-rate coding, is it feasible to construct a $(2, 7; 2, 3)$ code?

3.3.3: Design a state-dependent fixed-length $(2, 7; 2, 5)$ code.

3.4.1: Find a $(1, 4; 1, 2)$ variable-length fixed-rate code. Use this code to encode the source sequence 1 1 0 1 0 0 0 0 0 1 0 0 1 (read left to right).

3.5.1: Find a $(2, 5; 1, 3)$ look-ahead code. Specify its state table.

3.5.2: Using the $(1, 7; 2, 3)$ code of Table 3.5.1, encode the following source sequence (read left to right), assuming that the encoder's initial state is A:

$$00\ 10\ 11\ 11\ 00\ 01\ 11\ 01$$

3.6.1: Using the MFM encoder of Figure 3.3.1, encode the following source sequence (read left to right), assuming that the encoder's initial state is A:

$$1\ 1\ 0\ 1\ 1\ 0\ 1\ 1\ 0$$

Calculate the RDS of the coded sequence.

3.6.2: What is the relationship between the number of states in Figure 3.6.2 and the digital sum variation of the code?

3.6.3: Find the k-constraint and the capacity of a code of the form of Figure 3.6.2 having a DSV of 7.

CHAPTER 4

Linear Block Error-Correcting Codes

4.1 GENERAL CONSIDERATIONS

4.1.1 Channel Coding for Error Correction

Information theory and the theory of error-correcting codes arose almost simultaneously in the years shortly after World War II. The former was the creation of Shannon, while the latter is usually credited to Richard W. Hamming. Shannon and Hamming were familiar with each other's work during those early days, although Hamming's paper on the subject of error-correcting codes was published a few years after Shannon's now famous paper. (Hamming's paper had to wait for some patents to be filed before appearing.)

Shannon's theory deals with measures of information and fundamental limits to its transmission. Through Shannon's second theorem, we know it is possible to send information with an arbitrarily small probability of error, provided the information rate is kept less than the channel's capacity. However, Shannon's theory does not tell us *how* to accomplish this. That is where error-correcting codes come in.

Error-correcting codes are often called *channel codes*. This is not to be confused with data translation codes (which are also used because of the channel). Indeed, source codes, error-correcting codes, and data translation codes are often used simultaneously, as illustrated in Figure 4.1.1. The source encoder performs data compression on data from the information source. The error-correcting code (ECC) encoder adds controlled redundancy to this data sequence, so that up to some limit, data errors that occur during transmission and reception can be corrected and the original information sequence recovered. Finally, the data translation code (DTC) encoder prepares the information sequence for physical transmission by applying any constraints on run length or power spectrum that the channel might require.

This process is essentially repeated in reverse at the receiver. The DTC decoder decodes the DTC to recover the encoded data stream. The ECC decoder checks this stream for errors and, if possible, corrects errors that have occurred during transmission. Finally, the source decoder decompresses the data sequence to provide the original information sequence to the information sink.

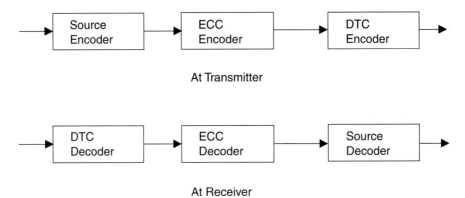

Figure 4.1.1: Encoding and Decoding Processes

A block error-correcting code accepts as its input a serially-transmitted block of symbols m_t drawn from a source alphabet M. If $M = \{0, 1\}$, the error-correcting code is called a *binary* error-correcting code. While non-binary error-correcting codes are important, in this introductory textbook we will confine ourselves to binary codes. The source data is organized into *blocks* of k bits each, $\overline{m} = (m_0, m_1, \cdots, m_{k-1})$. By standard notational convention, the bit m_{k-1} is taken as the first bit to be transmitted and the other bits follow in descending order of their subscripts. The ECC encoder maps this block into a *coded* sequence of n bits, $\overline{c} = (c_0, c_1, \cdots, c_{n-1})$, with $n > k$. The block $\overline{m}$ is called the *message* word, and $\overline{c}$ is called the *code word*. For a binary code, the code bits are drawn from the binary alphabet M. If we assume the message is composed of statistically independent bits and each symbol is equally probable, then the entropy of the source message is k bits. The ECC encoder adds $r = n - k$ redundant bits to the block. As we saw in chapter one, the addition of redundant bits to a block of bits does not change the entropy. However, we now will use the channel n times to send k bits of information. Therefore, the entropy rate of the coded block will be $R = k/n$. This is called the *code rate*. If the code rate is less than the channel capacity, Shannon's second theorem tells us there exists a code such that the probability of error at the receiver may be made arbitrarily small. Therefore, R is always chosen to be less than the channel capacity.

4.1.2 Error Rates and Error Distributions for the Binary Symmetric Channel

The ECC system views everything between its encoder and decoder as "the channel." (For this reason, DTCs are sometimes called "noiseless channels.") In most of what we will do in this text, we will assume the channel can be modeled as the discrete memoryless channel of chapter two. (One exception to this will be when we consider burst error-correcting codes.) For serial transmission of binary data, the appropriate DMC is the binary symmetric channel (BSC) with its crossover probability p. The BSC can be viewed graphically as a mapping from M to M with probabilities as shown in Figure 4.1.2. From chapter two, the transition probability matrix for the BSC is

$$P_{v|c} = \begin{bmatrix} 1 - p & p \\ p & 1 - p \end{bmatrix}, \qquad 4.1.1$$

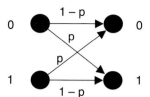

Figure 4.1.2 The BSC

where c is the channel input bit and v is the resulting channel output bit. The capacity of the BSC was given in chapter two as

$$C_c = 1 + p \log_2(p) + (1-p)\log_2(1-p). \tag{4.1.2}$$

The channel capacity and its cutoff rate

$$R_o = -\log_2(0.5 + \sqrt{p(1-p)}) \tag{4.1.3}$$

will be figures of merit against which we will compare the efficiencies of the codes we will be looking at.

There is a limit to the number of errors that may be corrected in any finite block $\bar{c}$. Under the binary symmetric channel model, we assume the probability of error in any particular bit is independent of any errors that might have been made in any other bit. In this case, the probability of having exactly t errors in a block of n bits with the BSC is

$$\Pr(t;p,n) = \binom{n}{t} p^t (1-p)^{n-t}, \tag{4.1.4}$$

where

$$\binom{n}{t} = \frac{n!}{(n-t)!\,t!} \tag{4.1.5}$$

is the binomial coefficient. Equation 4.1.4 is the binomial probability mass function.

The performance of the code and the system is governed by the statistics associated with Equation 4.1.4, so some review of its properties are in order. The probability that a block of n bits will have *fewer* than t errors during transmission is given by

$$\Pr(<t) = \sum_{j=0}^{t-1} \Pr(j;p,n) = \sum_{j=0}^{t-1} \binom{n}{j} p^j (1-p)^{n-j}. \tag{4.1.6}$$

Likewise, the probability of t or more errors is

$$\Pr(\geq t) = 1 - \Pr(<t). \tag{4.1.7}$$

The average number of errors in a block of n bits is

$$\bar{t} = \sum_{j=0}^{n} j \binom{n}{j} p^j (1-p)^{n-j} = np, \tag{4.1.8}$$

and the variance in the number of errors is

$$\sigma_t^2 = E[(t-\bar{t})^2] = \sum_{j=0}^{n} (j-\bar{t})^2 \binom{n}{j} p^j (1-p)^{n-j} = np(1-p). \tag{4.1.9}$$

The standard deviation, σ_t, is simply the square root of Equation 4.1.9 or $\sigma_t = \sqrt{np(1-p)}$.

The *raw error rate* of a channel is defined as the average number of errors in a block, divided by the number of bits in the block. From Equation 4.1.8, we see that the raw error rate and the channel crossover probability are the same in the BSC. A typical telecommunication channel is considered "good" if its raw error rate is better than about $p \approx 10^{-5}$ and is considered by many to be unusable for $p < 10^{-3}$. The current industry practice for hard disk drives requires the raw error rate to be better than $p = 10^{-9}$. The *unrecoverable* error rate, which is the error rate after error-correcting codes are applied by the disk drive, is typically specified in the range from $p < 10^{-12}$ to 10^{-15}, depending on the manufacturer and model.

EXAMPLE 4.1.1

Calculate the "three-sigma" error range $t_{3\sigma} = \bar{t} + 3\sigma_t$ for

$$p \in \{.1, .01, 10^{-3}, 10^{-4}, 10^{-5}\},$$

$$n \in \{7, 15, 31, 63, 127, 255\}.$$

Solution: Applying the equations above, the results are

			n			
p	7	15	31	63	127	255
0.1	3.081	4.986	8.111	13.444	22.842	39.872
0.01	0.86	1.306	1.972	2.999	4.634	7.317
0.001	0.258	0.382	0.559	0.816	1.196	1.769
10^{-4}	0.080	0.118	0.17	0.244	0.351	0.505
10^{-5}	0.025	0.037	0.053	0.076	0.108	0.154

It is interesting to note in this example that, although p changes by four orders of magnitude, the "three-sigma" range displays only about two orders of magnitude variation for a given value of n. On the other hand, as a function of n the "three-sigma" range varies only about one order of magnitude for a given p for slightly more than a one order of magnitude change in n.

EXAMPLE 4.1.2

Calculate the probability that the number of errors in a block will equal or exceed the value of $t_{3\sigma}$ from Example 4.1.1 for n up to 127.

Solution: The number of errors must always be an integer. The function $\lceil x \rceil$ is called the "ceiling function" and is defined as the smallest integer greater than or equal to x. The probability we seek is given by Equation 4.1.7, which can be written as

$$\Pr(t \geq t_{3\sigma}) = \sum_{j=\lceil t_{3\sigma}\rceil}^{n} \binom{n}{j} p^j (1-p)^{n-j} = 1 - \sum_{j=0}^{\lceil t_{3\sigma}\rceil -1} \binom{n}{j} p^j (1-p)^{n-j},$$

depending on which form is more convenient for calculating. Using the results from Example 4.1.1, the probabilities are

			n		
p	7	15	31	63	127
0.1	.0027	.0127	.0026	.0033	.0037
0.01	.068	.01	.038	.025	.0093
0.001	.007	.015	.031	.061	.0074
10^{-4}	$7 \cdot 10^{-4}$	.0015	.0031	.0063	.0126
10^{-5}	$7 \cdot 10^{-5}$	$15 \cdot 10^{-5}$	$31 \cdot 10^{-5}$	$63 \cdot 10^{-5}$	$1.27 \cdot 10^{-3}$

Notice that for those cases where $t_{3\sigma} > 1$, the probability of exceeding it is a few percent. This means that out of 100 n-bit blocks, only a few of these blocks will exceed the "three-sigma" value for numbers of errors. For those cases where $t_{3\sigma} < 1$, the probability of exceeding it in a block approaches the average number of errors per block. That is, if $p = .001$ and $n = 15$, the expected number of errors in a 15-bit block is .015 and the probability that a block will have greater than $t_{3\sigma}$ errors is also .015. This somewhat convenient result occurs because as $t_{3\sigma}$ falls below 1, the first term in the probability series comes to dominant the others and

$$\Pr(t \geq t_{3\sigma}) = \sum_{j=\lceil t_{3\sigma} \rceil}^{n} \binom{n}{j} p^j (1-p)^{n-j} \approx \binom{n}{1} p(1-p)^{n-1},$$

which is approximately for $\bar{t}$ for $p \ll 1$.

4.1.3 Error Detection and Correction

The raw error rate properties of the channel determine the amount of error control capability required of the channel code. Generally, a given code can be characterized in terms of its amount of *error detection* capability and its amount of *error correction* capability. Error detection refers to the ability of the decoder to tell if an error has been made in transmission. Error correction refers to the ability of the decoder to not only tell if one or more errors have been made, but, in addition, to be able to tell which bits are in error. A code may have the ability to do both. For example, a code might be designed to detect up to two bits of error in the block, but may only be able to correct errors if no more than one occurs in the block.

EXAMPLE 4.1.3 Repetition Codes

A repetition code simply repeats the transmitted information bits r additional times where r is called the redundancy of the code. Consider an encoder function G that performs the following mapping.

$$G(0) \rightarrow 000,$$
$$G(1) \rightarrow 111.$$

This code has a block length $n = 3$, a message length $k = 1$, and redundancy $r = 2$. This code can be used either to *correct* a single bit error or to *detect* up to two bits of error. It can not simultaneously do both. When used to detect errors, the decoder simply ascertains whether or not the received word is a valid code word. The operation can be tabulated as follows.

Received Word	Decoded Word	Error Flag
000	0	No
001	?	Yes
010	?	Yes
011	?	Yes
100	?	Yes
101	?	Yes
110	?	Yes
111	1	No

If the decision is made (by the system's designers) that having two bits of error in a block is very unlikely, the code can be used to correct single bit errors. Since it is assumed that any error that occurs will consist only of a single bit being in error, the logical decoding rule is to assign the received word to the legal code word "closest" to it. The "distance" between the received word and the two legal code words is measured by counting the number of bit positions in which the two words disagree. For example, 010 and 000 disagree in one bit position so the "distance" between them is 1. Words 010 and 111 disagree in two bit positions so the distance between them is 2. This measure of distance is called the *Hamming distance* between the two words.

Using the rule that assigns the received word to the code word at least Hamming distance, the error correction rule would be as follows:

Received Word	Decoded Word
000	0
001	0
010	0
011	1
100	0
101	1
110	1
111	1

In this case, each of the eight possible received words have a decoded word assigned to it. If a received word has two bits in error, it is not possible to detect this because the decoder will assume it is seeing a single bit error and base its "correction" on that assumption. Hence, we conclude that the $r = 2$ repetition code can detect up to two bits in error *or* it can correct single bit errors, but it can not do *both*. Furthermore, if we use the single-bit-correction decoding rule, not only will we not detect a two-bit error, we will in fact *miscorrect* the received word. For example, suppose the transmitted block was 111 and the received block was 001. This is a two-bit error. The decoder will "correct" this and output $m = 0$. The original message bit was $m = 1$.

EXAMPLE 4.1.4

A repetition code with $r = 3$ can correct single-bit errors *and* detect two-bit errors. Let the encoding rule be

$$G(0) \rightarrow 0000,$$

$$G(1) \rightarrow 1111,$$

and the decoding rule be

Received Word	Decode	Received Word	Decode
0000	0	1000	0
0001	0	1001	Error
0010	0	1010	Error
0011	Error	1011	1
0100	0	1100	Error
0101	Error	1101	1
0110	Error	1110	1
0111	1	1111	1

Under this decoding rule, if the received word has a Hamming distance of less than or equal to 1 from a valid code word, the decoder assumes that code word was originally transmitted and decodes the received word accordingly. If the Hamming distance between the received word and the two code words is 2, the decoder can not tell which code word to assign to the received word and indicates the occurrence of a two-bit error. An error of three or four bits will be miscorrected.

4.1.4 The Maximum Likelihood Decoding Principle

In the previous two examples, our decoding rule has been based on taking as the assumed transmitted word that code word lying at the least Hamming distance from the received word. Let us introduce some notation and then discuss why and when this is a good decoding rule. Let $\bar{v}$ be the received word, and let $\bar{c}_1$ and $\bar{c}_2$ be two valid code words. Let the notation $d_H(\bar{v},\bar{c}_1)$ denote the Hamming distance between $\bar{v}$ and $\bar{c}_1$. If $\bar{c}_1$ was the actual transmitted code word, then the number of errors which must have occurred during transmission is $t_1 = d_H(\bar{v},\bar{c}_1)$. If, on the other hand, the actual transmitted word was $\bar{c}_2$, then the number of errors during transmission must be $t_2 = d_H(\bar{v},\bar{c}_2)$. Suppose the decoder must decide which of $\bar{c}_1$ and $\bar{c}_2$ is most likely based on $\bar{v}$. Which code word should it choose?

The most likely code word is, by definition, the one with the greatest probability of occurring with the received word, $\bar{v}$. If the joint probabilities are written $p_{\bar{v},\bar{c}_1}, p_{\bar{v},\bar{c}_2}$, then we should choose $\bar{c}_1$ if $p_{\bar{v},\bar{c}_1} > p_{\bar{v},\bar{c}_2}$ and choose $\bar{c}_2$ if the opposite is true. The concise notation for this type of testing is sometimes written

$$p_{\bar{v},\bar{c}_1} \gtrless p_{\bar{v},\bar{c}_2}.$$

Suppose we assume $\bar{c}_1$ is the correct choice. If this assumption is correct, then it must be true that

$$\frac{p_{\bar{v},\bar{c}_1}}{p_{\bar{v},\bar{c}_2}} > 1.$$

If we take the logarithm of each side of this, we get the equivalent test condition

$$\ln(p_{\bar{v},\bar{c}_1}) - \ln(p_{\bar{v},\bar{c}_2}) > 0. \qquad 4.1.10$$

The individual terms in this expression are sometimes called the *log likelihoods*.

For the BSC with independent errors, the joint probabilities can be written

$$p_{\bar{v},\bar{c}_j} = p_{\bar{v}|\bar{c}_j} \cdot p_{\bar{c}_j}, j \in \{1,2\},$$

with the joint probabilities given as

$$p_{\bar{v}|\bar{c}_j} = \Pr(t_j) = p^{t_j}(1-p)^{n-t_j},$$

where n is the blocklength of the code word and p is the crossover probability of the BSC. The binomial coefficient does not appear in this expression because $\bar{v}$ has a *specific* error pattern in it for the given conditioning. Inserting these expressions into Equation 4.1.10 gives us

$$t_1 \ln(p) + (n - t_1)\ln(1-p) + \ln(p_{\bar{c}_1}) - t_2 \ln(p)$$
$$- (n - t_2)\ln(1-p) - \ln(p_{\bar{c}_2}) > 0.$$

Rearranging these terms gives us

$$(t_1 - t_2)\ln\left(\frac{p}{1-p}\right) > \ln\left(\frac{p_{\bar{c}_2}}{p_{\bar{c}_1}}\right). \qquad 4.1.11$$

If every possible message $\bar{m}$ is equally likely, then assuming that each $\bar{m}$ gets its own unique code word, then the quantity on the right-hand side of Equation 4.1.11 is equal to zero because all code words would have the same probability. This is the first condition that must be met for decoding based on minimum Hamming distance to be maximum likelihood.

If, in addition, the channel crossover probability satisfies $p < 0.5$, then

$$\ln\left(\frac{p}{1-p}\right) < 0,$$

from which we get the condition $t_1 < t_2$ if $\bar{c}_1$ is the most likely code word. This is equivalent to the Hamming distances satisfying the relation

$$d_H(\bar{v},\bar{c}_1) < d_H(\bar{v},\bar{c}_2).$$

This demonstrates why the decoding rules used in the previous examples were used and tells us the conditions under which these rules are correct in the probabilistic sense.

4.1.5 Hamming Distance and Code Capability

We are now in a position to understand the relationship between Hamming distance and the detection/correction capability of a code. Let M be a set of equally likely messages $\bar{m}$ of k bits each. We will assume $|M| = 2^k$. Let G be an encoding rule that assigns to every $\bar{m}_i \in M$ its own *unique* code word, $G(\bar{m}_i) \mapsto \bar{c}_i$, of n bits. Let C denote the set of legal code words. We require that C contain *only* code words in one-to-one correspondence with the elements of M and no others. In terms of mutual information, this restriction guarantees that $I(M;C) = H(M)$, so given any code word $\bar{c}_i$, we can uniquely identify its associated message word $\bar{m}_i$.

For every pair of code words, $\bar{c}_i, \bar{c}_j \in C (i \neq j)$, we can calculate a non-zero Hamming distance $d_H(\bar{c}_i,\bar{c}_j)$. There will be at least one pair of code words for which this distance is the least. We call this the *minimum Hamming distance of the code* and give it the symbol $d_{\min}$. The ability of the code to detect or correct errors is governed by the code's minimum Hamming distance. The relationships are:

1. A code can *detect* up to t errors if and only if $d_{min} \geq t + 1$;
2. A code can *correct* up to t errors if and only if $d_{min} \geq 2t + 1$;
3. A code can correct up to t_c errors *and* detect up to $t_d > t_c$ errors if and only if

$$d_{min} > 2t_c + 1 \quad and \quad d_{min} \geq t_c + t_d + 1.$$

EXAMPLE 4.1.5

The $r = 3$ binary repetition code of example 4.1.4 has a minimum Hamming distance of 4. Since $4 = 1 + 2 + 1$, this code can correct one-bit errors *and* detect two-bit errors. Since $4 < 2(2) + 1$, this code can not correct two-bit errors. Since $4 = 0 + 3 + 1$, this code can detect three-bit errors provided it does not attempt to correct one-bit errors ($1 + 3 + 1 = 5 > 4$).

The minimum Hamming distance of a code is related to the number of redundant bits, r, in the code word. In general, this relationship is not a simple one and is highly dependent on the rules by which the code is constructed. In general, however, the minimum distance is upper bounded by the relationship

$$d_{min} \leq r + 1. \qquad 4.1.12$$

This is called the *Singleton bound*.

Consider a set of 2^k unique binary messages of k bits each. For this set, we clearly have $d_{min} = 1$, since the set includes messages $(0\,0\cdots 0)$ and $(1\,0\cdots 0)$ and and all messages are unique. Since we have not added any redundancy, $r = 0$ and Equation 4.1.12 is satisfied with equality.

Now suppose we encode this set into a set of 2^k code words of length $n = k + 1$ by appending a *parity-check* bit to the end of each message. We define this check bit to be "1" if the message contains an odd number of "1" bits and "0" if the message contains an even number of "1" bits. This will increase d_{min} to 2. To prove this, notice that two messages $\overline{m}_1, \overline{m}_2$ are at minimum distance from each other only if they differ in exactly one bit position. Therefore, if one of them contains an even number of "1" bits, the other must have an odd number of "1" bits. The parity check bit for the "even" message will be "0," while that of the other will be a "1." Therefore, their respective code words will now have $d_{min} = 2 = r + 1$.

To see that it is possible to have $d_{min} < r + 1$, suppose our encoder merely copies the first bit of the message as its "check bit." If we did this, two minimum-distance messages having the same first bit will get the same value of check bits and therefore will still be at unit distance from each other. Since $1 < r + 1$, we see Equation 4.1.12 is an upper bound.

EXAMPLE 4.1.6

Suppose we wish to transmit seven-bit code words over a BSC with crossover probability $p = 0.05$, and we wish the probability of error in a block at the receiving end to be less than 10^{-3}. What is the maximum possible code rate we could achieve, and how does this rate compare to the channel capacity and the cutoff rate?

Solution: Suppose we use an error-correcting code capable of correcting t_c errors. The probability of having an uncorrectable error in a block then becomes

$$P_u = \sum_{j=t_c+1}^{7} \binom{7}{j} p^j (1-p)^{7-j} < 10^{-3}.$$

Trying different values for t_c, we find that $t_c = 2 \Rightarrow P_u = .0038$, $t_c = 3 \Rightarrow P_u = 1.936 \cdot 10^{-4}$. We would therefore need to correct three errors to achieve the goal. This would require that $d_{\min} \geq 7$. Since this exceeds the block size, we can not achieve the goal through error correction alone.

Suppose, instead, that we decide to use a code that can correct one error and detect up to three errors. If an error is detected, we could request the source to repeat the message. (This is a legitimate form of error control.) The minimum distance now required is $d_{\min} = 1 + 3 + 1 = 5$, which, in turn, requires $r \geq 4$. A direct inspection of the seven-bit block reveals that the only code that can meet the required $d_{\min}$ is a seven-bit repetition code, so $k = 1$. This code can correct one and detect five errors. This will give us a block error rate of $1.05 \cdot 10^{-7}$. We must *retransmit* any blocks having from two to five errors. The probability of retransmission is

$$P_{rx} = \sum_{j=t_c+1}^{t_d} \binom{7}{j} p^j (1-p)^{7-j} = 0.0444.$$

Retransmission slows down our information bits per channel use, and on the average, we must send each block $1/(1 - P_{rx})$ times. This gives us an information rate of

$$R = \frac{1 \cdot (1 - P_{rx})}{7} = .1365 \text{ bits per channel use.}$$

From Equations 4.1.2 and 4.1.3, we have channel capacity and cutoff rate, respectively, of

$$C_c = 0.7136; \quad R_o = 0.4781.$$

We see that this short block can not achieve the cutoff rate.

It is worthwhile at this point to remind ourselves of Shannon's second theorem back in Chapter two, Section 2.3. Recall from our discussion there that Shannon's theorem does *not* say we can achieve the channel capacity for any old block size n we may care to choose. The example above illustrates an important theme in information theory, namely, that longer block lengths are required to achieve higher rates and come closer to the channel capacity. The example also illustrates that the Singleton bound sets a *lower limit* on the redundancy, r. A number of powerful and important codes, such as the binary BCH codes, do not achieve the Singleton bound.

4.2 BINARY FIELDS AND BINARY VECTOR SPACES

4.2.1 The Binary Field

To do anything meaningful with the concepts discussed in the previous section, we must have a mathematical way to describe our codes. This is provided by a mathematical structure called a *field*. A field is defined to be a set of elements A and two arithmetic operations called *addition* ($+$) and *multiplication* ($\cdot$) such that the following set of properties hold:

1. *Closure:* For any two elements, $a, b \in A$, addition and multiplication are defined such that $a + b \in A$ and $a \cdot b \in A$;
2. *Associative properties of addition and multiplication:* For any three elements, $a, b, c \in A$, addition and multiplication are associative, i.e.,

$$a + b + c = (a + b) + c = a + (b + c)$$

and

$$a \cdot b \cdot c = (a \cdot b) \cdot c = a \cdot (b \cdot c);$$

3. *Identity elements:* The set A contains an element 0 called the *additive identity* and an element 1 called the *multiplicative identity* such that for any $a \in A$

$$a + 0 = 0 + a = a$$

and

$$a \cdot 1 = 1 \cdot a = a;$$

4. *Additive inverses:* For any element $a \in A$ there is some element called the additive inverse of a such that $a + b = 0$ (note: an element a is allowed to be its own additive inverse, i.e., $a + a = 0$); the additive inverse is usually written as $-a$;
5. *Multiplicative Inverses:* For any element $a \in A$ except 0, there is some element $b \in A$ called the multiplicative inverse of a such that $a \cdot b = 1$; the multiplicative inverse is usually written as a^{-1};
6. *Addition is commutative:* For any elements $a, b \in A$, the addition operation is commutative, i.e., $a + b = b + a$;
7. *Distributive property:* For any three elements $a, b, c \in A$,

$$a \cdot (b + c) = a \cdot b + a \cdot c$$

and

$$(a + b) \cdot c = a \cdot c + b \cdot c.$$

These seven properties are sufficient to build an algebra of amazing power and flexibility. We will usually follow the convention of writing multiplication as ab rather than $a \cdot b$. Notice that multiplication is *not* required to be commutative. We may have $ab \neq ba$ in our algebra.

EXAMPLE 4.2.1 Boolean arithmetic

The field composed of the set $A = \{0, 1\}$ is of fundamental importance to what we're doing. In much of the literature, this field is given the symbol GF(2) which stands for "Galois field with two elements." Addition and multiplication are defined for this field as

$$0 + 0 = 0; \quad 0 + 1 = 1 + 0 = 1; \quad 1 + 1 = 0;$$
$$0 \cdot 1 = 1 \cdot 0 = 0 \cdot 0 = 0; \quad 1 \cdot 1 = 1.$$

In digital logic, this addition operation is known as the "exclusive-or," and the multiplication operation is known as the "and" function. Notice that "1" is its own additive inverse, i.e., $-1 = 1$.

We can use the idea of the binary field GF(2) to construct binary vectors. Define A^n to be a set with elements $\bar{a} = (a_0, a_1, \cdots, a_{n-1})$ with each $a_i \in A = \{0,1\}$. We also need to define two arithmetic operations:

1. *Vector addition:* If $\bar{a}, \bar{b} \in A^n$, then vector addition (+) is defined as

$$\bar{a} + \bar{b} \equiv ((a_0 + b_0), (a_1 + b_1), \cdots, (a_{n-1} + b_{n-1}))$$

where the scalar addition terms are defined as in Example 4.2.1;

2. *Scalar multiplication:* If $\bar{a} \in A^n$ and $b \in A$ is a binary scalar, then

$$b \cdot \bar{a} = \bar{a} \cdot b = (ba_0, ba_1, \cdots, ba_{n-1}).$$

The two arithmetic operations just defined carry with them the assumption that these operations make sense. It seems like a safe assumption to make, but under what conditions is it really okay to do this? For example, what if $\bar{a} + \bar{b} \notin A^n$? Does this addition make sense if the result is outside of our set of defined vectors?

To properly set up a system of vector algebra using binary-valued numbers taken from GF(2), we need to establish a "universe" within which our arithmetic and algebraic operations work and make sense. This universe is called a *vector space*. A vector space is a structure made up of a set of vectors, A^n, a set of scalars, A, and the two arithmetic operations defined above. The definitions of the sets and these arithmetic operations are also subject to the following constraints:

1. *Closure:* For every $\bar{a}, \bar{b} \in A^n$, the sum $\bar{a} + \bar{b} \in A^n$;
2. Addition is commutative;
3. Addition is associative;
4. A^n contains a vector $\bar{0}$ such that for any element of $A^n, \bar{a} + \bar{0} = \bar{a}$;
5. *Additive Inverses:* For every $\bar{a} \in A^n$ there is some vector $\bar{b} \in A^n$ such that $\bar{a} + \bar{b} = \bar{0}$;
6. For every scalar $a \in A$ and every vector $\bar{b} \in A^n$ there is a vector $a\bar{b} \in A^n$;
7. Scalar multiplication is associative;
8. Scalar multiplication is distributive with respect to vector addition;
9. Scalar multiplication is distributive with respect to scalar addition;
10. If $1 \in A$ is the scalar multiplicative identity, then for every $\bar{a} \in A^n, 1\bar{a} = \bar{a}$.

These ten constraints plus the sets and arithmetic operations defined above seem innocent enough. We will find, however, that these modest rules and restrictions will suffice to let us develop a very powerful set of mathematics for describing error-detecting and error-correcting codes. Using these rules, we can construct matrices, define the transpose of a vector or a matrix, and generally do all of the things we are used to doing in ordinary linear algebra. The only trick (and experience shows it to be a tough one to catch on to) is *we must remember that* $1 + 1 = 0$. Our scalar field, A, does not contain a "2." Everything else is the same as what we are already used to.

EXAMPLE 4.2.2   A 2-D Code

A 2-D code is a simple single-error-correcting code with good additional error detection capability. In this example, we will look at a rate 9/16 2-D code and learn how to represent it using a binary vector space approach.

Section 4.2 Binary Fields and Binary Vector Spaces

Toward this end, suppose we have a 9-bit message vector

$$\bar{m} = [m_0 \ m_1 \ \cdots \ m_8], \quad m_i \in \{0,1\}.$$

By standard notational convention, we assume bit m_8 is the first bit transmitted. We wish to encode this message into a 16-bit binary vector

$$\bar{c} = [c_0 c_1 \cdots c_{15}] = [c_0 c_1 c_2 m_0 c_4 m_1 c_6 m_2 c_8 m_3 m_4 m_5 c_{12} m_6 m_7 m_8].$$

The bits labeled c_i are parity check bits calculated from the message bits. The encoding rule can be represented using the table shown below.

m_8	m_7	m_6	c_{12}
m_5	m_4	m_3	c_8
m_2	m_1	m_0	c_2
c_6	c_4	c_1	c_0

The check bits at the end of the rows are generated by taking the sum of the message bits of that row. For instance, $c_{12} = m_8 + m_7 + m_6$. (Remember: $1 + 1 = 0$.) Similarly, the bits at the bottom of each column are produced by summing down the column, i.e., $c_6 = m_8 + m_5 + m_2$. The peculiar order of the bits in the code vector is because the message is transmitted bit serially and the parity check bits are transmitted as soon as they become available from the message bit stream. The bit c_0 is defined as $c_0 = m_8 + m_7 + \cdots + m_1 + m_0$.

During transmission, bit errors may occur. We can represent the received word as a vector

$$\bar{v} = \bar{c} + [e_0 e_1 \cdots e_{15}] = \bar{c} + \bar{e},$$

where $e_i = 1$ if there is an error in the ith bit position and $e_i = 0$ if the ith bit is transmitted without error. (Again, remember that $1 + 1 = 0$.) At the receiver, we calculate seven error check bits, which we will call *syndrome bits*, according to the following table:

v_{15}	v_{14}	v_{13}	v_{12}	s_6
v_{11}	v_{10}	v_9	v_8	s_5
v_7	v_5	v_3	v_2	s_4
v_6	v_4	v_1	v_0	
s_3	s_2	s_1		s_0

The syndrome bits s_4, s_5, s_6 are calculated by summing across their respective rows. For instance, $s_6 = v_{15} + v_{14} + v_{13} + v_{12}$. Bits s_1 through s_3 are calculated by summing down their respective columns. Bit s_0 is given by

$$s_0 = v_0 + v_1 + \cdots + v_{15} = \sum_{i=0}^{15} v_i,$$

where the summation symbol implies modulo-2 summation (i.e., $1 + 1 = 0$, again).

This code can correct any single-bit error that occurs in the transmitted block. It can detect all two-bit errors. It can not be guaranteed to detect any larger number of errors (although it can detect *some* of them), and it can not be guaranteed that we will not *miscorrect* some cases where there is an odd number of multiple errors (although we can make it correct *some* of them). To see this, note the following properties of this code:

1. If no errors occur in the block, all the s_i bits equal zero;
2. If an *odd* number of errors occur, $s_0 = 1$;

3. If a single error occurs among the coded bits corresponding to one of the m_i message bits, one of the row *and* one of the column syndrome bits will be equal to 1; in this case, we can correct the erroneous bit by adding a 1 to the v_i bit corresponding to the intersection of the row and column belonging to the non-zero syndrome bits;
4. If a single error occurs among the coded bits corresponding to one of the parity bits, either one row syndrome *or* one column syndrome bit (but not both!) will be equal to one and the s_0 bit will be a 1; in this case, no correction is necessary because none of the message bits are in error;
5. If two errors occur, $s_0 = 0$, but *multiple* syndrome bits in a row *or* a column will be 1; if the multiple non-zero syndrome bits occur among the "column" bits, the "row" syndrome bits may or may not all be zero (and vice versa); we can not guarantee to correct this error, but we can *detect* it because of this state of the syndrome bits;
6. If an odd number of multiple errors occur (3, 5, etc.), we cannot guarantee that we will not *miscorrect* the error, because we can get syndrome bits that look like situation #4; we *can* guarantee to detect this kind of error, but if we do so, we will not be able to correct single-bit errors;
7. If an even number of multiple errors (more than two) occurs, we can not guarantee to detect the error.

It is left to you as an exercise to verify all of the previous statements.

4.2.2 Representing Linear Codes in a Vector Space

In describing the generation of the parity-check bits and the syndrome bits, we have used a lot of words. The description can be made much more concise by using the mathematics of the vector space. For the previous example, the code-word generation is given by

$$\bar{c} = \bar{m} \begin{bmatrix} 1 & 1 & 1 & 1 & 0 & 0 & 0 & 0 & 0 & 0 & 0 & 0 & 0 & 0 & 0 \\ 1 & 0 & 1 & 0 & 1 & 1 & 0 & 0 & 0 & 0 & 0 & 0 & 0 & 0 & 0 \\ 1 & 0 & 1 & 0 & 0 & 0 & 1 & 1 & 0 & 0 & 0 & 0 & 0 & 0 & 0 \\ 1 & 1 & 0 & 0 & 0 & 0 & 0 & 0 & 1 & 1 & 0 & 0 & 0 & 0 & 0 \\ 1 & 0 & 0 & 0 & 1 & 0 & 0 & 0 & 1 & 0 & 1 & 0 & 0 & 0 & 0 \\ 1 & 0 & 0 & 0 & 0 & 0 & 1 & 0 & 1 & 0 & 0 & 1 & 0 & 0 & 0 \\ 1 & 1 & 0 & 0 & 0 & 0 & 0 & 0 & 0 & 0 & 0 & 0 & 1 & 1 & 0 & 0 \\ 1 & 0 & 0 & 0 & 1 & 0 & 0 & 0 & 0 & 0 & 0 & 1 & 0 & 1 & 0 \\ 1 & 0 & 0 & 0 & 0 & 0 & 1 & 0 & 0 & 0 & 0 & 1 & 0 & 0 & 1 \end{bmatrix},$$

or in more concise form, $\bar{c} = \bar{m}G$, with G given by the preceding matrix. The G matrix is given by directly implementing the equations for the parity-check bits described earlier. It is worth your while to verify for yourself that the matrix equation above does, indeed, generate the proper parity-check bits. Notice that there is one column in G for each bit in the code word and one row in G for each bit in the message vector.

In a similar fashion, we can define a *syndrome vector* $\bar{s} = [s_0 s_1 \cdots s_6]$, which can be computed from an equation of the form $\bar{s} = \bar{c} H^T$, where H^T is a matrix having sixteen rows and seven columns. (The superscript T stands for matrix transpose; it is written this way to be consistent with some later notation; for now, you don't need to worry about "why transpose?".) It is left to you as an exercise to find the correct H^T matrix for this code.

EXAMPLE 4.2.3

What is d_{min} for the code in Example 4.2.2? Also, find the probability of a block error if this code is used with the BSC of Example 4.1.6, and find the code rate if we correct single errors and retransmit when we detect double errors.

Solution: This code corrects $t_c = 1$ errors and detects $t_d = 2$ errors. Its minimum distance between code words is therefore $d_{min} = t_c + t_d + 1 = 4$. (Note: $r = 7$).

Using the method described in the previous example, we find that the probability of an undetected error in a block is $P_u = 0.0429$. This is really no improvement over sending uncoded messages over this channel. The probability of having to retransmit is $P_{rx} = 0.1463$ and so the code's rate is $R = 0.4802$. Notice this is *above* the cutoff rate of the BSC (from Example 4.1.6). The poor performance of this code and the fact its code rate exceeds the cutoff rate in the BSC is *not* a coincidence.

By the way, the fact we have to retransmit almost one in every seven blocks at this poor error rate has an adverse effect on our probability of an undetected block. We will show later in this text that the probability of an uncorrected block when we retransmit blocks with detected errors is degraded by the fact that we now have more opportunities to accept a block with undetected errors. (We detected an error the first time, but we might miss detecting the error the second time!) The probability of accepting blocks with undetected errors when we retransmit will later be shown to be

$$P_B = \frac{P_u}{1 - P_{rx}}, \qquad 4.2.1$$

which is 0.0503 for this example. (The code is actually *worse* than sending uncoded messages for this channel!) If nothing else, this example serves to illustrate the need to understand the channel capacity characteristics when deciding on an error-correcting code.

4.3 LINEAR BLOCK CODES

4.3.1 Elementary Properties of Vector Spaces

Linear block codes are one of the most important classes of error-correcting codes. This importance is due in no small respect to the fact that such codes can be treated mathematically using the mathematics of vector spaces. The example code in the previous section is a linear block code. In this text, we will be dealing with binary linear block codes. There are also nonbinary linear block codes, of which the most important example are the Reed-Solomon codes, but we must save the nonbinary codes for a more advanced course in error-correcting codes.

One of the most important defining features of a linear block code is *superposition*. Suppose that the set of legal code words is C. For a linear block code, if $\bar{c}_1, \bar{c}_2, \in C$, then $\bar{c}_1 + \bar{c}_2 \in C$. The sum of any two code words is also a code word. (This can be taken as the *definition* of a linear code.) For binary codes, one obvious consequence of this is that the all-zero code word, $\bar{0}$, must be an element of the set of code words C.

This is easy to see because, for any binary-valued code word, $\bar{c}_1 + \bar{c}_1 = \bar{0}$, and this sum must also be a code word. (As it turns out, *nonbinary* linear block codes must also have $\bar{0} \in C$.) Furthermore, for the scalars 0 and 1, we have $0 \cdot \bar{c} = \bar{0}$ and $1 \cdot \bar{c} = \bar{c}$ from our previous definition of scalar multiplication. If you think about it for a little while, you should be able to conclude that these facts add up to the statement that a linear block code satisfies all of the conditions required for a *vector space*. Thus, all the rules for vector spaces we discussed in the previous section apply to linear block codes. From this point on, whenever we say "code" we will mean a linear block code of this type.

An (n, k) linear block code encodes k-bit message vectors into n-bit code vectors. The code rate is $R = k/n$. If our set of legal messages is the set of all possible k-bit vectors, then our code word set will have $|C| = 2^k$ distinct legal code vectors. (We know that they are distinct because it would be a bad move to assign the *same* code word to *two different* messages; we are only interested in error-correcting codes that are information lossless!) The vector space of our code therefore consists of 2^k vectors of n bits each.

It is always possible to define our code words using a matrix G having k rows and n columns, as we did for the example code in the previous section. Suppose that we let the n-element vectors $\bar{g}_0, \bar{g}_1, \cdots, \bar{g}_{k-1}$ represent the rows of this G matrix. Our code words are then defined as

$$\bar{c} = \bar{m}G = \bar{m} \begin{bmatrix} \bar{g}_0 \\ \hline \bar{g}_1 \\ \hline \vdots \\ \hline \bar{g}_{k-1} \end{bmatrix} \equiv m_0 \bar{g}_0 + m_1 \bar{g}_1 + \cdots + m_{k-1} \bar{g}_{k-1}. \qquad 4.3.1$$

The code vectors are merely linear combinations of the rows of the G matrix. Since we must have a set of *unique* code words, this means the k rows of the G matrix must be *linearly independent*. This means that no row vector $\bar{g}_i$ can be expressed as a linear combination of the *other* rows of the G matrix. These are very special vectors. They are called the *basis vectors* of the vector space C. If you look at Equation 4.3.1 for a minute, you should be able to see the $\bar{g}_i$ vectors must all be legal code vectors and none of them can be the all-zero vector.

The *dimension* of a vector space is defined to be the number of basis vectors it takes to describe it. The dimension of a code is defined to be equal to the dimension of its vector space. From Equation 4.3.1, we see the dimension of the vector space C is k. We therefore have a k-dimensional vector space made up of 2^k n-element vectors with $n > k$. It is possible that this is the first time you have encountered a vector space where the number of dimensions is less than the number of elements in the vector. If so, you should not let this bother you. If it does, it may be because you are used to equating the number of elements in the vector with the dimension of the space. If so, you are presented here with the opportunity to unlearn something that isn't true[1]. The

---

[1] In the words of the great American "philosopher," Will Rogers, "It ain't what we don't know that gets us into trouble. It's what we know that ain't so."

dimension of a space equals the number of basis vectors required to describe it. The dimension doesn't (necessarily!) have anything to do with how many elements the vectors have. If you are still uncomfortable, have patience. We'll have more to say about this distinction in a little while.

4.3.2 Hamming Weight, Hamming Distance, and the Hamming Cube

In Example 4.1.3 we defined the Hamming distance between two code words to be the number of positions in which they differed. We can apply this definition to vectors as well. Let us represent the Hamming distance between two vectors with the notation $d_H(\bar{c}_1, \bar{c}_2)$. Let us also define the *Hamming weight* of a vector as the number of nonzero elements in that vector. Our notation for Hamming weight will be $w_H(\bar{c})$. For instance,

$$\bar{c} = [1\,0\,0\,1\,1\,0\,1] \Rightarrow w_H(\bar{c}) = 4.$$

Since $1 + 1 = 0$, the Hamming distance $d_H(\bar{c}_1, \bar{c}_2)$ between two vectors must be equal to the Hamming weight of their sum,

$$d_H(\bar{c}_1, \bar{c}_2) = w_H(\bar{c}_1 + \bar{c}_2). \qquad 4.3.2$$

We can use this to learn something very useful about linear block codes. Since it is clear that

$$d_H(\bar{c}, \bar{0}) = w_H(\bar{c}), \qquad 4.3.3$$

it must also be true that

$$w_H(\bar{c}_1 + \bar{c}_2) = d_H(\bar{c}_3 = \bar{c}_1 + \bar{c}_2, \bar{0}). \qquad 4.3.4$$

Since $\bar{c}_3 \in C$, Equations 4.3.2 and 4.3.4 tell us that *the Hamming distance between any two code words in* **C** *must equal the Hamming distance between some other code word and the all-zero code word, which, in turn, is merely the Hamming weight of that code word*. This has an important corollary. Suppose that $\bar{c}_1$ and $\bar{c}_2$ are at the minimum Hamming distance of the code, i.e.,

$$d_{\min} = d_H(\bar{c}_1, \bar{c}_2).$$

Then there exists a code word $\bar{c}_3 = \bar{c}_1 + \bar{c}_2$ with Hamming weight $w_H(\bar{c}_3) = d_{\min}$. Therefore, *the minimum Hamming distance of a linear block code is equal to the minimum Hamming weight of the nonzero code vectors.*

Among other things, this tells us something useful about our basis vectors. Since each $\bar{g}_i \in C$, we must require that $w_H(\bar{g}_i) \geq d_{\min}$ for each basis vector. This condition is not *sufficient* to define a code for a particular $d_{\min}$, but it *is* a *necessary* condition.

Now let us consider the set V of possible received vectors defined by

$$\bar{v} = \bar{c} + \bar{e}. \qquad 4.3.5$$

Since the error vector $\bar{e}$ may be *any* vector of n bits, it is clear that V is the set of 2^n n-bit binary vectors and is an n-dimensional vector space. Further, since $\bar{e} = \bar{0}$ is one of the permitted "error" vectors (the "zero-error vector"), we clearly have $C \subset V$. The vector space defined by C is called a vector *subspace* of the vector space V.

The relationship between C and V may be understood with the help of a *Hamming cube*. Consider the (3, 1) repetition code of Example 4.1.3. This is a 1-dimensional code ($k = 1$) with a single basis vector $\bar{g}_0 = [1\,1\,1]$. The Hamming cube of Figure 4.3.1 illustrates

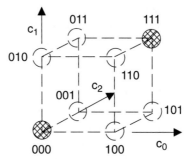

Figure 4.3.1: Hamming Cube

how C is embedded in the three-dimensional space V. The cross-hatched circles represent the two legal code vectors. The open dashed circles are points in V which are not legal code vectors. The minimum Hamming distance of the (3, 1) repetition code is 3. This is shown clearly in the figure. It is also clear why this code can correct a single-bit error. Such an error produces a vector in V that is closer to the transmitted code vector than to the other one. It is also clear why this code must give up error correction if it is to detect double-bit errors. A double-bit error produces a vector in V that is closer to the wrong legal code vector.

The two legal code vectors define a kind of "line" within the three-dimensional space of V. In a real sense, our code space is analogous to restricting ourselves to one line (or one plane in the case of a higher-dimensional code in a higher dimension V-space) within the higher dimensional space defined by V. This "line" does not lie along one of the "coordinate" axes labeled c_i in the figure[2]. The code's basis vector is [1 1 1] while the space V uses [1 0 0], [0 1 0], and [0 0 1] as its basis vectors.

So far, we've been a bit loose in our use of words such as "distance," and you've probably been willing to accept the term "Hamming distance" as just another colorful technical term (perhaps mentally viewing it as "Hammingdistance"). With the introduction of the Hamming cube in Figure 4.3.1, we're starting to sound like we're serious about the notion that the Hamming distance really is a distance in a real sense. What sense could we mean, though? We all know from elementary analytic geometry that the "distance" between [0 0 0] and [1 1 1] in Figure 4.3.1 is "really" $\sqrt{1^2 + 1^2 + 1^2} = \sqrt{3}$, not 3 (!). What are we talking about?

In mathematics, the term "distance" is derived from a more abstract concept called a "metric space." A metric space is defined as a set (call it A^n for our purposes) and a function ρ called a "metric function." ρ is a function of two elements from the set A^n and the function's result is an element of the set of real numbers. It is a generalized measure of the "distance" between these two elements in the space defined by A^n. If $\bar{a}, \bar{b}, \bar{c}$ are elements of the set A^n, a metric function is any function that satisfies the following three properties:

1. $\rho(\bar{a}, \bar{b}) = 0 \Rightarrow \bar{a} = \bar{b}$;
2. $\rho(\bar{a}, \bar{b}) = \rho(\bar{b}, \bar{a})$; and
3. $\rho(\bar{a}, \bar{c}) \leq \rho(\bar{a}, \bar{b}) + \rho(\bar{b}, \bar{c})$ (This is called the triangle inequality.)

---

[2] It would be more correct to label these axes as v_i in the figure but I wanted to make the correspondence with the code bits in the code vector clear.

Observe that our familiar, everyday "distance" $\sqrt{1^2 + 1^2 + 1^2} = \sqrt{3}$ is one example of such a function and is called the "Euclidean distance" metric function. The Hamming distance is another example of a metric function and its result (while always an integer) is a legitimate measure of "how close or how far apart" two things are in our vector space. Thus, from a slightly abstract point of view, the Hamming distance really *is* a distance. Other abstract "distances" are going to be coming up later in the text, so it's just as well we prepare ourselves for them now.

4.3.3 The Hamming Sphere and Bounds on Redundancy Requirements

Let's get back to codes. Suppose that $\bar{e} \in C$ in Equation 4.3.5. Then $\bar{v} \in C$, too. However, unless $\bar{e} = \bar{0}$, an error has been made. Since $\bar{v}$ is a legal code vector, there is no way to know that the error has occurred. On the other hand, if $\bar{e} \notin C$, then $\bar{v} \notin C$. How do we know this? If $\bar{e} \notin C$, it cannot be represented by any linear combination of our code's basis vectors. Therefore, $\bar{v}$ must contain a term that cannot be represented by the basis vectors, and consequently, $\bar{v}$ cannot be represented by the code's basis vectors. Therefore, $\bar{v}$ cannot be an element of C. In this case, we can detect the occurrence of the error by noticing that $\bar{v}$ is not a legal code vector. We conclude that the set of *undetectable errors, U,* is simply *the set of all errors that are legal code vectors.*

We can use the geometric interpretation given by the Hamming cube to gain some more useful insights into error-correcting codes. Suppose that V is an n-dimensional space within which lie our 2^k valid code vectors. We define a *Hamming sphere* of radius t as the set of all possible vectors $\bar{v}$ that are at a Hamming distance less than or equal to t from a code word. Without any loss of generality, we may assume that this code word is $\bar{0}$. The *volume* of the Hamming sphere is defined to be the number of vectors $\bar{v}$ contained within it. For our binary codes, it is relatively easy to show that the volume of the Hamming sphere is

$$\zeta(n, t) = \sum_{j=0}^{t} \binom{n}{j}. \qquad 4.3.6$$

We can associate each of the 2^k code vectors with a Hamming sphere of radius t. Suppose that these spheres do not overlap. Since the total volume of all of these spheres cannot exceed the total number of vectors in V, it follows that

$$2^k \zeta(n, t) \leq 2^n$$

Taking logarithms to the base two of each side and rearranging terms, we have

$$n - k = r \geq \log_2 \left(\zeta(n, t) \right). \qquad 4.3.7$$

Equation 4.3.7 is known as the *Hamming Bound.* It sets a lower limit on the number of redundant bits, r, required to correct t errors in an (n, k) linear block code.

There is another useful bound, called the *Gilbert Bound,* which places an *upper* limit on the number of redundant bits required to correct t errors. The proof of the Gilbert bound is a little more abstract than the proof of the Hamming bound (although it, too, uses the Hamming sphere idea) and we shall not give this proof here. We will, however, state the result. The Gilbert bound is

$$r \leq \log_2 \left(\zeta(n, 2t) \right). \qquad 4.3.8$$

Between Equations 4.3.7 and 4.3.8, we can rather quickly get a range on the value of r needed. Note, however, that the Gilbert bound does *not* say any code picked at random with r redundant bits satisfying Equations 4.3.8 and 4.3.7 will be able to correct t errors. The Gilbert bound merely says that *there exists* a code that can do this.

A code that satisfies the Hamming bound with equality is called a "perfect code." There are only three kinds of perfect codes. These are the binary repetition codes, the Hamming codes, and the Golay code. If you were wondering, "perfect" does *not* mean "best." There are a number of much more powerful codes than these that are not "perfect."

4.4 DECODING LINEAR BLOCK CODES

4.4.1 Complete Decoders and Bounded-Distance Decoders

We have seen that an encoder for an ECC can be implemented in the form $\bar{c} = \bar{m}G$. We now turn to the decoder. The job of the decoder is to take as its input the received vector $\bar{v} = \bar{c} + \bar{e}$ and produce as its output the best estimate of the transmitted message vector $\bar{m}$. Since the correspondence between $\bar{m}$ and $\bar{c}$ is one-to-one, the decoder's task can also be viewed as producing the most likely code vector $\bar{c}$ given $\bar{v}$ and the statistical properties of the channel's error vectors As discussed in Section 4.1.4, the best choice for $\bar{c}$ is the one that maximizes the joint probability $p_{\bar{v},\bar{c}}$. The joint probability can be written as $p_{\bar{v}|\bar{c}} p_c$, and if all message vectors are equally probable, the code vectors are also equally probable. In this case, the best decoder strategy is the one which maximizes the conditional probability $p_{\bar{v}|\bar{c}}$. However, given $\bar{v}$ and $\bar{c}$, this is equivalent to picking that $\bar{c}$ for which the error probability $p_{\bar{e}}$ is maximized.

For the BSC with crossover probability p, the expected number of errors in a block was seen earlier to be np. If the ECC is designed to correct t errors and if $np < t$, then the expected number of errors will fall inside the Hamming sphere of the code word for which $d_H(\bar{v}, \bar{c})$ is least. In this case, the decoder is called a "minimum distance" decoder. Its selected code word will be that code vector $\bar{c}$ which minimizes $d_H(\bar{v}, \bar{c})$. This is the strategy most often employed in selecting an ECC, since as our previous examples have indicated, having $np < t$ is mandatory if the code is to help improve the corrected error rate over the channel. (It is also often a good idea to include additional error detection capability up to $t_d \approx np + 3\sigma_t$ where $\sigma_t = \sqrt{np(1-p)}$.)

For the present, let us focus on the problem of error correction. ECC decoders may be classified into two categories. A *complete* decoder is a decoder that selects that $\bar{c}$ which produces the minimum $d_H(\bar{v}, \bar{c})$. A *bounded-distance* decoder is one that does the same thing if there is a $\bar{c}$ such that $d_H(\bar{v}, \bar{c}) \leq t$ and declares a *decoder failure* otherwise. The reasoning behind this strategy is simply that if $d_H(\bar{v}, \bar{c}) > t$, the probability of making a decoding error is significantly higher than if $d_H(\bar{v}, \bar{c}) \leq t$. Declaration of a decoder failure, therefore, can be viewed as a form of error detection: The decoder is recommending that the block be retransmitted. A bounded distance decoder would therefore be one that adopts the combined error correction plus error-detection strategy mentioned earlier.

The simplest, least clever, and often most expensive strategy for implementing error correction is to simply look up $\bar{c}$ in a decoding table that contains all possible $\bar{v}$. This is called a *standard-array decoder,* and the lookup table is called the standard array. The standard array must obviously contain 2^n table entries and is economically practical only for small block lengths n. The lookup table is constructed as a matrix with 2^k columns and $2^{n-k} = 2^r$ rows. The first row contains all the valid code vectors. The remaining entries in each column are given by adding to the code vector in the first row of that column all of the possible error vectors, beginning with the error vectors of highest probability of occurrence, until all 2^n table entries have been made. An example will illustrate this.

EXAMPLE 4.4.1

Construct the standard array for a $(6, 2)$ repetition code having code vectors of the form

$$\bar{c} = [m_0 \, m_1 \, m_0 \, m_1 \, m_0 \, m_1].$$

Solution: The first row of the standard array consists of the four legal code vectors

$$[0\,0\,0\,0\,0\,0], \; [0\,1\,0\,1\,0\,1], \; [1\,0\,1\,0\,1\,0], \; [1\,1\,1\,1\,1\,1].$$

Since this is a linear block code, it is clear that we have $d_{\min} = 3$, since the smallest Hamming weight of the nonzero code vectors is three. This code therefore has correction capability of $t = 1$. This implies that we have a channel in which single-bit errors are the most probable kind of error. There are six different single-bit error patterns, so the first $6 + 1 = 7$ rows of the standard array are found by adding the six possible single-error patterns to the legal code vectors as follows:

000000	010101	101010	111111
000001	010100	101011	111110
000010	010111	101000	111101
000100	010001	101110	111011
001000	011101	100010	110111
010000	000101	111010	101111
100000	110101	001010	011111

This does not complete the table, since there are still nine more rows to be added. If we were implementing a bounded-distance decoder, we would stop here. If $\bar{v}$ did not appear in this table, we would declare a decoding failure.

To implement a complete decoder, we have to add the remaining nine rows. We have now exhausted all of the single-error cases, since, as you may observe from the above, we have generated all the possible $\bar{v}$ that can result from single-error error vectors. The next most likely error patterns would be weight 2 (contain two errors). There are

$$\binom{6}{2} = \frac{6 \cdot 5}{2!} = 15$$

of these errors. We do not have 15 more rows left to fill, so we will not be able to use them all. The ones we must use are ones that do not result in *duplicate* entries in the table. If you have an eye for patterns, you can see (and if you don't, you can verify) the following weight-2 error patterns generate duplicate table entries: [1 0 1 0 0 0], [0 1 0 1 0 0], [0 0 1 0 1 0], [0 0 0 1 0 1], [0 1 0 0 0 1], and [1 0 0 0 1 0]. The remaining nine weight-2 patterns do not generate duplicate table entries, so we use them.

000000	010101	101010	111111
000011	010110	101001	111100
000110	010011	101100	111001
001100	011001	100110	110011
011000	001101	110010	100111
110000	100101	011010	001111
001001	011100	100011	110110
010010	000111	111000	101101
100100	110001	001110	011011
100001	110100	001011	011110

These additional nine rows plus the seven we did earlier complete the standard array. Notice how the error pattern we are using always appears in the first column of the standard array. Note also that the six weight-2 patterns we did not use are every bit as likely as the ones we did. We will miscorrect those error patterns should they occur.

When decoding with the standard array, we identify the column of the array where the received vector appears. The decoded vector is the vector in the first row of that column. In mathematical language, the rows of the standard array are called *cosets*. The elements in the first column (which are the error patterns being "corrected") are called *coset leaders*. These are terms that occasionally are used in the technical literature, so it can be handy to know them.

4.4.2 Syndrome Decoders and the Parity-Check Theorem

The standard-array method of decoding very quickly becomes impractically inefficient with increasing block length. It is also not a very clever approach because it fails to exploit the tremendous amount of *structure* built into a linear block code. We will now take a look at a much more efficient method known as the *syndrome decoder* method.

Recall that the rows of the code's *generator matrix*, G, are basis vectors in the code space C. The basis vectors of a space are not unique. Given any G with k linearly independent rows, we can always transform G using simple row and column operations to obtain another generator matrix. The rows of this new matrix are also a set of linearly independent basis vectors (just not the same ones we began with), and this matrix can generate a code that is equivalent, in terms of all of its important properties, to the original code. In particular, it is always possible to find an equivalent *systematic* code. A systematic code is one for which the generator matrix is of the form

$$G = [P_{k \times r} \vdots I_{k \times k}], \qquad 4.4.1$$

where $I_{k \times k}$ is a $k \times k$ identity matrix and $P_{k \times r}$ is a $k \times r$ matrix called the *parity-bit generator*. A code generated by a generator matrix of the form in Equation 4.4.1 is called a systematic code because its code vectors are of the form

$$\bar{c} = [c_0 \, c_1 \cdots c_{r-1} \, m_0 \, m_1 \cdots m_{k-1}]. \qquad 4.4.2$$

The original message bits appear in unaltered form in the code vector. This has a number of cost advantages when it comes to building hardware implementations of encoders and decoders, and since a systematic code is always possible (and there are no disadvantages to having one), the vast majority of linear block codes are usually implemented in systematic form. Bits $c_0 \to c_{r-1}$ of the code are called *parity bits*.

Assume we have a code with a generator matrix of the form in Equation 4.4.1. Let us define another matrix

$$H = [I_{r \times r} \mid -P^T], \qquad 4.4.3$$

where $I_{r \times r}$ is an $r \times r$ identity matrix and P^T is the transpose of the P matrix in Equation 4.4.1. H is called the *parity-check matrix*. For any code vector $\bar{c}$, we have

$$\bar{c}H^T = \bar{m}GH^T = \bar{m}[P_{k \times r} \mid I_{k \times k}]\begin{bmatrix} I_{r \times r} \\ -P_{k \times r} \end{bmatrix} = \bar{m}(P_{k \times r} - P_{k \times r}) \equiv \bar{0}. \qquad 4.4.4$$

Equation 4.4.4 is an important theorem. It says that the product of a valid code vector with the transpose of the parity-check matrix is always zero. Since by elementary row and column operations we can always transform G and H to equivalent nonsystematic codes, the parity-check theorem is true for any arbitrary linear block code.

There is a wealth of useful stuff resulting from the parity-check theorem. The first good thing is syndrome decoding. Let us define the syndrome vector

$$\bar{s} = [s_0 \; s_1 \cdots s_{r-1}] \qquad 4.4.5$$

in terms of the received vector $\bar{v}$ as

$$\bar{s} \equiv \bar{v}H^T. \qquad 4.4.6$$

Substituting for $\bar{v}$, we have

$$\bar{s} = (\bar{c} + \bar{e})H^T = \bar{c}H^T + \bar{e}H^T = \bar{0} + \bar{e}H^T = \bar{e}H^T. \qquad 4.4.7$$

The syndrome vector is a function only of the error vector $\bar{e}$ and is independent of $\bar{c}$. It is zero if and only if $\bar{e}$ is a valid code vector. (Remember: no errors = all-zero $\bar{e}$ vector = valid code vector.) We can detect the entire set of *detectable* errors by checking to see if $\bar{s}$ is zero or not.

We can say still more. Since $\bar{s}$ is independent of the code vector, all of the elements in the same row of the standard array must have the same syndrome. We can take this even further. We have seen that $GH^T = \bar{0}$. The only way this could be so is if the rows of H are vectors that are orthogonal to the rows of G. Since the rows of G are linearly independent, the rows of H must also be linearly independent, and therefore, the columns of H^T are linearly independent. Therefore, each of the 2^r rows of the standard array must generate a unique syndrome. Thus, the possible syndrome vectors are in one-to-one correspondence with the error patterns in the standard array, and a particular syndrome *uniquely identifies* a particular error pattern in the standard array!

This means that we do not ever need to store the standard array. It is sufficient to store in a table the $2^r - 1$ nonzero error patterns corresponding to the $2^r - 1$ possible nonzero syndromes. This is a very significant savings! We then perform error correction as follows. If the syndrome is zero, we assume that there is no error. If the syndrome is nonzero, we look up the error pattern associated with our syndrome and subtract it from the received vector. (Since we're doing binary codes, every vector is its own additive inverse, and therefore, subtraction and addition are the same thing in this vector space; we can't even make a sign mistake!)

If we are doing a complete decoder, our syndrome lookup table will contain all of the possible syndromes. If we are doing a bounded-distance decoder, we need store only those error patterns which satisfy our distance criterion. If we get a nonzero syndrome for which there is no table entry, we declare a decoder failure. This is how we might do a "correct t_c, detect t_d" ECC strategy.

There is one more useful item stemming from the parity-check theorem. We have seen that the rows of the H matrix are linearly independent and there are r rows. These r linearly independent rows form a set of basis vectors for an r-dimensional vector space. Every vector in this space is orthogonal to the vectors in C. The vector space defined by H is called the *dual space* of C and is usually given the symbol $C^\perp$. Since H has n columns and defines a vector space, H can be used as the generator matrix for an (n, r) linear block code. This code is called the *dual code* of C. (When you find one code, you automatically get another; that's kind of handy, isn't it?) The parity-check matrix for the dual code is simply G^T.

4.5 HAMMING CODES

4.5.1 The Design of Hamming Codes

The Hamming codes were the first major class of linear block error-correcting codes. They were developed by Richard Hamming at essentially the same time Shannon was developing information theory. Initial publication of these codes was delayed for patent reasons, and Hamming's initial paper on them did not appear until 1950, although Shannon makes reference to Hamming's work in his landmark paper.

The Hamming codes are a family of single-error-correcting codes. They are "perfect" codes, i.e., their redundancy meets the Hamming bound with equality. A Hamming code exists for any $r \geq 3$. The block length of a Hamming code is given by

$$n = 2^r - 1, r \geq 3.$$

The rate of a Hamming code is therefore given by

$$R = \frac{2^r - r - 1}{2^r - 1}.$$

The first few (n, k) parameters of the Hamming codes are $(7, 4), (15, 11), (31, 26)$, etc.

One distinguishing feature of Hamming codes is that they are among the easiest error-correcting codes to construct. To specify a Hamming code of length $2^r - 1$, we begin with the parity-check matrix H for the systematic Hamming code. This matrix has r rows and n columns. The first r columns are specified as the $r \times r$ identity matrix. For the remaining k columns, we simply choose the columns so that the columns consist of all nonzero binary vectors of length r. An example will illustrate this.

EXAMPLE 4.5.1

Construct the parity-check matrix for the (7, 4) systematic Hamming code.

Solution:

$$H = \begin{bmatrix} 1 & 0 & 0 & 1 & 1 & 0 & 1 \\ 0 & 1 & 0 & 1 & 0 & 1 & 1 \\ 0 & 0 & 1 & 0 & 1 & 1 & 1 \end{bmatrix}.$$

Note how the set of columns in H make up all possible three-bit binary vectors.

Recall from the previous section that the form of the parity-check matrix for a systematic code is

$$H = [I \mid -P^T].$$

Therefore, the generator matrix G for the systematic code can be written down immediately from the H matrix as

$$G = [P \mid I].$$

EXAMPLE 4.5.2

Write down the generator matrix for the Hamming code of Example 4.5.1.

Solution:

$$G = \begin{bmatrix} 1 & 1 & 0 & 1 & 0 & 0 & 0 \\ 1 & 0 & 1 & 0 & 1 & 0 & 0 \\ 0 & 1 & 1 & 0 & 0 & 1 & 0 \\ 1 & 1 & 1 & 0 & 0 & 0 & 1 \end{bmatrix}.$$

Note how the rows of the submatrix P make up the set of all r-bit binary vectors of weight greater than 1.

For all Hamming codes, $d_{min} = 3$. They may be used to correct single errors or they may be used to detect double errors. Error correction is by far the most common use of these codes. The codes may be decoded using a syndrome table. The table is particularly easy to calculate, since all of its entries correspond to error vectors $\bar{e}$ of unity Hamming weight. For the error vector with its error in the ith column of $\bar{e}$, the corresponding syndrome is simply the ith row of H^T.

EXAMPLE 4.5.3

Construct the syndrome table of the Hamming code in Example 4.5.1.

Solution:

$\bar{e}$	$\bar{s}$
1000000	100
0100000	010
0010000	001
0001000	110
0000100	101
0000010	011
0000001	111

There is an interesting relationship between these syndromes and the position of the error in the received code vector. If we count bit positions in the error vector from left to right, starting with 1, this position corresponds to the number obtained from reading the syndrome from *right* to *left*, with the exception of syndromes 001 and 011. If we were to swap these two rows of H^T and the corresponding two columns of G, we would have a nonsystematic Hamming code in which the syndromes, when read from right to left, gave the column position of the error in the received vector (numbering the columns left to right as 1 through 7). For the code in this example, the corresponding matrices are

$$G = \begin{bmatrix} 1 & 1 & 1 & 0 & 0 & 0 & 0 \\ 1 & 0 & 0 & 1 & 1 & 0 & 0 \\ 0 & 1 & 0 & 1 & 0 & 1 & 0 \\ 1 & 1 & 0 & 1 & 0 & 0 & 1 \end{bmatrix}, \quad H = \begin{bmatrix} 1 & 0 & 1 & 0 & 1 & 0 & 1 \\ 0 & 1 & 1 & 0 & 0 & 1 & 1 \\ 0 & 0 & 0 & 1 & 1 & 1 & 1 \end{bmatrix}.$$

Note how the columns of the H matrix, when read from the bottom row to the top row, form a simple counting sequence $1, 2, 3, \ldots, 7$. Also note that we have exchanged the third and fourth columns of this H matrix relative to Example 4.5.1 and that we have exchanged the same columns of the G matrix relative to Example 4.5.2. The code vectors formed by G are of the form

$$\bar{c} = [c_0 \ c_1 \ m_0 \ c_3 \ m_1 \ m_2 \ m_3].$$

While this is not the systematic form, we see that it is a trivial column permutation of the systematic form and, therefore, equally easy to generate. The advantage of this form of the Hamming code is in permitting a very simple implementation of the error correction function in the decoder. Since the syndrome itself actually "points" at the erroneous bit in the received vector, a syndrome lookup table is not actually required, or rather, the syndrome table merely becomes a 1-of-r decoder.

Since all Hamming codes are single-error-correcting codes, the "trick" discussed above can be used for Hamming codes of any length. Construction of the G and H is first done in systematic form. The columns of H are then permuted to obtain a simple counting sequence from 1 to n when the columns are read from bottom row to top row. The columns of G are permuted in the same manner to produce the appropriate code word.

In the case of long Hamming codes this "permutation" procedure may be undesirable from the point of view of the costs of the encoder if additional buffering is incurred.

In this case, it may be cheaper to use a systematic form for the code and to implement the error column location in the decoder using standard logic design techniques.

EXAMPLE 4.5.4

Construct the systematic parity-check matrix for the (15, 11) Hamming code.

Solution:

$$H = \begin{bmatrix} 1 & 0 & 0 & 0 & 1 & 1 & 0 & 1 & 1 & 0 & 1 & 0 & 1 & 0 & 1 \\ 0 & 1 & 0 & 0 & 1 & 0 & 1 & 1 & 0 & 1 & 1 & 0 & 0 & 1 & 1 \\ 0 & 0 & 1 & 0 & 0 & 1 & 1 & 1 & 0 & 0 & 0 & 1 & 1 & 1 & 1 \\ 0 & 0 & 0 & 1 & 0 & 0 & 0 & 0 & 1 & 1 & 1 & 1 & 1 & 1 & 1 \end{bmatrix}$$

4.5.2 The Dual Code of a Hamming Code

The Hamming codes are simple to use and very straightforward. It might seem that not much more remains to be said about them, but there are some interesting variations on a theme of occasional use. These come in the forms of modifications to the Hamming code and the dual code of a Hamming code.

Let us first consider the dual codes. In the previous section, we discovered that the parity-check matrix of a code consists of r linearly independent basis vectors that are orthogonal to the basis vectors that make up the G matrix. Therefore, the H matrix of an (n, k) Hamming code is also the generator matrix of an (n, r) dual code. To what use could we put this knowledge?

Consider the (7, 4) Hamming code. It's dual code is a (7, 3) code. Since the parity-check matrix of the (7, 4) Hamming code is the generator matrix of the (7, 3) dual code, the H matrix in Example 4.5.1 is a generator for the (7, 3) code. However, this generator is not in systematic form. It can be put in systematic form by simple permutation of its columns. The result is

$$G^\perp = \begin{bmatrix} 1 & 1 & 0 & 1 & 1 & 0 & 0 \\ 1 & 0 & 1 & 1 & 0 & 1 & 0 \\ 0 & 1 & 1 & 1 & 0 & 0 & 1 \end{bmatrix}.$$

Since this is now in systematic form, its parity-check matrix $H^\perp$ can be written down directly from Equation 4.4.3. The result is

$$H^\perp = \begin{bmatrix} 1 & 0 & 0 & 0 & 1 & 1 & 0 \\ 0 & 1 & 0 & 0 & 1 & 0 & 1 \\ 0 & 0 & 1 & 0 & 0 & 1 & 1 \\ 0 & 0 & 0 & 1 & 1 & 1 & 1 \end{bmatrix}.$$

The set of code vectors can be found by multiplying $G^\perp$ by the eight possible three-bit message vectors. The resulting set of code words is

[0000000], [1101100], [1011010], [0111001], [0110110], [1010101], [1100011], [0001111].

The minimum Hamming weight of the nonzero code vectors is 4. Therefore, the minimum distance of this code is 4. It can correct single-bit errors and detect double errors $(1 + 2 + 1 = 4)$. The dual code therefore provides us with an easy design for a bounded-distance decoder.

Now let us consider the dual code to the $(15, 11)$ Hamming code. This is a $(15, 4)$ code with $r = 11$. What are the capabilities of this code? Since it has only 16 code words, it is relatively easy to generate them. If we do so, we find that the minimum Hamming weight of the nonzero code words is 7, so this code can correct up to three errors.

There is an interesting way to arrive at this result by using an approximation without generating the code vectors first, by using the Hamming sphere. We know that the dual code is not a "perfect" code (since the only perfect codes are the repetition, Hamming, and Golay codes). Consequently, we know that the total volume of the 16 Hamming spheres must be less than the number of vectors in the 15-bit vector space V. Therefore,

$$2^4 \, \zeta(15, t) < 2^{15} \Rightarrow \zeta(15, t) < 2048.$$

Applying Equation 4.3.6, we have

$$\zeta(15, 2) = 121, \quad \zeta(15, 3) = 576, \quad \zeta(15, 4) = 1941.$$

The Hamming sphere volume $\zeta(15, 4)$ is only about 5% below the limit of 2048 vectors per sphere. Since we know this is not a perfect code, this is cutting it a bit close. Therefore, we can tentatively conclude that this code has a correction capability of $t = 3$. We must, of course, confirm this by confirming the code's minimum Hamming distance, but if we're just trying to "ballpark" the capability of the code to see if it fits our needs, the Hamming sphere volume provides a "back of the envelope" way to proceed.

4.5.3 The Expanded Hamming Code

We now turn to the other variation on the Hamming code theme. We are interested in slightly modifying a Hamming code to add double-error-detection capability, while maintaining the same k as our original Hamming code. This method is called *expanding* the Hamming code.

Let C be a systematic (n, k) Hamming code with code words

$$\bar{c} = [c_0 \, c_1 \cdots c_{r-1} \, m_0 \, m_1 \cdots m_{k-1}].$$

Let us define a new code, C', having code words

$$\bar{c}' = [c_0' \, c_0 \cdots c_{r-1} \, m_0 \cdots m_{k-1}] = [c_0' \mid \bar{c}].$$

C' is an $(n + 1, k)$ systematic code. It is obtained by adding an extra parity bit to the (n, k) Hamming code. We call C' an expanded Hamming code. Let the new parity bit be given by

$$c_0' = \sum_{i=0}^{n-1} c_i,$$

bearing in mind that this summation is under addition in GF(2), i.e., $1 + 1 = 0$. Therefore, $c_0' = 0$ if $w_H(\bar{c})$ is even, and $c_0' = 1$ if $w_H(\bar{c})$ is odd.

The expanded Hamming code has $d_{min} = 4$. To see this, observe that $w_H(\bar{c}') = w_H(\bar{c}) + 1$ if $w_H(\bar{c})$ is odd and that $w_H(\bar{c}') = w_H(\bar{c})$ if $w_H(\bar{c})$ is even. Consequently, the minimum Hamming weight of the nonzero code words is now 4, since d_{min} is odd for code words in C. Also notice that nothing in this argument depends on the fact that C is a systematic Hamming code. Therefore, the construction for c_0' defined above works equally well for expanding a nonsystematic Hamming code. Since we now have $d_{min} = 4$ for the expanded Hamming code, we can correct single errors *and* detect double errors. We have added some additional error detection capability to the basic Hamming code.

Let G be the generator matrix for C, and let $g_{i,j}$ be the element in the ith row and j^{th} column of G. It is easy to verify (and, therefore, left as a homework problem) the generator matrix G' for the expanded code is given by

$$G' = \begin{bmatrix} \sum_{j=0}^{n-1} g_{0,j} & g_{0,0} & g_{0,1} & \cdots & g_{0,n-1} \\ \sum_{j=0}^{n-1} g_{1,j} & g_{1,0} & g_{1,1} & \cdots & \vdots \\ \vdots & \vdots & & \ddots & \\ \sum_{j=0}^{n-1} g_{k-1,j} & g_{k-1,0} & \cdots & & g_{k-1,n-1} \end{bmatrix}. \qquad 4.5.1$$

If C is a systematic code, then so is the expanded code and Equation 4.5.1 is of the form

$$G' = [P' \mid I].$$

On the other hand, construction of the parity-check matrix is governed by two considerations. First, we would like to construct H' so that in the event of a single error, we retain the simple error correction method of the underlying Hamming code. Second, we would like an easy method to detect the occurrence of a double error.

Let the received word be $\bar{v}' = \bar{c}' + \bar{e}'$ with

$$\bar{e}' = [e_0'\, e_0\, e_1 \cdots e_{n-1}] = [e_0' \mid \bar{e}],$$

where $\bar{e}$ is an error vector denoting errors that occur in the code bits associated with the underlying Hamming code. The associated syndrome at the receiver is

$$\bar{s}' = \bar{v}' H'^T = \bar{e}' H'^T = [s_0'\, s_1' \cdots s_r'].$$

If H is the parity-check matrix of the underlying Hamming code, we would like $\bar{s}'$ to maintain the original Hamming code syndrome

$$\bar{e} H^T = \bar{s} = [s_0\, s_1 \cdots s_{r-1}].$$

Therefore, we require that

$$[s_1'\, s_2' \cdots s_r'] = [s_0\, s_1 \cdots s_{r-1}].$$

Also, since c'_0 is a parity bit generated from the underlying code vector $\bar{c}$, we would like the occurrence of a single error to be detected as a parity error $s'_0 = 1$. Both of these conditions are satisfied if

$$H' = \begin{bmatrix} 1 & 1 & 1 & \cdots & 1 & 1 \\ \hline 0 & & & & & \\ \vdots & & & H & & \\ 0 & & & & & \end{bmatrix}, \qquad 4.5.2$$

since

$$[e'_0 \mid \bar{e}]H'^T = \left[\sum_{i=0}^{n} e'_i \mid \bar{e}H^T \right].$$

EXAMPLE 4.5.5

Construct the generator and parity-check matrices for expanding the systematic (7,4) Hamming code.

Solution: Using the generator from Example 4.5.2, we have

$$G' = \begin{bmatrix} 1 & 1 & 1 & 0 & 1 & 0 & 0 & 0 \\ 1 & 1 & 0 & 1 & 0 & 1 & 0 & 0 \\ 1 & 0 & 1 & 1 & 0 & 0 & 1 & 0 \\ 0 & 1 & 1 & 1 & 0 & 0 & 0 & 1 \end{bmatrix}.$$

From Equation 4.5.2 and Example 4.5.1, we have

$$H' = \begin{bmatrix} 1 & 1 & 1 & 1 & 1 & 1 & 1 & 1 \\ 0 & 1 & 0 & 0 & 1 & 1 & 0 & 1 \\ 0 & 0 & 1 & 0 & 1 & 0 & 1 & 1 \\ 0 & 0 & 0 & 1 & 0 & 1 & 1 & 1 \end{bmatrix}.$$

As a check, note that $G'H'^T = \bar{0}$.

How do we interpret the syndrome for the expanded Hamming code? Let the error vector for the received word be $\bar{e}' = [e'_0 \, e_0 \cdots e_{n-1}] = [e'_0 \mid \bar{e}]$. If no errors occur, then clearly $\bar{s}' = \bar{0}$. Now suppose that $w_H(\bar{e}') = 1$, but $e'_0 = 0$. Then $s'_0 = 1$ and $\bar{s} \neq \bar{0}$. We would proceed to correct a single-bit error in $\bar{c}$. Now suppose $w_H(\bar{e}') = 1$ and $e'_0 = 1$. We would then have $\bar{s} = \bar{0}$ and $s'_0 = 1$. Therefore, this syndrome condition tells us that only the expanded parity bit is in error. Finally, suppose $w_H(\bar{e}') = 2$. In this case, we must have $\bar{s} \neq \bar{0}$, since at least one of the errors must occur in $\bar{e}$, and the original Hamming code has a minimum Hamming distance of three, which allows it to *detect* (i.e., generate a nonzero syndrome $\bar{s}$) double errors. On the other hand, an error with an even Hamming weight produces $s'_0 = 0$. Therefore, $\bar{s}' = [0 \mid \bar{s} \neq \bar{0}]$ denotes the *detection* of an *uncorrectable* error.

The form of Equation 4.5.2 of the parity-check matrix has the virtue of being independent of whether or not the underlying Hamming code is in systematic form.

Consequently, we may still permute columns of G and H to obtain the simple error correction strategy previously discussed for Hamming codes for application to the expanded Hamming code.

4.6 ERROR RATE PERFORMANCE BOUNDS FOR LINEAR BLOCK ERROR-CORRECTING CODES

4.6.1 Block Error Rates

When selecting an error-correcting code for use over a given channel, our primary objective is typically the achievement of some specified error rate performance given the error properties of that channel. In this section, we will explore some of the bounds on error rate performance for linear block error-correcting codes over the binary symmetric channel. We will begin with decoders that only perform error correction and explore error detection in Section 4.7.

A complete decoder will attempt to perform error correction in all cases. This is often called *forward error correction* or FEC. A FEC decoder does not attempt to detect errors without correction. Let us assume that the BSC has a crossover probability of p and the code is an (n, k) code with the ability to correct up to t errors in a block. If the minimum Hamming distance of the code is d_{min}, then t is the largest integer such that $t \leq (d_{min} - 1)/2$.

We begin by examining the probability of an uncorrectable error in a block. This is simply equal to the probability of the number of errors exceeding t so

$$P_B = 1 - \sum_{j=0}^{t} \binom{n}{j} p^j (1-p)^{n-j} \qquad 4.6.1$$

Given p and t, we can, in principle, calculate P_B exactly. However, for large values of n, this calculation may be difficult due to the presence of $n!$ in the binomial coefficient. It is therefore useful to consider an asymptotic approximation to Equation 4.6.1 for large n. We can express the term under the summation as

$$\binom{n}{j} p^j (1-p)^{n-j} = \frac{n! p^j}{(n-j)!} \frac{(1-p)^{n-j}}{j!}.$$

For $p \ll 1$,

$$(1-p)^{n-j} \approx (e^{-p})^{n-j} \approx e^{-np} \quad \text{if } j \ll n.$$

Likewise,

$$\frac{n! p^j}{(n-j)!} = n(n-1) \cdots (n-j+1) p^j \approx n^j p^j = (np)^j$$

for $j \ll n$. Therefore,

$$P_B \approx 1 - e^{-np} \sum_{j=0}^{t} \frac{(np)^j}{j!}. \qquad 4.6.2$$

It is worth noticing that $\bar{t} = np$ is the expected number of errors in a block of n bits for the BSC.

Equation 4.6.2 illustrates that the block error probability is a function only of the expected number of errors per block and the correction capability t. For $np \ll 1$, P_B approaches np for $t = 0$ and $(np)^2$ for $t = 1$. For $np > 1$, P_B is of the order of unity until t surpasses some threshold correction capability. This is illustrated in Figures 4.6.1. These figures plot P_B vs. the correction ratio t/np on a log-log scale.

As seen in Figures 4.6.1(a) through (d), the correction capability of the code must be several times the expected number of errors per block in order to obtain even reasonable error rate performance when $np \geq 1$. This tends to require block lengths to be kept short enough to hold the expected number of errors per block below one if a reasonably high code rate is to be obtained. Unfortunately, high code rates also tend to be difficult to obtain with short block lengths. This might lead us to expect that it is very difficult to obtain the BSC channel capacity or even the channel cutoff rate with simple block codes. Experience bears this out. The known linear block codes are not good enough to achieve a code rate near the channel capacity. (This does not make Shannon's theorem false; it just means that more sophisticated coding methods must be used; in these more sophisticated approaches, block codes still have a part to play, as we shall see much later in this text.)

4.6.2 Bit Error Rate

Equations 4.6.1 and 4.6.2 describe the probability of a block being uncorrectable. This is *not* the same as the bit error rate. An exact bit error rate analysis of the complete

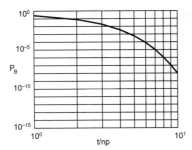

Figure 4.6.1(a): $np = 1$

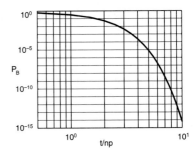

Figure 4.6.1(b): $np = 2$

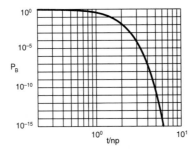

Figure 4.6.1(c): $np = 5$

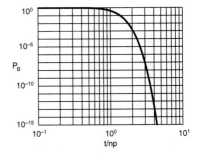

Figure 4.6.1(d): $np = 10$

Block error probabilities as a function of code correction capability

Section 4.6 Error Rate Performance Bounds for Linear Block Error-Correcting Codes

decoder is very difficult to achieve, and the analysis must be carried out individually for each particular code. It is possible, however, to place upper and lower bounds on the bit error rate performance. One reason that exact analysis is difficult is because the decoder itself may introduce more errors in the information bits in an uncorrectable block. We can, however, reasonably expect that, on the average, at least *some* of the information bits will be correct after decoding. Therefore, the bit error rate of the code is upper bounded by

$$p_b \leq P_B. \qquad 4.6.3$$

It is also reasonable to expect *at least* one information bit to be incorrect after "decoding" an uncorrectable block. Therefore, the bit error rate is lower bounded by

$$\frac{P_B}{k} \leq p_b. \qquad 4.6.4$$

Depending on the size of the block, the upper and lower bounds given by Equations 4.6.3 and 4.6.4 may be one or more orders of magnitude apart. This can make analysis during the code selection process frustrating. Is the bit error rate more likely to be near the lower bound, the upper bound, or somewhere in the middle? And what *is* the "middle"? Since an exact analysis is likely to prove elusive for many codes, there is some merit in having some sort of estimate of the "middle" that guide us in the early stages of the code selection process. One such reasonable estimate of the middle is the expected number of errors in an uncorrectable block prior to decoding.

Given a block that is uncorrectable and a code that corrects up to t errors per block, the error probabilities are described by a "truncated" probability distribution

$$\Pr[j|j > t] = \frac{1}{P_B}\binom{n}{j}p^j(1-p)^{n-j},$$

where the factor $1/P_B$ is a normalizing factor, which guarantees that

$$\sum_{j=t+1}^{n} \Pr[j|j > t] = 1.$$

The expected number of errors in an uncorrectable block before decoding is then given by

$$E[j|j > t] = \frac{1}{P_B}\sum_{j=t+1}^{n} j\binom{n}{j}p^j(1-p)^{n-j} = \frac{1}{P_B}\left[np - \sum_{j=1}^{t} j\binom{n}{j}p^j(1-p)^{n-j}\right].$$

If we assume that the decoder, on the average, produces $E[j|j>t]$ information bit errors in an uncorrectable block, the overall expected number of information bit errors per block becomes

$$E[j] = 0 \cdot (1 - P_B) + E[j|j > t]P_B = np - \sum_{j=1}^{t} j\binom{n}{j}p^j(1-p)^{n-j},$$

and the bit error rate then becomes

$$p_b \approx \frac{E[j]}{k} = \frac{np}{k} - \frac{1}{k}\sum_{j=1}^{t} j\binom{n}{j}p^j(1-p)^{n-j}. \qquad 4.6.5$$

For $t \ll n$, $p \ll 1$, we can employ our approximation used previously in Equation 4.6.2. After a small amount of algebra, Equation 4.6.5 becomes

$$p_b \approx \frac{np}{k}\left[1 - e^p e^{-np} \sum_{j=0}^{t-1} \frac{(np)^j}{j!}\right]. \qquad 4.6.6$$

If $np \ll 1$ and $t = 1$, this expression evaluates to approximately $p_b \approx (np)^2(1+p)/k$, which is slightly greater than the lower bound calculated from Equation 4.6.4 using approximation 4.6.2. For small values of p and $np > 1$, approximation 4.6.6 yields results that tend to track the lower bound of Equation 4.6.4 at a slightly larger error rate. Therefore, it is often safe to assume that the actual system performance will tend to be closer to the lower bound than the upper bound. However, this should always be verified by simulation before making a final commitment to the selection of a particular code. The assumption that went into the foregoing discussion was that, on the average, the decoder will not create additional errors in the information bits. This is approximately true of decoder failures in a bounded-distance decoder that does not employ error detection, but it is frequently untrue for complete decoders. Unfortunately, an exact analysis of the complete decoder requires a full understanding of the distribution of Hamming weights in its code words, and for a number of important linear block codes, this weight distribution is not known.

We can get a qualitative understanding of this from geometrical considerations. Figure 4.6.2 illustrates the legal code vectors as points in the space V of possible received vectors.

Valid code vectors are represented as solid dots in Figure 4.6.2, and the bounded-distance decoding spheres (for a bounded-distance decoder) are shown as cross-hatched regions. The transmitted code vector is located in the center of the figure. Also shown is a sphere of radius j representing possible received vectors containing errors with Hamming weight j.

If the received vector falls inside the cross-hatched region centered around the transmitted code vector, it will be properly decoded without error. If the received vector falls into some other cross-hatched region, it will be decoded as the code vector

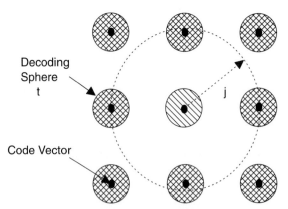

Figure 4.6.2: Vector space model of the Hamming sphere decoding strategy

Section 4.6 Error Rate Performance Bounds for Linear Block Error-Correcting Codes

centered in that region and one or more bit errors will result. The number of information bit errors that results depends on the Hamming distance between the actual and decoded code vectors.

In the case of the bounded-distance decoder, a decoding failure occurs when the received vector does not fall within any cross-hatched region. If the bounded-distance decoder does not request retransmission of the transmitted block, it typically will not change the received vector, and the number of information bit errors will be the equal to the number of information bit errors in the received vector, assuming a systematic code.

In the case of a complete decoder, the received vector will be decoded as the nearest valid code vector. In this case, the decoding spheres in the figure are ignored (or can be thought of as filling the entire space of the figure). If $j > (d_{min} - 1)/2$, the received vector will be erroneously decoded and the number of resulting information bit errors is a complicated function of the Hamming distance between the actual and decoded code vectors. Since the complete decoder attempts to decode every received vector, it will sometimes actually change correct information bits to incorrect ones, generating more errors in information bits than were actually present in the original received vector.

EXAMPLE 4.6.1

A (31, 26) Hamming code is used over a BSC having a crossover probability $p = .001$. Calculate the bit error rate bounds. Compare the results obtained using Equation 4.6.1 with those obtained from Equation 4.6.2. Then calculate and compare the bit error rates estimated by Equations 4.6.5 and 4.6.6.

Solution: Using Equation 4.6.1, we get

$$P_B = 4.561 \cdot 10^{-4}.$$

Since $k = 26$, the bit error rate bounds are

$$1.754 \cdot 10^{-5} \leq p_b \leq 4.561 \cdot 10^{-4}.$$

Using the approximation 4.6.2, we get

$$P_B \approx 4.707 \cdot 10^{-4},$$

so

$$1.81 \cdot 10^{-5} \leq p_b \leq 4.707 \cdot 10^{-4}.$$

Finally, from Equation 4.6.5 we have

$$p_b \approx 3.526 \cdot 10^{-5},$$

while from Equation 4.6.6 we get

$$p_b \approx 3.524 \cdot 10^{-5}.$$

EXAMPLE 4.6.2

Repeat Example 4.6.1 for the (127, 120) Hamming code. Compare the code rate with the channel capacity and the cutoff rate.

Solution: Repeating the method of the previous example, Equation 4.6.1 yields

$$6.137 \cdot 10^{-5} \leq p_b \leq 0.007,$$

while Equation 4.6.2 gives us

$$6.178 \cdot 10^{-5} \leq p_b \leq 0.007.$$

From Equation 4.6.5, we have

$$p_b \approx 1.253 \cdot 10^{-4},$$

while Equation 4.6.6 gives us

$$p_b \approx 1.253 \cdot 10^{-4}.$$

From Equation 2.2.4, the channel capacity is $C_c = 0.989$. The cutoff rate is given by Equation 2.3.5 as $R_o = 0.912$. For this code, we have $R = k/n = 0.945$, which exceeds the cutoff rate, but does not achieve the channel capacity.

In Example 4.6.1, the code rate was $R = 0.839$ and is below the cutoff rate. Note, however, that the upper error rate bound in this example was less than p, while in Example 4.6.2 the upper bound exceeds p. Note, too, that the differences between the lower and "middle" error rate estimates for these two examples. Using the "middle" estimates, the code in Example 4.6.1 improves the error rate over that of the raw channel by approximately 28 times, while the code of Example 4.6.2 achieves only about an eight-times improvement.

4.7 PERFORMANCE OF BOUNDED-DISTANCE DECODERS WITH REPEAT REQUESTS

4.7.1 Approximate Error Performance

The decoders discussed in Section 4.6 do not attempt error detection without correction. Complete decoders always do this. A bounded-distance decoder, on the other hand, attempts to correct the received block if it can find a legal code word within a Hamming distance $d \leq (t_c - 1)/2$. If no such code word exists, the decoder declares a decoding failure. It should be noted that some bounded-distance decoders, such as Hamming code decoders, are both complete decoders *and* bounded-distance decoders at the same time. (This is because Hamming codes are "perfect" codes; their decoding spheres "touch.") Such decoders do not detect without correction. (We saw in Section 4.5 how to expand a Hamming code to allow for uncorrected error detection; the resulting code is no longer "perfect.")

Provided that the communication channel allows two-way communication between the transmitter and receiver, the receiver can request the block to be retransmitted. (If this capability does not exist, or if the receiver chooses not to employ repeat requests, then the performance of the code is the same as discussed in the previous section.) In this section we will look at the performance of bounded-distance decoders using codes that can correct up to t_c and detect up to $t_d > t_c$ errors. This requires a code with minimum Hamming distance $d_{\min} = t_c + t_d + 1$, with $d_{\min} > 2t_c + 1$. Systems using this strategy are called *automatic repeat request* or ARQ systems.

Section 4.7 Performance of Bounded-Distance Decoders with Repeat Requests

We have three possible outcomes for any given transmitted block. The block may be received either with no errors or with correctable errors. Therefore, the probability of correctly receiving a block is

$$P_c = \sum_{j=0}^{t_c} \binom{n}{j} p^j (1-p)^{n-j}. \qquad 4.7.1$$

For large n, $p \ll 1$, and $t_c \ll n$, we may employ our approximation and get

$$P_c \approx e^{-np} \sum_{j=0}^{t_c} \frac{(np)^j}{j!}. \qquad 4.7.2$$

The second possibility is that the block may be received with uncorrectable, but detectable errors. This probability is lower-bounded by[3]

$$P_d = \sum_{j=t_c+1}^{t_d} \binom{n}{j} p^j (1-p)^{n-j}, \qquad 4.7.3$$

which can be approximated under the large n conditions above as

$$P_d \approx e^{-np} \sum_{j=t_c+1}^{t_d} \frac{(np)^j}{j!}. \qquad 4.7.4$$

Finally, we have the possibility that the block may be received with undetectable errors. This probability is upper-bounded[4] Equation 4.6.1 or 4.6.2 with t replaced by t_d. In this case, the block will be erroneously *accepted* by the receiver and P_B is then typically called the *undetected block error rate*. Since these three cases are all the possible cases, we have $P_c + P_d + P_B = 1$.

In ARQ systems, we are typically concerned with the *accepted packet error rate*, P_e, which is the percentage of transmitted blocks containing errors accepted by the receiver. P_e is not the same as P_B because a block with a detectable error, when retransmitted, may have an undetectable error when the retransmitted block is received. This system is a Markov process, and we may represent this process using a state diagram as shown in Figure 4.7.1.

We can find the accepted packet error rate by summing the probabilities of all paths from the transmit state to the "accept with errors" state. From inspection of Figure 4.7.1, we see that this probability is

$$P_e = P_B + P_B P_d + P_B P_d^2 + P_B P_d^3 + \cdots.$$

This is a geometric sequence, and, since $P_d < 1$, the solution is given by

$$P_e = P_B \sum_{j=0}^{\infty} P_d^j = \frac{P_B}{1 - P_d} \qquad 4.7.5$$

We can compute upper and lower bit error rate bounds for this system using Equations 4.6.3 and 4.6.4 by replacing P_B with P_e in these expressions.

---

[3]This is a lower bound because some cases with a larger number of errors may also be detectable.
[4]This is an upper bound because only errors that are legal codewords are undetectable.

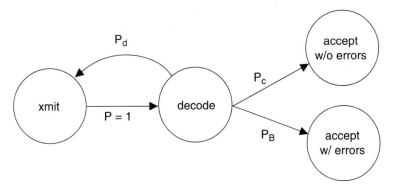

Figure 4.7.1: Markov process state diagram of an automatic repeat request system

We can gain some additional insight into the performance of this system by slightly re-arranging Equation 4.7.5. Since $P_c + P_d + P_B = 1$,

$$P_e = \frac{P_B}{P_c + P_B} = \frac{1}{1 + P_c/P_B}. \qquad 4.7.6$$

From this, we see that the accepted packet error rate is a function of the ratio of P_c to P_B. If $P_B \ll P_c$, then the accepted packet error rate approaches $P_e \to P_B/P_c$.

EXAMPLE 4.7.1

Calculate the accepted packet error rate for the (32, 26) expanded Hamming code for a BSC with crossover probability $p = 0.001$. Calculate the upper and lower bounds for bit error rate achieved by this system.

Solution: For this system, $t_c = 1$ and $t_d = 2$. Using Equation 4.7.1, we have

$$P_c = 0.99951.$$

Equation 4.6.1 gives us

$$P_B = 4.853 \cdot 10^{-6}.$$

Applying Equation 4.7.6 then gives us

$$P_e = 4.855 \cdot 10^{-6}.$$

For upper and lower bounds, Equations 4.6.3 and 4.6.4 give us

$$0.187 \cdot 10^{-6} \le p_b \le 4.855 \cdot 10^{-6}.$$

4.7.2 Effective Code Rate of ARQ Systems

Comparing the results of Example 4.7.1 with those of Example 4.6.1, we see that the former system has achieved a very significant improvement over the simple Hamming code. The upper bound in Example 4.7.1 is less than the lower bound of the previous example. This improvement is obtained at the expense of code rate. First, the rate for

Section 4.7 Performance of Bounded-Distance Decoders with Repeat Requests

the (32, 26) expanded Hamming code is 0.8125. Additionally, some of the blocks must be transmitted more than once, and this also decreases the overall system's code rate. We will now look at the effect of repeat requests on code rate.

We first look at the average number of times a block must be transmitted before it is accepted by the receiver. Tracing the paths in Figure 4.7.1 from the transmit block to either of the received blocks and recognizing the number of times a block is repeated is equal to the number of times we go through the retransmit loop, the average number of times a block must be transmitted is given by

$$T_x = (P_c + P_B) + 2(P_c + P_B)P_d + 3(P_c + P_B)P_d^2 + \cdots.$$

Since $P_c + P_B = 1 - P_d$, this expression becomes

$$T_x = (1 - P_d) \sum_{j=1}^{\infty} j P_d^{j-1}.$$

By observing that

$$j P_d^{j-1} \equiv \frac{d}{dP_d}(P_d^j)$$

we can rewrite this as

$$T_x = (1 - P_d)\frac{d}{dP_d} \sum_{j=0}^{\infty} P_d^j = (1 - P_d)\frac{d}{dP_d}\left(\frac{1}{1 - P_d}\right) = \frac{1}{1 - P_d}. \qquad 4.7.7$$

Applying Equation 4.7.7 to Example 4.7.1, the average number of times a block must be transmitted in Example 4.7.1 is $T_x = 1.0005$ times.

We are now ready to look at a simple ARQ system. We will assume that the transmitter sends one block and then waits for the receiver to acknowledge the receipt of this block. The receiver will require a certain amount of time to process the received block and then send its acknowledgment or its request to retransmit the block. Let us measure the average amount of time this requires in terms of the equivalent number of bits the transmitter *could* have sent if it hadn't stopped and waited for acknowledgment. We will use the symbol Γ to denote this idle time in *bits*. Γ must account for: 1) The number of bit cycles in excess of n required for the receiver to determine whether to accept the block or request retransmission, and; 2) the number of bit cycles required for the receiver to send its acknowledgment to the transmitter and for the transmitter to receive and decode this acknowledgment. The *throughput* (equivalent rate) of this system is then given by

$$\eta \equiv \frac{k}{T_x(n + \Gamma)} = \frac{k/n}{T_x(1 + \Gamma/n)} = R\frac{1 - P_d}{1 + \Gamma/n}, \qquad 4.7.8$$

where $R = k/n$ is the code rate of the base code.

EXAMPLE 4.7.2

As an example, let us assume the same system as in Example 4.7.1. We will also assume that the receiver requires one-bit time after receiving the block to recognize from the syndrome whether or not to request retransmission. Let us further assume that the receiver acknowledges using an (8, 4) expanded Hamming code. In addition we will assume that there is a one-bit delay from the time the receiver sends its acknowledgment to the time this acknowledgment arrives at the transmitter. And we will assume that the transmitter requires one-bit time to decode whether the acknowledgment is a request for retransmission. We will further assume that the transmitter has buffered up the next block to be transmitted and also saved the previous block so it can begin transmitting immediately. Finally, we will assume that one more bit time is required for the next block to reach the receiver. This gives us a total of $\Gamma = 1 + 8 + 1 + 1 + 1 = 12$ bit times. Therefore, $\Gamma/n = 0.375$, and using our previously computed value for T_x, $\eta = 0.591$.

4.7.3 ARQ Protocols

The type of ARQ system described in the preceding section is known as stop-and-wait ARQ (SW-ARQ) protocol. There are two other basic retransmission protocols that are widely used. These are the go-back-N ARQ (GBN-ARQ) and the selective-repeat ARQ (SR-ARQ) protocols. In the GBN-ARQ protocol, the transmitter keeps N previous blocks in its buffer and transmits continuously. The basic idea behind this protocol is that repeat requests are infrequent, and it is better to transmit continuously. Whenever a repeat request is received, however, the transmitter must go back N blocks and repeat everything beginning with that block. The receiver, after requesting a retransmission, ignores the $N-1$ blocks it has received since requesting the retransmit. The parameter N is the smallest integer such that $N \geq \Gamma/n$. It can be shown that the throughput of the GBN-ARQ protocol is

$$\eta_{\text{GBN}} = R\left(\frac{1 - P_d}{1 + P_d(N - 1)}\right). \qquad 4.7.9$$

In the selective-repeat ARQ protocol, both the transmitter and the receiver maintain block buffers. As in the GBN-ARQ protocol, the transmitter sends continuously. When the receiver requests a retransmission, it identifies the block it wishes to have retransmitted. The transmitter then responds by sending the requested block, which it has stored in its buffer. It then resumes sending blocks from the point at which it was interrupted by the repeat request.

At the receiver, the receiver processes and buffers the received blocks as they come in. When a retransmit is necessary, it ceases to output to the information sink and saves the following blocks in its buffer until the repeated block is received. It then outputs the repeated block to the information sink, followed by the other blocks in their proper time sequence. The throughput of this system can be shown to be

$$\eta_{\text{SR}} = R(1 - P_d). \qquad 4.7.10$$

This equation indicates the SR-ARQ protocol achieves the highest possible rate. However, both Equations 4.7.9 and 4.7.10 are mildly deceptive. In a practical system using either GBN- or SR-ARQ strategies, it is usually necessary to include in the transmitted block a data header that identifies which block this is. This header is usually of no interest to the information sink and is merely overhead in the communication system. Therefore, the *effective* rate R should be expressed as the number of information bits in the block minus the number of these bits which are "overhead" (which gives us an "effective" or nonoverhead k) divided by the blocklength n.

SUMMARY

We have covered quite a bit of ground in this chapter. In Section 4.1, we defined the basic concepts of linear block error-correcting codes and the probabilities associated with transmission over the binary symmetric channel. We looked at some examples of these calculations and then introduced binary repetition codes. From this, we derived the maximum-likelihood decoding rule and discussed the relationship between the minimum Hamming distance of a code and its error correction and error detection capabilities.

In Section 4.2 we introduced the mathematical idea of binary fields and binary vector spaces. This provides the mathematical toolset we need for working with linear block codes. It is important to remember that when working with binary vector spaces, $1 + 1 = 0$. We illustrated the basic concepts by introducing simple 2-D codes.

Section 4.3 introduced us to the general theory of linear block codes. We discussed the mathematical form of these codes and a number of important properties. One such important property is superposition: the sum of any two code words is itself a code word. From this property, we found the relationship between the minimum Hamming distance of a code and the minimum Hamming weight of the nonzero code vectors. We also introduced the idea of the Hamming cube, with its geometrical interpretation of linear vector spaces, and the generalized mathematical notion of a metric function. We saw that the Hamming distance was an example of such a metric function.

Decoding of linear block codes was the topic of Section 4.4. We introduced the standard array for decoding block codes and then went on to show how decoding can be simplified considerably by the use of syndrome tables. We defined the syndrome of a received vector in terms of the parity-check matrix. We also defined systematic codes, which are codes of the highest practical importance, and discussed how any linear block code can be expressed in systematic form.

Section 4.5 introduced us to our first nontrivial class of codes, the Hamming codes. We discussed at length the design of these codes and their properties. We also saw how we could get to some other useful codes through modification of the basic Hamming code. The modifications we considered were the dual code of a Hamming code and the expanded Hamming code. We also saw how we could use the concept of the Hamming sphere to get quick approximations of the correction or detection capabilities of these modified codes.

We turned our attention in Section 4.6 to the problem of estimating upper and lower bounds for the error rate of the system. We saw how the performance of the system was related to the probability distribution of errors in the binary symmetric channel and looked at a useful way to approximate our error rate expressions for large block lengths.

We concluded this chapter with a look at systems that combine error correction with error detection and repeat requests. We saw how to model such a system as a Markov process and calculate its performance from that model. We discussed three standard ARQ system protocols and discussed their error rate performance and the effect of repeat requests on throughput.

In Chapter 5, we will extend our universe of known block codes by introducing the most important class of all linear block codes: the cyclic codes. It is hard to overstate the practical importance of these codes because of their many practical advantages.

REFERENCES

R. S. Hamming, "Error-detecting and error-correcting codes," *Bell System Technical Journal*, vol. 29, pp. 147–160, 1950.

D. A. Slepian, "A class of binary signaling alphabets," *Bell System Technical Journal*, vol. 35, pp. 203–234, 1956.

R. C. Singleton, "Maximum distance q-nary codes," *IEEE Transactions on Information Theory*, vol. IT-10, pp. 116–118, 1964.

G. Benelli, "An ARQ scheme with memory and soft error detectors," *IEEE Transactions on Communications*, vol. COM-33, no. 3, pp. 285–288, Mar., 1985

G. Benelli, "An ARQ scheme with memory and integrated modulation," *IEEE Transactions on Communications*, vol. COM-35, no. 7, pp. 689–697, July, 1987.

R. Comroe and D. J. Costello, Jr., "ARQ schemes for data transmission in mobile radio systems," *IEEE Journal on Selected Areas in Communications*, vol. SAC-2, no. 4, pp. 472–481, July, 1984.

EXERCISES

4.1.1: Calculate and plot the binomial distribution function for a block length of $n = 31$ with error probability $p = 0.05$.

4.1.2: Calculate the expected number of errors and the three-sigma error range for a block of 63 bits transmitted over a BSC with crossover probability $p = 10^{-3}$.

4.1.3: For the system of Exercise 4.1.2, what is the probability of two or more errors in any given block?

4.1.4: Find the Hamming distance between the following pairs of binary words:

a) 0 0 0 0, 0 1 0 1   b) 0 1 1 1 0, 1 1 1 0 0

c) 0 1 0 1 0 1, 1 0 1 0 0 1   d) 1 1 1 0 1 1 1, 1 1 0 1 0 1 1.

4.1.5: Using the repetition code of Example 4.1.4, decode the following code word sequences.

a) 0 0 0 1 1 0 0 0 1 1 1 1 1 1 0 1 0 0 0 0

b) 1 1 1 1 1 1 1 1 0 1 1 1 0 0 1 0 0 0 0 0

c) 0 1 0 0 0 1 0 1 1 1 1 1 0 0 0 0 0 1 1 1.

4.1.6: A sequence of 63 coded bits is transmitted over a BSC having $p = 0.01$. The bits are encoded using a three-bit repetition code. What is the probability of error in the decoded output?

4.1.7: A certain code has an alphabet consisting of the following code words:

0000000   1000111   0101011   0011101

1101100   1011010   0110110   1110001.

Find the minimum Hamming distance of this code and determine it error correction and error detection capabilities.

4.1.8: What is the code rate for the code in Exercise 4.1.7? How many check bits does it have?

4.2.1: Show that the code vectors for the code in Exercise 4.1.7 form a vector space.

4.2.2: Encode the message

$$\overline{m} = [0\,1\,1\,0\,0\,0\,1\,1\,1] = [m_0 \cdots m_8]$$

using the 2-D code of Example 4.2.2.

4.2.3: Assuming the 2-D code of Example 4.2.2, decode the following received block:
$$\bar{v} = [v_0 \cdots v_{15}] = [1 1 1 0 0 0 0 0 0 0 1 1 0 1 0 1]$$

4.2.4: Encode the message $\bar{m} = [0100]$ using a 2-D code with $n = 9$. What is the code rate?

4.3.1: An $(8,2)$ repetition code is defined to have code vectors of the form
$$\bar{c} = [m_0\, m_1\, m_0\, m_1\, m_0\, m_1\, m_0\, m_1],$$
where $\bar{m} = [m_0\, m_1]$ is the source vector. What is the generator matrix for this code? What are the basis vectors for this code?

4.3.2: For the repetition code of Exercise 4.3.1, find the minimum Hamming distance of the code. Determine its error detection and error correction capabilities.

4.3.3: Find the volume of the Hamming sphere of radius $t = 1$ for the code of Exercise 4.3.1. Show that this code exceeds the Hamming bound.

4.3.4: Does the code of Exercise 4.3.1 meet the Gilbert bound? Justify your answer.

4.3.5: Is the code in Exercise 4.3.1 a perfect code?

4.4.1: Construct the standard array for the $(n, k) = (3,1)$ repetition code.

4.4.2: Find the systematic G and H matrices of the code in Example 4.4.1, and construct the syndrome table for a bounded distance decoder for this code.

4.4.3: Using the results of Exercise 4.4.2, identify all of the two-bit errors that will be miscorrected by a bounded-distance decoder for the code of Exercise 4.4.2.

4.5.1: Using a systematic (7,4) Hamming code, generate the code words for the following information words:
  a) $\bar{m} = [0100]$ b) $\bar{m} = [0101]$ c) $\bar{m} = [1110]$ d) $\bar{m} = [1001]$.

4.5.2: Assuming a systematic (7,4) Hamming code, decode the following received vectors:
  a) $\bar{v} = [1101001]$ b) $\bar{v} = [0010111]$
  c) $\bar{v} = [1111100]$ d) $\bar{v} = [1011001]$.

4.5.3: Construct the syndrome table for a bounded-distance decoder for the systematic (7,3) dual code of the (7,4) Hamming code.

4.5.4: Using the results of Exercise 4.5.3, decode the following received vectors for the systematic (7,3) dual code:
  a) $\bar{v} = [1010001]$ b) $\bar{v} = [1101001]$ c) $\bar{v} = [1101011]$.

4.5.5: Show that Equation 4.5.1 is the generator matrix of an $(n+1, k)$ expanded Hamming code.

CHAPTER 5

Cyclic Codes

5.1 DEFINITION AND PROPERTIES OF CYCLIC CODES

Cyclic codes are a class of linear block error-correcting codes. These codes, as a class, are probably the most widely used form of error-correcting and error-detecting codes. The popularity of cyclic block codes arises primarily from the fact that these codes can be implemented with extremely cost-effective electronic circuits. The codes themselves also possess a great deal of structure and regularity, which gives rise to their cost-effective implementability, and there is a certain beauty and elegance in the theory that describes them.

The defining feature of cyclic codes, which distinguishes them from the other linear block codes, is the *cyclic shift* property. Let $\bar{v} = [v_0 \; v_1 \; \cdots \; v_{n-1}]$ be an n-element binary vector. A vector $\bar{v}'$ is a cyclic shift of $\bar{v}$ if $\bar{v}' = [v_{n-1} \; v_0 \; \cdots \; v_{n-2}]$. A *cyclic code* is a linear block code C with code vectors $\bar{c} = [c_0 \; c_1 \; \cdots \; c_{n-1}]$ such that for every $\bar{c} \in C$, the vector given by the cyclic shift of $\bar{c}$ is also a code vector.

EXAMPLE 5.1.1

The $(6, 2)$ repetition code

$$C = \{[000000], [010101], [101010], [111111]\}$$

is a cyclic code, since a cyclic shift of any of its code vectors results in a vector that is an element of C.

EXAMPLE 5.1.2

The $(5, 2)$ linear block code defined by the generator matrix

$$G = \begin{bmatrix} 1 & 0 & 1 & 1 & 1 \\ 0 & 1 & 1 & 0 & 1 \end{bmatrix}$$

Section 5.1 Definition and Properties of Cyclic Codes

is a single-error-correcting code, but is *not* a cyclic code. Its code vectors are

$$00000$$
$$10111$$
$$01101$$
$$11010.$$

The cyclic shift of [1 0 1 1 1] is [1 1 0 1 1] and is not an element of C. Similarly, the cyclic shift of [0 1 1 0 1] is [1 0 1 1 0], which is also not a code vector.

Since cyclic codes are linear block codes, all the properties of linear block codes discussed in Chapter 4 apply to cyclic codes. However, because of the special structure imparted to cyclic codes by the fact that the cyclic shift of any code vector is also a code vector, cyclic codes have some additional properties not found in general in linear block codes. These properties give rise to encoding and decoding procedures that are algebraic and computationally efficient. Rather than attempt to list these properties now, we will develop them gradually over the course of this chapter. In that way, the methods by which we can take advantage of these properties will hopefully be more readily apparent.

In Chapter 4, we discussed the generator matrix G of a linear block code and saw that the rows of the generator matrix comprised a set of basis vectors in a k-dimensional linear space. Because a cyclic code contains as a code vector the cyclic shift of some other code vector, the generator matrix of a cyclic code reflects this property. In particular, the generator matrix of a non-systematic (n, k) cyclic code can always be expressed in the form

$$G = \begin{bmatrix} g_0 & g_1 & \cdots & g_{n-k-1} & g_{n-k} & 0 & 0 & \cdots & & 0 \\ 0 & g_0 & g_1 & \cdots & g_{n-k-1} & g_{n-k} & 0 & 0 & \cdots & 0 \\ 0 & 0 & g_0 & g_1 & \cdots & g_{n-k-1} & g_{n-k} & 0 & 0 & \cdots & 0 \\ \vdots & & & & & & & & & \vdots \\ 0 & 0 & 0 & 0 & \cdots & 0 & g_0 & g_1 & \cdots & g_{n-k-1} & g_{n-k} \end{bmatrix}. \qquad 5.1.1$$

The rows of this generator matrix are merely cyclic shifts of the $1 \times n$ basis vector

$$\bar{g} = [g_0 \, g_1 \cdots g_{n-k-1} \, g_{n-k} \, 0 \, 0 \cdots 0]. \qquad 5.1.2$$

Since cyclic codes are linear codes, it is also always possible to express the generator matrix in systematic form

$$G = [P \mid I], \qquad 5.1.3$$

as we did in Chapter 4. The systematic form can always be obtained from elementary row and column operations on the G matrix of Equation 5.1.1. However, we will see a bit later in this chapter that a simpler method for generating systematic cyclic codes exists.

The code vectors for the non-systematic code generated by Equation 5.1.1 are given by

$$\bar{c} = \bar{m}G = [m_0 \, m_1 \cdots m_{k-1}]G,$$

where $\overline{m}$ is the message vector. Applying Equation 5.1.1 to this, the individual elements of the code vector can readily be seen to be

$$c_0 = m_0 g_0$$
$$c_1 = m_0 g_1 + m_1 g_0$$
$$c_2 = m_0 g_2 + m_1 g_1 + m_2 g_0$$
$$\vdots$$
$$c_{n-k} = m_0 g_{n-k} + m_1 g_{n-k-1} + \cdots + m_{n-k} g_0$$
$$\vdots$$
$$c_{n-1} = m_{k-1} g_{n-k}.$$

If we adopt the convenient notation convention $m_j = 0$ if $j < 0$ or $j > k - 1$ and $g_j = 0$ for $j < 0$ or $j > n - k$, the set of equations given above can be concisely represented as

$$c_\ell = \sum_{j=0}^{k-1} m_j g_{\ell-j}. \qquad 5.1.4$$

Equation 5.1.4 is known as a *convolution sum*. (Remember: $1 + 1 = 0$.) Convolution sums play an extremely important role in the theory of ordinary linear difference equations with constant coefficients. Their role there is so important that a number of efficient techniques have been developed for working with convolution sums. The appearance of a convolution sum in Equation 5.1.4 suggests that these efficient mathematical tools can also be applied to cyclic block codes, and as luck would have it, this is indeed the case.

In the theory of ordinary linear difference equations with constant coefficients, one of the most important tools is the use of *linear transforms,* particularly the z-transform. Assuming you have some familiarity with z-transforms, it is worth recalling that z-transforms have the property of mapping convolution sums into algebraic products of z-transforms. Since it is frequently easier to deal with algebraic equations than to work directly with convolution sums, the transform approach is very popular. In a similar vein, we will find it easier to work with cyclic codes if we introduce an analogous type of transform. This we will do in Section 5.2.

5.2 POLYNOMIAL REPRESENTATION OF CYCLIC CODES

The basic notion behind classical transform methods (Fourier, Laplace, z, and so on) is the introduction of an *operator,* sometimes called an *indeterminate,* that describes something useful about the quantity being transformed. For example, in Chapter 2 we introduced the idea of the power spectrum of a code. This is an example of a transform and its operator was the digital frequency θ.

For cyclic codes, a very convenient operator is the *bit position operator*. If v_j is the jth element of a vector $\overline{v}$, its bit position is j, and we can describe this using the bit position operator x^j. The indeterminate variable x is a sort of "place holder" and is *not* necessarily an element of our scalar field GF(2). We may multiply x^j by a scalar, and we may perform arithmetic on functions of x^j. The superscript j is restricted to be an integer, while any scalar multiplying x^j is restricted to be an element of our base field GF(2). The rules are very simple. If a and b are binary elements of GF(2),

$$ax^j + bx^j = (a + b)x^j,$$
$$(ax^j) \cdot (bx^k) = (a \cdot b)x^{j+k}, \quad 5.2.1$$

where addition or multiplication of the scalars is defined under the rules for GF(2) and addition of the bit position superscripts j and k uses ordinary integer addition. If $j \neq k$, the operation

$$ax^j + bx^k$$

cannot be simplified and must be left in this polynomial form.

These rules are simply the familiar, everyday rules for polynomial arithmetic. The only aspect that may seem a little strange at first is the restriction of the scalar multipliers to GF(2). However, we have been working with binary vector spaces for some time now, so perhaps the notion of binary polynomials is not too strange.

With these basic ideas firmly in place, we can now define a polynomial transform representation for the vectors in our linear vector space. The bit position (or polynomial) transform of a vector

$$\bar{v} = [v_0 \, v_1 \cdots v_{n-1}]$$

is defined to be

$$v(x) = v_0 x^0 + v_1 x^1 + \cdots + v_{n-1} x^{n-1} = \sum_{j=0}^{n-1} v_j x^j. \quad 5.2.2$$

If you are familiar with the z-transform, you might notice the strong similarity between the z-transform of a sequence and the polynomial transform defined by Equation 5.2.2[1]. (If you are not familiar with the z-transform, you don't need to be worried; just think "polynomials.")

Polynomial arithmetic for our polynomial transform works in the usual way (except the scalar coefficients obey the rules for GF(2)). Polynomial addition and multiplication are defined term by term. For example, if $m(x) = m_0 x^0 + m_1 x^1 + m_2 x^2$ and $g(x) = g_0 x^0 + g_1 x^1$, then

$$m(x) + g(x) = (m_0 + g_0)x^0 + (m_1 + g_1)x^1 + (m_2 + 0)x^2$$

and

$$m(x)g(x) = m_0 g_0 x^0 + (m_0 g_1 + m_1 g_0)x^1 + (m_1 g_1 + m_2 g_0)x^2 + m_2 g_1 x^3.$$

Let us compare the multiplication example above with the code vector generation defined by

$$\bar{c} = [m_0 \, m_1 \, m_2] \begin{bmatrix} g_0 & g_1 & 0 & 0 \\ 0 & g_0 & g_1 & 0 \\ 0 & 0 & g_0 & g_1 \end{bmatrix}.$$

By direct multiplication, we have

$$\bar{c} = [m_0 g_0 \, (m_0 g_1 + m_1 g_0) \, (m_1 g_1 + m_2 g_0) \, m_2 g_1].$$

---

[1] The z-transform and this polynomial transform are very similar in many respects. Probably the most important difference is x should not be interpreted as being the same as z^{-1}. It is more accurate to interpret x as being equivalent to z.

The elements of $\bar{c}$ are identical to the coefficients in $m(x)g(x)$. The matrix approach to code vector generation for a nonsystematic cyclic code is identical to the polynomial transform representation $c(x) = m(x)g(x)$. Thus, we can exchange our matrix algebra representation of cyclic codes for a polynomial representation. The polynomial $g(x)$ is called the *generator polynomial* of the cyclic code. It will play a fundamental role in the theory that follows. We will also see later how this representation of cyclic codes leads directly to simple circuits for implementing these codes.

There are a couple additional observations we can make about our polynomial representations. The *degree* of a polynomial is defined to be the highest power of x in the polynomial for which the coefficient of x is not zero. The degree of a polynomial $m(x)$ is often denoted by $\deg(m(x))$. If M is our set of k-bit source vectors, the highest degree polynomial in the set of polynomials $m(x)$ that describe our source vectors is $k-1$. If $\deg(g(x)) = r$, then the highest degree of our set of code polynomials $c(x)$ is $k-1+r = n-1$. Therefore, $\deg(g(x))$ tells us the number of check bits in our code.

Since $x^0 \cdot x^j = x^j$, the operator x^0 acts like a multiplicative identity element. For that reason, it is common to use the abbreviation $x^0 \equiv 1$. You should not take this to mean that x must have a "value" of either 0 or 1. x is an indeterminate and is quite independent of the elements in our scalar field GF(2). Once this convention is adopted, it is a natural step to use the notational abbreviation $v_0 x^0 \equiv v_0$ and let the x^0 operator be "understood". In other words, the absence of an x term in a polynomial implies x^0. It is also common and notationally convenient to use the abbreviation $x^1 \equiv x$. We will adopt these conventions for the remainder of our discussions.

5.3 POLYNOMIAL MODULO ARITHMETIC

5.3.1 Polynomial Rings

The polynomial $p(x) = x^n - 1$ plays a key role in the theory of cyclic codes. (In this text we are interested in binary codes; for such codes, $-1 = 1$; however, we will often use the "additive inverse" notation $-a$ to maintain consistency with the theory for non-binary codes.) As an illustration of its role, consider a code-word polynomial $c(x)$ and its cyclic shift:

$$c(x) = c_0 + c_1 x + \cdots + c_{n-1} x^{n-1},$$
$$c'(x) = c_{n-1} + c_0 x + \cdots + c_{n-2} x^{n-1}.$$

We can write $c'(x)$ as

$$c'(x) = xc(x) - c_{n-1} x^n + c_{n-1} = xc(x) - c_{n-1}(x^n - 1).$$

This expression gives us our first hint of the role $p(x) = x^n - 1$ is to play in cyclic codes. It is, however, not the only role for this very special polynomial. While the expression for $c'(x)$ above is perfectly true, it turns out not to be the most *convenient* algebraic expression for *systematic* cyclic codes.

As an alternative, let us consider the expression $xc(x)$ *modulo* $(x^n - 1)$ which we will denote using either of two forms,

$$xc(x) \bmod (x^n - 1) \equiv xc(x)/(x^n - 1).$$

The modulo operation a mod b in ordinary arithmetic is defined as the *remainder* that results from dividing integer a by integer b. In polynomial arithmetic, the modulo operation is defined the same way, except now a and b are both polynomials. We calculate the remainder using long division. Let us do this for $xc(x)$ mod $(x^n - 1)$. We have

$$
\begin{array}{r}
c_{n-1} \phantom{x^n + c_{n-2}x^{n-1} + \cdots + c_0 x} \\
x^n - 1 \overline{)\, c_{n-1}x^n + c_{n-2}x^{n-1} + \cdots + c_0 x \phantom{+ c_{n-1}}} \\
\underline{c_{n-1}x^n \phantom{+ c_{n-2}x^{n-1} + \cdots + c_0 x} - c_{n-1}} \\
c_{n-2}x^{n-1} + c_{n-3}x^{n-2} + \cdots + c_0 x + c_{n-1}
\end{array}.
$$

Since $x^n - 1$ does not divide $c_{n-2}x^{n-1}$ using nonnegative powers of x, the remainder left by this operation is

$$xc(x) \bmod (x^n - 1) = c_{n-2}x^{n-1} + c_{n-3}x^{n-2} + \cdots + c_0 x + c_{n-1} \equiv c'(x).$$

Therefore, a cyclic shift of $c(x)$ is equivalent to $xc(x)$ mod $(x^n - 1)$.

The set of polynomials modulo some polynomial $p(x)$ forms a mathematical structure called a *ring*. If the polynomials have scalar coefficients drawn from GF(2), the structure is referred to as "the ring $GF(2)[x]/p(x)$," where this complicated-looking notation tells us: 1) the scalars are from GF(2); 2) the $[x]$ denotes polynomials, and; 3) "$/p(x)$" indicates the polynomials are modulo $p(x)$. The term "ring" arises from the periodic character of the structure. This can be illustrated by considering the set of integers modulo 3. This structure has the form

        Integer: 0 1 2 3 4 5 6 7 ...

        Integer mod 3: 0 1 2 0 1 2 0 1 ...

and we see that the integers "wrap around" from 0 to 2 and back to zero (forming a "ring").

Polynomial rings do not necessarily follow this same simple ordering of the elements but the main characteristic, namely higher degree polynomials being mapped back into lower degree polynomials, is the same (as we saw in the cyclic shift example above). The set of polynomials belonging to $GF(2)[x]/p(x)$ and the operations of polynomial addition and polynomial multiplication are said to form a "commutative ring with identity" and possess the following properties:

1. For the set of polynomials $P = GF(2)[x]/p(x)$ and the operation polynomial addition, $+$,
   a) polynomial addition is closed [i.e., for $p_1(x), p_2(x) \in P, p_1(x) + p_2(x) \in P$];
   b) the zero polynomial, $f(x) = 0$, is the additive identity;
   c) every $f(x) \in P$ is its own additive inverse;
   d) polynomial addition is associative;
   e) polynomial addition is commutative. (Properties 1 form what is called an "Abelian group.")
2. Polynomial multiplication
   a) is associative;
   b) is commutative;
   c) is such that the polynomial $f(x) = 1$ is the multiplicative identity.
3. Polynomial multiplication is distributive over polynomial addition.

These are simple, but very useful and powerful properties with which our system of binary polynomials is endowed.

Consider a polynomial ring $GF(2)[x]/(x^n - 1)$. For $n > 1$, the polynomial $x^n - 1 = x^n + 1$ can always be factored into the product of some number of lower degree polynomials. For example, $x^7 + 1 = (x + 1)(x^3 + x + 1)(x^3 + x^2 + 1)$. These three lower order polynomials can *not* be factored into even lower degree polynomials. A polynomial that cannot be factored into lower degree polynomials is said to be *irreducible*. *Our cyclic code generator polynomials will be constructed* from the factors of $x^n + 1$. In other words, if $g(x)$ is the generator polynomial of a cyclic (n, k) code, then there will exist some other polynomial $h(x)$ such that $g(x)h(x) = x^n + 1$.

If we have two polynomials, $f(x)$, $g(x)$, such that $f(x)$ mod $g(x) = 0$, we say that $g(x)$ *divides* $f(x)$. (Obviously, for this to be meaningful, we must have $g(x) \neq 0$.) We may always express $f(x)$ in terms of $g(x)$ in the form

$$f(x) = Q(x)g(x) + \rho(x), \qquad 5.3.1$$

where $\deg[\rho(x)] < \deg[g(x)]$. If $g(x)$ does not divide $f(x)$, then we will have $\rho(x) \neq 0$. The polynomial $Q(x)$ is called the quotient, while the polynomial $\rho(x)$ is called the remainder. Equation 5.3.1 is the basis for our definition of polynomial division. Given $f(x)$ and $g(x)$, we may always find $Q(x)$ and $\rho(x)$ by long division (as we did earlier in this section to prove the cyclic shift relationship).

5.3.2 Some Important Algebraic Identities

For working with cyclic codes, the remainder $\rho(x)$ is usually much more interesting to us than the quotient $Q(x)$. We will use the notation

$$\rho(x) = f(x)/g(x)$$

to designate the remainder resulting from division of $f(x)$ by $g(x)$. There are two algebraic identities which arise from Equation 5.3.1 that are very useful in working with polynomials. Suppose that we have two polynomials

$$f_1(x) = Q_1(x)g(x) + \rho_1(x),$$
$$f_2(x) = Q_2(x)g(x) + \rho_2(x).$$

Adding these two equations together, we have

$$f_1(x) + f_2(x) = [Q_1(x) + Q_2(x)]g(x) + \rho_1(x) + \rho_2(x).$$

Since $\deg[\rho_1(x)] < \deg[g(x)$ and $\deg[\rho_2(x)] < \deg[g(x)]$, we have

$$\deg[\rho_1(x) + \rho_2(x)] < \deg[g(x)],$$

and therefore,

$$[f_1(x) + f_2(x)]/g(x) = \rho_1(x) + \rho_2(x), \qquad 5.3.2$$

since it is clear that $g(x)$ divides $[Q_1(x) + Q_2(x)]g(x)$.

There is a similar useful identity for the remainder of the product of two polynomials. We may write

$$f_1(x) \cdot f_2(x) = [Q_1(x)Q_2(x)g(x) + Q_1(x)\rho_2(x) + Q_2(x)\rho_1(x)]g(x) + \rho_1(x) \cdot \rho_2(x).$$

Section 5.3 Polynomial Modulo Arithmetic

The first term on the right-hand side of this equation is clearly a multiple of $g(x)$, so its remainder resulting from division by $g(x)$ is zero. Since

$$\deg[p_1(x) \cdot p_2(x)] = \deg[p_1(x)] + \deg[p_2(x)], \qquad 5.3.3$$

we have

$$[f_1(x) \cdot f_2(x)]/g(x) = [p_1(x) \cdot p_2(x)]/g(x). \qquad 5.3.4$$

Equations 5.3.2 and 5.3.4 are often very useful in analyzing cyclic codes.

EXAMPLE 5.3.1

Divide $f(x) = x^6$ by $g(x) = 1 + x + x^3$.

Solution: We use the long division algorithm to express $f(x)$ in the form of Equation 5.3.1.

$$
\begin{array}{r}
x^3 + x + 1 \\
x^3 + x + 1 \overline{) x^6 } \\
\underline{x^6 + x^4 + x^3} \\
x^4 + x^3 \\
\underline{x^4 + x^2 + x} \\
x^3 + x^2 + x \\
\underline{x^3 + x + 1} \\
x^2 + 1
\end{array}
$$

Therefore, $x^6 = (x^3 + x + 1)(x^3 + x + 1) + x^2 + 1$. In carrying out the long division algorithm, notice that we always write the polynomials with the x terms arranged in order of descending powers of x from left to right. The division ceases when the degree of the remainder is less than the degree of the divisor. In this example, $Q(x) = g(x)$ and $p(x) = x^2 + 1$.

EXAMPLE 5.3.2

The ring $GF(2)[x]/(x^7 - 1)$ contains $2^n = 2^7$ distinct polynomials. Suppose we wished to build a division table containing all the remainders for all $f(x) \in GF(2)[x]/(x^7 - 1)$ divided by $g(x) = 1 + x + x^3$. Because of Equation 5.3.2, we only need to compute remainders for the terms $x^j, j = 1, \ldots, n - 1$, where $n = 7$. The remainders for all other polynomials can then be obtained through addition of the remainders in this table. Use this method to calculate the remainder for

$$f(x) = 1 + x^3 + x^5 + x^6.$$

Solution: By long division, we obtain the remainder table

x^j	$p(x)$
x^6	$1 + x^2$
x^5	$1 + x + x^2$
x^4	$x + x^2$
x^3	$1 + x$
x^2	x^2
x	x

Also note the remainder of $f(x) = 1$ is 1. So,

$$1 + x^3 + x^5 + x^6 \Rightarrow \rho(x) = 1 + 1 + x + 1 + x + x^2 + 1 + x^2 = 0.$$

EXAMPLE 5.3.3

Prove that $1 + x$ is a factor of $x^n + 1$ for any $n > 0$ in $GF(2)[x]/(x^n - 1)$.

Solution: Using the long division algorithm and induction,

$$x^n + 1 = (x + 1) \sum_{j=0}^{n-1} x^j.$$

Since $\rho(x) = 0$, we see that $1 + x$ divides $x^n + 1$.

EXAMPLE 5.3.4 Meggitt's Theorem

Meggitt's theorem is the key to simple hardware implementation for decoders for cyclic codes that correct up to a few errors. The theorem states

Theorem 5.3.1: Suppose that $g(x)h(x) = x^n - 1$ and $v(x)/g(x) = \rho(x)$. Then

$$[xv(x) \bmod (x^n - 1)]/g(x) = [x\rho(x)]/g(x).$$

(Notice that this expression is the remainder with respect to $g(x)$ of the cyclic shift of $v(x)$.)
Proof: $v(x)$ can be uniquely expressed in the form of Equation 5.3.1 as

$$v(x) = Q_1(x)g(x) + \rho(x), \quad \deg[\rho(x)]/g(x).$$

Therefore,

$$xv(x) = xQ_1(x)g(x) + x\rho(x).$$

We can express the cyclic shift of $v(x)$ as

$$xv(x) \bmod (x^n - 1) = xv(x) - (x^n - 1)v_{n-1} = xv(x) - v_{n-1}h(x)g(x).$$

We also know that we can uniquely express $x\rho(x)$ in the form

$$x\rho(x) = Q_2(x)g(x) + t(x), \quad \deg[t(x)] < \deg[g(x)].$$

Therefore,

$$\begin{aligned} xv(x) \bmod (x^n - 1) &= xQ_1(x)g(x) + x\rho(x) - v_{n-1}h(x)g(x) \\ &= [xQ_1(x) + Q_2(x) - v_{n-1}h(x)]g(x) + t(x) \\ &= Q_3(x)g(x) + t(x). \end{aligned}$$

But the remainder $t(x) \equiv [x\rho(x)]/g(x)$. QED

5.4 GENERATION AND DECODING OF CYCLIC CODES

5.4.1 Generator, Parity-Check, and Syndrome Polynomials

We are now ready to put the mathematics of binary polynomial algebra to work. In Section 5.2 we saw that we could represent a nonsystematic (n, k) cyclic code word in polynomial form as

$$c(x) = m(x)g(x), \qquad 5.4.1$$

where the message word is given by the polynomial $m(x)$ having

$$\deg[m(x)] \leq k - 1 \qquad 5.4.2$$

and the generator polynomial $g(x)$ has

$$\deg[g(x)] = r = n - k. \qquad 5.4.3$$

Our code polynomials are elements of the ring $GF(2)[x]/(x^n - 1)$, and we pick our generator polynomial so that

$$x^n - 1 = x^n + 1 = h(x)g(x). \qquad 5.4.4$$

With this selection of our generator polynomial, we also have the identity

$$c(x)h(x)/(x^n - 1) = m(x)g(x)h(x)/(x^n - 1) = 0. \qquad 5.4.5$$

Note the similarity between this expression and the expression from Chapter 4.

$$\bar{c}H^T = \bar{m}GH^T = \bar{0}$$

By analogy, the polynomial $h(x)$ is called the *parity-check polynomial*. Let us represent the received block vector $\bar{v}$ in polynomial form as

$$v(x) = c(x) + e(x), \qquad 5.4.6$$

where $e(x)$ is the *error polynomial*. We can then define a syndrome polynomial

$$v(x)h(x)/(x^n - 1) = c(x)h(x)/(x^n - 1) + e(x)h(x)/(x^n - 1) = e(x)h(x)/(x^n - 1).$$

The syndrome polynomial depends only on the error polynomial and not the transmitted code polynomial.

5.4.2 Systematic Cyclic Codes

In practice, there are many advantages in using *systematic* block codes. We now turn our attention to how to define systematic cyclic codes. In polynomial form, a systematic code word is of the form

$$c(x) = x^r m(x) - d(x) = x^r m(x) + d(x), \qquad 5.4.7$$

where $r = n - k$ and the check bits polynomial $d(x)$ has $\deg[d(x)] \leq r - 1$. Remember that for binary polynomials, $d(x) = -d(x)$. Let us define $d(x)$ as the remainder term

$$d(x) \equiv x^r m(x)/g(x). \qquad 5.4.8$$

Since $g(x)$ is constrained by Equation 5.4.3, this satisfies the systematic form of Operation 5.4.7. In addition, the syndrome polynomial defined by

$$s(x) = c(x)/g(x) = x^r m(x)/g(x) + d(x)/g(x) = d(x) + d(x) = 0,$$

since $\deg[d(x)] \leq r - 1 \Rightarrow d(x)/g(x) = d(x)$. This syndrome check has the same desirable property as Equation 5.4.5, that is, the syndrome of a code polynomial is zero.

Does the set of code polynomials defined by the Operations 5.4.7 and 5.4.8 still satisfy the property that for every $c(x) \in C$ there is also a polynomial $c'(x) = xc(x)/(x^n - 1) \in C$? This condition is required for the code to be a cyclic code. We already know for $g(x)$ satisfying Equation 5.4.4, any polynomial $q(x)$ with $\deg[q(x)] \leq k - 1$ forms a systematic code word

$$c(x) = q(x)g(x).$$

Now consider the polynomial $x^r m(x)$ with $\deg[m(x)] \leq k - 1$. By the definition of polynomial division, we know we can express

$$x^r m(x) = q(x)g(x) + d(x)$$

for some $q(x)$ with $\deg[q(x)] \leq k - 1$ and $d(x) = x^r m(x)/g(x)$. This implies that

$$q(x)g(x) = c(x) = x^r m(x) - d(x) = x^r m(x) + d(x).$$

Since $c(x)$ is a code word in a cyclic code defined by $g(x)$, it follows that Operations 5.4.7 and 5.4.8 do indeed define a cyclic code. Is Equation 5.4.4 a necessary condition for the code to be cyclic? The answer is "yes." For a code to be cyclic, Meggitt's theorem (Theorem 5.3.1) must be satisfied, but this theorem requires Equation 5.4.4 to be true. (This argument does not constitute a rigorous proof of the assertion, but it would be possible for us to prove it rigorously if only we knew a little more abstract algebra at this point.) We raise Equation 5.4.4 to the status of a theorem.

Theorem 5.4.1: A linear (n, k) block code defined by a generator polynomial $g(x)$ is a cyclic code if and only if $g(x)$ satisfies Equation 5.4.4.

This theorem places some restrictions on the form of our generator polynomials $g(x)$. If $g(x)$ is of degree r, it must have the form

$$g(x) = 1 + g_1 x + \cdots + g_{r-1} x^{r-1} + x^r,$$

that is, $g_0 = g_r = 1$. The condition $g_r = 1$ comes about because we have defined $g(x)$ to be of degree r. (Such a polynomial is sometimes called a *monic* polynomial.) The condition $g_0 = 1$ is because $g(x)$ must be a factor of $x^n - 1 = x^n + 1$ and *all* such factors must have $g_0 = 1$. The proof of this statement is simple and is left to you as an exercise.

EXAMPLE 5.4.1

Construct a systematic (7, 4) cyclic code.

Solution: We previously found that the factorization $x^7 + 1 = (x+1)(x^3 + x + 1)(x^3 + x^2 + 1)$. The generator polynomial must be of degree $r = n - k = 7 - 4 = 3$. Let our generator polynomial be

$$g(x) = x^3 + x + 1.$$

The code words are the 16 polynomials defined by

$$c(x) = x^3(m_0 + m_1 x + m_2 x^2 + m_3 x^3)/g(x) + x^3 m(x) = d(x) + x^3 m(x).$$

In Example 5.3.2, we found the remainders for this $g(x)$ for the terms x^3, x^4, x^5, and x^6. Using these results and Equation 5.3.2, we get the following code table:

$m(x)$	$c(x)$	$m(x)$	$c(x)$
0	0	x^3	$1 + x^2 + x^6$
1	$1 + x + x^3$	$1 + x^3$	$x + x^2 + x^3 + x^6$
x	$x + x^2 + x^4$	$x + x^3$	$1 + x + x^4 + x^6$
$1 + x$	$1 + x^2 + x^3 + x^4$	$1 + x + x^3$	$x^3 + x^4 + x^6$
x^2	$1 + x + x^2 + x^5$	$x^2 + x^3$	$x + x^5 + x^6$
$1 + x^2$	$x^2 + x^3 + x^5$	$1 + x^2 + x^3$	$1 + x^3 + x^5 + x^6$
$x + x^2$	$1 + x^4 + x^5$	$x + x^2 + x^3$	$x^2 + x^4 + x^5 + x^6$
$1 + x + x^2$	$x + x^3 + x^4 + x^5$	$1 + x + x^2 + x^3$	$1 + x + x^2 + x^3 + x^4 + x^5 + x^6$

By inspection, we can see that this code satisfies the cyclic code property of having the cyclic shift of each code word as a code word. Also notice that the minimum Hamming weight of the nonzero code words is 3, and therefore, this code has a minimum Hamming distance of 3. We can construct the systematic generator matrix G of this code easily, since the P submatrix of this code is defined by the message words of weight one, i.e., [1 0 0 0], [0 1 0 0], [0 0 1 0], and [0 0 0 1]. The G matrix for this code is

$$G = \begin{bmatrix} 1 & 1 & 0 & 1 & 0 & 0 & 0 \\ 0 & 1 & 1 & 0 & 1 & 0 & 0 \\ 1 & 1 & 1 & 0 & 0 & 1 & 0 \\ 1 & 0 & 1 & 0 & 0 & 0 & 1 \end{bmatrix}.$$

From G we can write down the parity-check matrix H using Equation 4.4.3 with the result

$$H = \begin{bmatrix} 1 & 0 & 0 & 1 & 0 & 1 & 1 \\ 0 & 1 & 0 & 1 & 1 & 1 & 0 \\ 0 & 0 & 1 & 0 & 1 & 1 & 1 \end{bmatrix}.$$

This is the parity-check matrix of a Hamming code! Compare this matrix to the H matrix of Example 4.5.1. The two matrices are the same, except for a trivial permutation of the last three columns. This amounts to nothing more than a permutation of the check bits, which, of course, leaves the basic code properties unaltered. The Hamming codes turn out to be cyclic codes.

5.4.3 Hardware Implementation of Encoders for Systematic Cyclic Codes

The mechanics of the long division process used in Equation 5.4.8 have a simple implementation for binary polynomials. We assume that the bits are transmitted serially with the highest power of x being transmitted first. Since we know from Equation 5.3.2 that we can calculate the remainder term in Equation 5.4.8 for individual powers of x and sum the results, we can derive a simple expression for computing the remainder of x^t modulo $g(x)$. Consider the calculation of $x^{n-1}/(x^r+g_{r-1}x^{r-1}+\cdots+g_1x+1)$. We can illustrate the mechanics of this by example using $n=7, r=3$. For the first step in the long division process, we have

$$
\begin{array}{r}
x^3 \\
x^3+g_2x^2+g_1x+1 \overline{)\, x^6 } \\
x^6+g_2x^5+g_1x^4+x^3 \\ \hline
g_2x^5+g_1x^4+x^3
\end{array}.
$$

We can represent the remainder after the first cycle of the division as a vector

$$S_1^T = [g_2\ g_1\ 1] \equiv [g_{r-1}\ g_{r-2}\cdots g_1\ g_0].$$

For the next division cycle, we have

$$
\begin{array}{r}
g_2x^2 \\
x^3+g_2x^2+g_1x+1 \overline{)\, g_2x^5+g_1x^4+x^3 } \\
g_2x^5+g_2x^4+g_1g_2x^3+g_2x^2 \\ \hline
(g_1+g_2)x^4+(1+g_1g_2)x^3+g_2x^2
\end{array},
$$

since $g_2^2 = g_2$ in GF(2). We can write the remainder after this cycle as

$$S_2 = \begin{bmatrix} g_2 & 1 & 0 \\ g_1 & 0 & 1 \\ 1 & 0 & 0 \end{bmatrix}\begin{bmatrix} g_2 \\ g_1 \\ 1 \end{bmatrix} = \begin{bmatrix} g_{r-1} & 1 & 0 & \cdots & 0 \\ g_{r-2} & 0 & 1 & \cdots & 0 \\ \vdots & 0 & & \cdots & 0 \\ g_1 & 0 & & \cdots & 1 \\ g_0 & 0 & 0 & \cdots & 0 \end{bmatrix} S_1.$$

This process continues two more times, for a total of k cycles. By induction, we find that

$$S_3 = \begin{bmatrix} g_{r-1} & 1 & 0 & \cdots & 0 \\ g_{r-2} & 0 & 1 & \cdots & 0 \\ \vdots & 0 & & \cdots & 0 \\ g_1 & 0 & & \cdots & 1 \\ g_0 & 0 & & \cdots & 0 \end{bmatrix} S_2 \equiv \Gamma S_2, \quad S_4 = \Gamma S_3.$$

The process for the term m_2x^5 is the same, except only $k-1=3$ cycles are involved. The same is true for each successive term in $x^t m(x)$ with one less shift for each decrease in the power of x. For a general (n, k) code, we can represent the long-division process for the remainder vector as

$$S_t = \Gamma S_{t-1} + \begin{bmatrix} g_{r-1} \\ g_{r-2} \\ \vdots \\ g_1 \\ g_0 \end{bmatrix} m_{k-t}, \quad t=1,2,\cdots,k, \qquad 5.4.9$$

where $S_0 \equiv \bar{0}$ and

$$\Gamma \equiv \begin{bmatrix} g_{r-1} \\ g_{r-2} \\ \vdots & & I_{(r-1)\times(r-1)} \\ g_1 \\ g_0 & 0 & \cdots & 0 \end{bmatrix}. \qquad 5.4.10$$

This process can be implemented in bit-serial form using the feedback shift register circuit shown in Figure 5.4.1. The addition blocks in this figure are implemented by logic EXCLUSIVE-OR gates, while the multiply blocks are a connection if $g_i = 1$ and no connection if $g_i = 0$. (We may also think of these multipliers as logic AND gates, but in practice it is rare to implement the connections this way.) The operation of this circuit is identical to Equation 5.4.9. This may be verified by direct analysis of the circuit.

After k shifts, the contents of the shift register contain the vector

$$S_k^T = [s_{r-1}\, s_{r-2}\, \cdots\, s_1\, s_0],$$

which defines the check field

$$d(x) = s_0 + s_1 x + \cdots + s_{r-1} x^{r-1}. \qquad 5.4.11$$

The circuit of Figure 5.4.1 can be used to implement the encoder of Equation 5.4.7 for an (n, k) cyclic code. Figure 5.4.2 illustrates the procedure. The details of the shift register circuitry are represented by the "divide by $g(x)$" block in the figure. The switches are set in the positions shown in the figure for the first k cycles of operation, while the message $x^r m(x)$ is shifted in. The switches are set to the second position for the remaining $n - k = r$ cycles. This breaks the feedback action of the divide circuit and shifts the remainder out of the shift register, appending it as $d(x)$ to the transmitted code word. The shift register is initialized to zero at the start of each block. (Notice that this happens automatically during the r shifts when switch 1 is set to the "zero" position.)

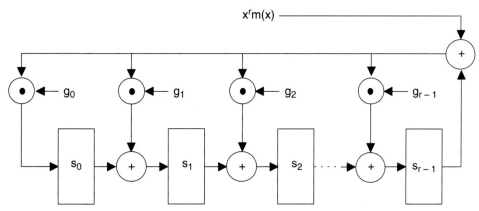

Figure 5.4.1: Divide-by-$g(x)$ Circuit

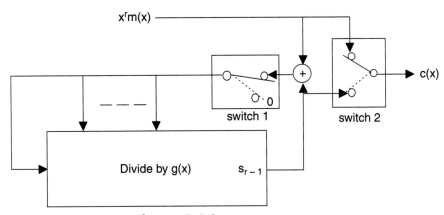

Figure 5.4.2: Systematic Encoder

5.4.4 Hardware Implementation of Decoders for Cyclic Codes

Syndrome calculation and decoding can be carried out using the same basic divide by $g(x)$ configuration. If we replace $m(x)$ by the received code polynomial $v(x)$ in Figure 5.4.1, the input to the divide circuit becomes $x^r v(x)$. Since $v(x)$ contains n bits instead of k, a total of n shifts are required to calculate the syndrome. The division equation becomes

$$x^r v(x) = Q(x)g(x) + s(x), \qquad 5.4.12$$

where $s(x)$ is the syndrome polynomial. The syndrome is given by

$$s(x) = x^r v(x)/g(x). \qquad 5.4.13$$

Assume that $v(x)$ is given by Equation 5.4.6. From Equation 5.3.2, we know that we can separate the syndrome calculation into a term involving $c(x)$ and a term involving $e(x)$. For the code word term, we have

$$[x^r c(x)]/g(x) = \left[(x^r/g(x))\bigl((x^r m(x) + d(x))/g(x)\bigr)\right]\bigg/ g(x)$$

by Equation 5.3.4. However,

$$(x^r m(x) + d(x))/g(x) = d(x) + d(x) = 0,$$

so the code word term is zero. Therefore,

$$s(x) = x^r e(x)/g(x). \qquad 5.4.14$$

Once again, the syndrome is a function only of the error term and not the code word.

Using Equation 5.4.14 as our syndrome definition, we can compile a table of syndromes for the correctable error patterns of the code. The size of this table is clearly less than or equal to $2^r - 1$ (since the zero syndrome denotes "no error"). It follows we can perform error correction using the circuit configuration of Figure 5.4.3. The divide circuit is initialized to zero at the start of the block and the n-bit received word is shifted into the syndrome calculator and an n-bit buffer shift register. At the end of the n bits, the output of the syndrome table equals the error pattern selected by the syn-

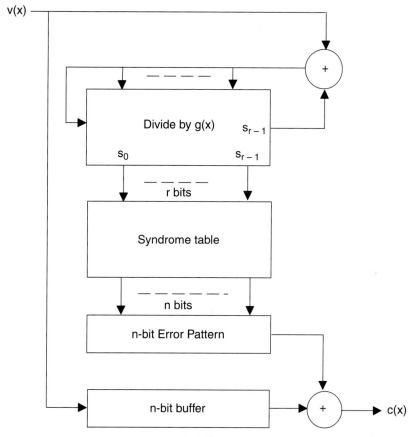

Figure 5.4.3: Error Correction

drome. This pattern is loaded into a buffer and shifted concurrently with the received-word buffer. The error pattern is corrected at the output of the received-word buffer. While this is taking place, the divide circuit is reset and begins calculating the syndrome for the next block.

5.4.5 The Meggitt Decoder

The error-correction circuit of Figure 5.4.3 is not a very efficient implementation and can be greatly improved on. In terms of its efficiency, we first note a t-error-correcting code typically has $t \ll n$. Consequently, most of the terms in the error pattern buffer are zero. Secondly, the number of possible error patterns of Hamming weight t in an n bit block is

$$n_e = \sum_{j=1}^{t} \binom{n}{j},$$

which is of order n^t for $t \ll n$.

Meggitt's theorem (Theorem 5.3.1) greatly simplifies the decoding problem. Let us categorize the possible error patterns into two sets

$$E_{\text{meg}} = \{e(x) | e_{n-1} = 1\} \qquad 5.4.15$$

and

$$E_{\text{shift}} = \{e(x) | e_{n-1} = 0\}.$$

The number of patterns in the Meggitt set is

$$n_{\text{meg}} = \sum_{j=0}^{t-1} \binom{n-1}{j},$$

which is on the order of $(n-1)^{t-1} \ll n^t$. In the correction circuit in Figure 5.4.3, correction takes place as the erroneous bits shift out the end of the buffer. If we store all of the syndromes for errors in the Meggitt set in our syndrome table, any syndrome that matches one of these patterns calls for correction of the received bit v_{n-1}.

Now consider the error patterns in the set E_{shift}. The degree of the error polynomial for an error in this set will be $\deg[e'(x)] = n - 1 - \ell$ for some $1 \le \ell \le n-1$, and after ℓ shifts of the buffer, the first erroneous term will be presented at the buffer output. The shifted error pattern is equivalent to ℓ cyclic shifts of $e'(x)$. If the syndrome of $e'(x)$ is $s'(x)$, then the syndrome associated with the error pattern in the Meggitt set corresponding to ℓ cyclic shifts of $e'(x)$ is given by Meggitt's theorem as

$$s(x) = [x^\ell e'(x) \bmod (x^n - 1)]/g(x) = [x^\ell s'(x)]/g(x). \qquad 5.4.16$$

Therefore, *we only need to store syndromes for the error patterns in the Meggitt set,* provided that we update the syndrome pattern as the buffer is shifted out.

How does the syndrome pattern get updated? This is quite easy. Consider the circuit in Figure 5.4.1, and assume there is a nonzero pattern stored in the shift register. Further, assume that the input $x^r m(x) = 0$. If we now shift the pattern in the shift register, the next syndrome contained in the shift register is given from Equation 5.4.9 as $S_{r+1} = \Gamma S_r$. This is equivalent to multiplying the syndrome by x and then dividing it by $g(x)$. If we shift ℓ times, we will have the syndrome of Equation 5.4.16.

We can implement this strategy by pipelining the error correction process and the syndrome calculation for the next incoming block using the circuit in Figure 5.4.4. The operation of this decoder is very simple. After the n bits of the received block have been shifted in, the syndrome $s(x)$ for this block is parallel loaded into a copy of the divide by $g(x)$ circuit. The shift register contents of this pipeline syndrome calculator are applied to a syndrome table that contains only the syndromes corresponding to an error pattern $e(x)$ in the set E_{meg}. If the syndrome matches one of these patterns, the bit at the output of the buffer is in error. The syndrome table outputs a "1" on its Match line, and the error is corrected.

If the initial syndrome does not match an entry in the table, the bit at the output of the buffer is correct and a "0" is output from the Match line. As the next received block is shifted in, the upper divide by $g(x)$ circuit is cleared and begins calculating the syndrome for the next block. The pattern in the buffer is shifted, and the pipeline syndrome calculator is updated for the shifted block. Whenever the updated syndrome matches a pattern stored in the syndrome table, the current bit coming out of

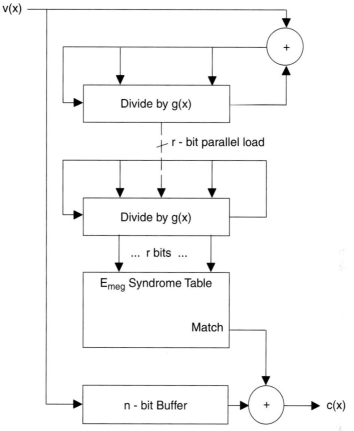

Figure 5.4.4: The Meggitt Decoder

the buffer is in error and is corrected by the Match line output. At the end of n shifts, the old block has been corrected and shifted out, the new received block is held in the buffer, and the operation begins all over again.

EXAMPLE 5.4.2

Using the (7, 4) Hamming code of Example 5.4.1, decode the received block

$$v(x) = 1 + x^3 + x^4 + x^5 + x^6.$$

Solution: Since $t = 1$ for a Hamming code, the syndrome table only contains the syndrome for the error pattern $e(x) = x^{n-1} = x^6$. The syndrome pattern is given by Equation 5.4.14 as

$$s(x) = (x^3 x^6)/(x^3 + x + 1).$$

We can calculate this by long division, but let's save ourselves a little labor by noting that this syndrome is equivalent to the syndrome obtained from applying Meggitt's theorem to the syndrome given by

$$x^6/(x^3 + x + 1).$$

In Example 5.3.1, we found this syndrome to be $s'(x) = x^2 + 1$. Therefore, the syndrome to be stored in the syndrome table is given by

$$S = \begin{bmatrix} 0 & 1 & 0 \\ 1 & 0 & 1 \\ 1 & 0 & 0 \end{bmatrix}^3 \begin{bmatrix} 1 \\ 0 \\ 1 \end{bmatrix} = \begin{bmatrix} 1 & 1 & 0 \\ 1 & 1 & 1 \\ 1 & 0 & 1 \end{bmatrix} \begin{bmatrix} 1 \\ 0 \\ 1 \end{bmatrix} = \begin{bmatrix} 1 \\ 0 \\ 0 \end{bmatrix},$$

or $s(x) = x^2$. The "syndrome table" is therefore merely a logic circuit that outputs a "1" if the most significant syndrome bit is a "1" and all other syndrome bits are "0". This is a trivial circuit. If we remember the flipflops that make up the shift register typically have both true and complementary outputs available, the syndrome "table" is merely a three-input logic AND circuit.

To decode the received block, we must calculate the block's syndrome. Using Equation 5.4.9 and replacing m_{k-t} by v_{n-t} and k by n, we have

$$S_1 = \begin{bmatrix} 0 \\ 1 \\ 1 \end{bmatrix}, \quad S_2 = \begin{bmatrix} 0 & 1 & 0 \\ 1 & 0 & 1 \\ 1 & 0 & 0 \end{bmatrix} \begin{bmatrix} 0 \\ 1 \\ 1 \end{bmatrix} + \begin{bmatrix} 0 \\ 1 \\ 1 \end{bmatrix} = \begin{bmatrix} 1 \\ 0 \\ 1 \end{bmatrix},$$

etc., until we arrive at

$$S_7 = \begin{bmatrix} 0 \\ 0 \\ 1 \end{bmatrix} \Rightarrow s(x) = 1.$$

This does not match our syndrome table, so bit v_6 is correct, and $c_6 = m_3 = 1$.

Next we shift the pattern in the buffer and update the syndrome to

$$S_8 = \begin{bmatrix} 0 & 1 & 0 \\ 1 & 0 & 1 \\ 1 & 0 & 0 \end{bmatrix} \begin{bmatrix} 0 \\ 0 \\ 1 \end{bmatrix} = \begin{bmatrix} 0 \\ 1 \\ 0 \end{bmatrix} \Rightarrow s(x) = x.$$

This also does not match our table, so v_5 is correct and $m_2 = 1$. Shifting once more, we get

$$S_9 = \begin{bmatrix} 0 & 1 & 0 \\ 1 & 0 & 1 \\ 1 & 0 & 0 \end{bmatrix} \begin{bmatrix} 0 \\ 1 \\ 0 \end{bmatrix} = \begin{bmatrix} 1 \\ 0 \\ 0 \end{bmatrix} \Rightarrow s(x) = x^2.$$

This matches our syndrome table, so v_4 is in error and is corrected to give us $m_1 = 0$. The remaining four shifts give us

$$S_{10} = \begin{bmatrix} 0 \\ 1 \\ 1 \end{bmatrix}, \quad S_{11} = \begin{bmatrix} 1 \\ 1 \\ 0 \end{bmatrix}, \quad S_{12} = \begin{bmatrix} 1 \\ 1 \\ 1 \end{bmatrix}, \quad S_{13} = \begin{bmatrix} 1 \\ 0 \\ 1 \end{bmatrix}.$$

None of these match our syndrome table, so all remaining bits are correct, and

$$m(x) = 1 + x^2 + x^3.$$

5.5 ERROR-TRAPPING DECODERS

5.5.1 Updating the Syndrome during Correction

In Example 5.4.2, the error pattern $e(x) = x^{n-1}$ gave us a syndrome $s(x) = x^{r-1}$. This is not an accident, and it can be used to our advantage to further reduce the size of the syndrome table for codes having $t > 1$. There are a couple of properties of the syndrome defined by Equation 5.4.14 we can exploit.

First, we can express any error pattern as

$$e(x) = e_{n-1}x^{n-1} + e'(x),$$

where $e'(x) \in E_{\text{shift}}$. We also know from Equation 5.3.2 that we can compute the syndrome for $e(x)$ by calculating the syndromes from each of the two terms above and adding the results. The syndrome for the first term is

$$e_{n-1}(x^r x^{n-1})/g(x) = e_{n-1}[x^{n+r-1} \bmod(x^n - 1)]/g(x) = e_{n-1}x^{r-1}/g(x) = e_{n-1}x^{r-1}. \quad 5.5.1$$

This is equal to the output of the Match line in the Meggitt decoder of Figure 5.4.4. Therefore, as we correct errors with the decoder, we can also remove the contribution of the corrected error from the syndrome calculation with the simple modification to the pipeline syndrome circuit shown in Figure 5.5.1. The initial syndrome is parallel loaded into the divide by $g(x)$ circuit at the beginning of the correction process. As each error is corrected coming out of the received block buffer (which is not in this figure), the contribution made by that error to the syndrome is "corrected" by adding Equation 5.5.1 to the syndrome. At the end of n shift cycles, the corrected syndrome contained in the divide by $g(x)$ shift register is given by the "corrected syndrome" outputs of the block. Assuming that the block contained only correctable errors, if all errors are corrected at the end of n shift cycles, the "corrected syndrome" output will be all zeroes.

There are two ways for the corrected syndrome output to be nonzero at the end of n shifts. The first way is if the decoder is a bounded-distance decoder with additional error detection capability. In this case, a nonzero corrected syndrome output would indicate detection of an uncorrectable error.

The second way to get a nonzero corrected syndrome output is if the syndrome table does not contain syndrome patterns for all possible error patterns of weight t or less. Why might we do this? Recall that the size of the syndrome table required for all possible errors in the Meggitt set is of order $(n-1)^{t-1}$. This can still be rather large for $t > 1$. The required size of the syndrome table can be further reduced for some codes by *multi-stage pipelining* of the error correction process. A decoder based on this approach is sometimes called an *error-trapping* decoder.

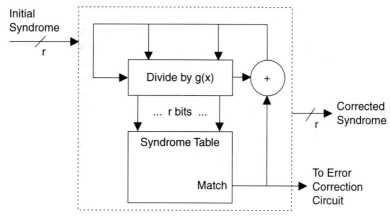

Figure 5.5.1: Syndrome Matcher

5.5.2 Burst Error Patterns and Error Trapping

The *burst length* of an error polynomial $e(x)$ is defined as the number of bits from the first error term in $e(x)$ to the last error term, inclusive. For example, the error polynomial

$$e(x) = x^3 + x^7$$

has a burst length

$$b = 7 - 3 + 1 = 5.$$

By definition, there can be only one burst in a block. In other words,

$$e(x) = x^3 + x^7 + x^{19} + x^{20}$$

has *one* burst of length $20 - 3 + 1 = 18$ and *not* two bursts of lengths 5 and 2.

An error-trapping decoder may be used if, for every correctable error $e(x)$, there is some cyclic shift of the error pattern

$$[x^\Delta e(x)]/(x^n - 1), \quad \Delta \geq 0,$$

for which $b \leq r$. A sufficient condition to guarantee this for double-error-correcting codes is if $r \geq k + 1$, i.e., the code rate is less than 1/2.

If the burst length criterion is met, then every correctable error can be represented as the cyclic shift of some error pattern $e'(x)$ by

$$e(x) = [x^\Delta e'(x)]/(x^n - 1),$$

where $\deg[e'(x)] \leq r - 1$.

Now suppose we store in our syndrome table only those syndromes associated with the polynomials $e'(x)$. The maximum number of terms needed in our syndrome table is then

$$\sum_{j=0}^{t-1} \binom{r-1}{j}.$$

This is on the order of $(r-1)^{t-1}$ entries and gives us a much smaller syndrome table. If $e(x)$ has a burst length $b \leq r$, its syndrome is in the table and the correction circuit of 5.5.1 will correct it. If the burst length of $e(x)$ exceeds r, this means the burst can be thought of as "wrapping around" the block. An example of this would be $e(x) = 1 + x^{n-1}$ (which, after one cyclic shift, would have a burst length of $b = 2$). In this case, the decoder of Figure 5.5.1 will "miss" the highest degree error bits but will find the lower degree error bits. It will do only a partial correction.

The error bits "missed" by the first error correction can be corrected by a follow-on error correction circuit. Since *at least* one error bit is corrected by the first-stage decoder, the syndrome table in the second-stage decoder only needs to deal with up to $t - 1$ errors, and so the size of the syndrome table required is reduced to

$$\sum_{j=0}^{t-2} \binom{r-1}{j},$$

which is of the order $(r-1)^{t-2}$. The error-trapping decoder is illustrated in Figure 5.5.2.

It is sometimes possible to use the error-trapping decoder even with codes that do not strictly satisfy the burst length criterion above. However, this requires the syndrome

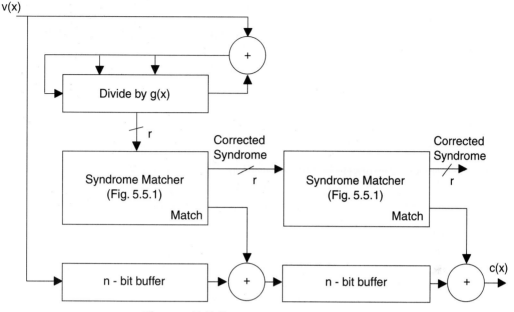

Figure 5.5.2: Error-Trapping Decoder

table in the first stage syndrome matcher to contain extra syndrome terms in addition to the syndromes for $e'(x)$ described above. The collection of syndromes used in the first stage must be sufficient to *guarantee* the correction of at least one error term in the first buffer, and it must deal with all correctable error cases which the subsequent syndrome matcher cannot. Likewise, it may sometimes be advantageous to add a third stage to the pipeline. Both of these cases, however, are rather infrequent in practice.

In the more usual case, for those errors (or their cyclic shifts) which meet the burst length criterion, the required syndromes are terms associated with error patterns

$$x^r e(x) = x^r x^k (e_0 + e_1 x + \cdots + e_{r-1} x^{r-1}) = x^n \sum_{j=0}^{r-1} e_j x^j.$$

Since $g(x)$ divides $x^n - 1$, it is clear that

$$x^n = Q(x)g(x) + 1.$$

So

$$s(x) = [x^r e(x)]/g(x) = \sum_{j=0}^{r-1} e_j x^j. \qquad 5.5.2$$

Therefore, the syndromes $s(x)$ stored in the syndrome table are identical to the correctable error patterns of Equation 5.5.2.

Equation 5.5.2 also suggests an alternative to having the decoder look up syndromes in a syndrome table. Suppose the code is designed to correct t errors. When the buffer register's output bit is in error, the number of "1" bits in the syndrome must be less than or equal to t by Equation 5.5.2. Furthermore, the highest order syndrome bit must be a "1". Therefore, if we *count* the number of "1" bits in the remaining syndrome bits and find less than or equal to $t-1$ of them, the syndrome must correspond to a correctable error pattern. This can be advantageous for larger codes to using a lookup table.

EXAMPLE 5.5.1

The $(15, 7)$ BCH code defined by the generator polynomial

$$g(x) = x^8 + x^7 + x^6 + x^4 + 1$$

can correct $t = 2$ errors. Decode the received block

$$v(x) x^{14} + x^{11} + x^8 + x^5 + x^4 + x^3 + x + 1.$$

Solution: Since $r = 8 = k + 1$, we may use an error-trapping decoder. The syndrome table for the first decoder must contain an entry for the single-bit error

$$e(x) = x^{14} \Rightarrow s(x) = x^{r-1} = x^7$$

and seven entries for the double-bit errors

$$e(x) = x^{n-1} + x^{k+\ell}, \ell = 0, \cdots, r - 2 = 6 \Rightarrow s(x) = x^7 + x^\ell.$$

The remaining k two-error cases have cyclic shifts that equal one of these seven cases and will be handled by the two-stage pipeline.

The initial syndrome is calculated from Equation 5.4.9 as

$$S_t = \Gamma S_{t-1} + [\overline{g}] v_{n-t}, \quad \overline{g} = [1\ 1\ 0\ 1\ 0\ 0\ 0\ 1]^T,$$

with

$$\Gamma = \begin{bmatrix} 1 & 1 & 0 & 0 & 0 & 0 & 0 & 0 \\ 1 & 0 & 1 & 0 & 0 & 0 & 0 & 0 \\ 0 & 0 & 0 & 1 & 0 & 0 & 0 & 0 \\ 1 & 0 & 0 & 0 & 1 & 0 & 0 & 0 \\ 0 & 0 & 0 & 0 & 0 & 1 & 0 & 0 \\ 0 & 0 & 0 & 0 & 0 & 0 & 1 & 0 \\ 0 & 0 & 0 & 0 & 0 & 0 & 0 & 1 \\ 1 & 0 & 0 & 0 & 0 & 0 & 0 & 0 \end{bmatrix}.$$

The initial syndrome calculation is shown in detail in the following table:

t	S_t^T	t	S_t^T
0	[00000000]	8	[01001000]
1	[11010001]	9	[10010000]
2	[01110011]	10	[00100000]
3	[11100110]	11	[10010001]
4	[11001100]	12	[00100010]
5	[01001001]	13	[01000100]
6	[10010010]	14	[01011001]
7	[00100100]	15	[01100011]

At the end of the $t=15$ cycle, the syndrome is loaded into the first syndrome matcher. This syndrome is not one of those in the table, so no correction takes place. The calculations for the first syndrome matcher are

$$S_{t+1} = \Gamma S_t + [1\ 1\ 0\ 1\ 0\ 0\ 0\ 1]^T \cdot match.$$

These calculations are given in the next table. A match is registered on the 28th cycle, which corresponds to bit v_1 being in error. At the 30th cycle, the syndrome in the first stage syndrome matcher is nonzero. This indicates that the error was a two-bit error with a burst length that exceeded r bits (without considering cyclic shifts). Therefore, one error was "missed" by the first-stage syndrome matcher.

The first-stage syndrome matcher's residual syndrome at the end of the 30th cycle is loaded into the syndrome register for the second stage and the first-stage syndrome matcher is loaded with the syndrome for the next incoming block. The calculations for the second-stage syndrome matcher are tabulated below. For the second stage, we are only concerned with matching a single-bit error (since $t=2$ for the code and the first stage is guaranteed to have corrected at least one error). The second-stage syndrome matcher is therefore looking for the syndrome for the single-bit error case.

First Syndrome Matcher Calculations

t	S_t^T	match	t	S_t^T	match
15	[0 1 1 0 0 0 1 1]	0	24	[0 1 1 1 1 1 0 0]	0
16	[1 1 0 0 0 1 1 0]	0	25	[1 1 1 1 1 0 0 0]	0
17	[0 1 0 1 1 1 0 1]	0	26	[0 0 1 0 0 0 0 1]	0
18	[1 0 1 1 1 0 1 0]	0	27	[0 1 0 0 0 0 1 0]	0
19	[1 0 1 0 0 1 0 1]	0	28	[1 0 0 0 0 1 0 0]	1
20	[1 0 0 1 1 0 1 1]	0	29	[0 0 0 0 1 0 0 0]	0
21	[1 1 1 0 0 1 1 1]	0	30	[0 0 0 1 0 0 0 0]	0
22	[0 0 0 1 1 1 1 1]	0	31	[x x x x x x x x]	
23	[0 0 1 1 1 1 1 0]	0			

Second Syndrome Matcher Calculations

t	S_t^T	match	t	S_t^T	match
30	[0 0 0 1 0 0 0 0]	0	39	[0 0 0 0 0 0 0 0]	0
31	[0 0 1 0 0 0 0 0]	0	40	[0 0 0 0 0 0 0 0]	0
32	[0 1 0 0 0 0 0 0]	0	41	[0 0 0 0 0 0 0 0]	0
33	[1 0 0 0 0 0 0 0]	1	42	[0 0 0 0 0 0 0 0]	0
34	[0 0 0 0 0 0 0 0]	0	43	[0 0 0 0 0 0 0 0]	0
35	[0 0 0 0 0 0 0 0]	0	44	[0 0 0 0 0 0 0 0]	0
36	[0 0 0 0 0 0 0 0]	0	45	[0 0 0 0 0 0 0 0]	0
37	[0 0 0 0 0 0 0 0]	0	46	[x x x x x x x x]	
38	[0 0 0 0 0 0 0 0]	0			

The second stage registers a match on the 33rd cycle, which corresponds to bit v_{11} being in error. The syndrome matcher will correct that bit. After correction, the syndrome is now zero. Provided that the error pattern was correctable (i.e., t was actually 2 and not some larger number of errors beyond the capability of the code), the final output sequence from the second stage is now correct.

The error polynomial corrected by this process was

$$e(x) = x^{11} + x,$$

so the final decoded message polynomial is

$$m(x) = x^6 + 1.$$

In this example, it did not happen that bit v_0 was in error. If it had been, this error would have been caught and corrected by the first-stage syndrome matcher. In that case, the syndrome loaded into the second stage would have been the *corrected* syndrome, i.e., the syndrome with the match correction added in. Otherwise, the second stage would have also missed the first error and would reinsert the error in v_0 corrected by the first stage.

5.6 SOME STANDARD CYCLIC BLOCK CODES

There are many known codes with good error correction capability. In this section, we will tabulate a number of codes that can be decoded with the techniques discussed in the previous two sections. The codes we will look at are not the most powerful of the known cyclic codes. Indeed, we do not have the theoretical background required to understand these powerful codes at this stage of our learning. However, the codes presented here, while not the most powerful available, enjoy reasonably widespread use in a number of applications and are important in their own right. In addition, there are some useful "variations on a theme" that can be practiced on these codes to tailor them for specific applications. We will discuss the base codes in this section and discuss some useful methods for modifying these codes in Section 5.7.

It is common to specify cyclic codes by means of the generator polynomial $g(x)$. It is also common to specify $g(x)$ in octal (base 8) format to save having to write out the entire polynomial. This notational convention is best illustrated by example. Suppose we wish to specify that

$$g(x) = x^4 + x + 1.$$

If we write down the coefficients g_i of this polynomial, starting with the highest power of x, we get

$$1\ 0\ 0\ 1\ 1.$$

Now group these binary digits in threes from right to left, and, if necessary, pad zeroes on at the left end to obtain a group of three binary digits. For the foregoing example, this gives us

$$010\ 011.$$

We now write this in octal notation as

$$23.$$

5.6.1 The Hamming Codes

We discussed Hamming codes at length in Chapter 4, and in this chapter we saw that Hamming codes can be expressed in cyclic form. The Hamming codes are single-error-correcting codes having block lengths defined by the degree of the generator polynomial as

$$n = 2^r - 1, \qquad r \geq 3.$$

The Hamming codes for several different block lengths are given in the following table:

TABLE 5.6.1 Hamming Codes

n	k	t	g(x) (octal)
7	4	1	13
15	11	1	23
31	26	1	45
63	57	1	103
127	120	1	211
255	247	1	435
511	502	1	1021
1023	1013	1	2011
2047	2036	1	4005
4095	4083	1	10123

5.6.2 BCH Codes

The Bose–Chaudhuri–Hocquenghem (BCH) codes are among the most important of all cyclic block codes. They were discovered by Hocquenghem in 1959 and independently discovered by Bose and Chaudhuri in 1960. These codes are described by an elegant algebraic theory that, unfortunately, is slightly beyond the scope of this text. (The theory of BCH codes is standard fare in a first graduate course on error-correcting codes.) The single-error-correcting BCH codes are none other than the Hamming codes. BCH codes can be designed to correct any number of errors although, of course, decoding complexity grows with the number of errors t the code can correct. The block length of a BCH code is always

$$n = 2^j - 1, j \geq 3,$$

but the parameter j does not correspond to r, except in the case of the single-error-correcting BCH codes. The following table lists some of the smaller BCH codes with $t > 1$:

TABLE 5.6.2 BCH Codes

n	k	t	g(x) (octal)
15	7	2	721
15	5	3	2467
31	21	2	3551
31	16	3	107657
31	11	5	5423325
63	51	2	12471
63	45	3	1701317
63	39	4	166623567
63	36	5	1033500423
63	30	6	157464165547
127	113	2	41567
127	106	3	11554743
255	239	2	267543
255	231	3	156720665

The first two of these BCH codes (with block lengths $n = 15$) can be decoded using the error-trapping decoder of the previous section. The other codes in this list cannot be corrected by the simple error-trapping decoder without the addition of a larger syndrome table. This obviously becomes more difficult to do as the block length increases or as t increases. A point is quickly reached where the simple Meggitt decoder is no longer a feasible approach to the decoding of BCH codes and another alternative must be sought.

The earliest solution to this problem was presented by Peterson in 1960. Peterson's "direct solution" algorithm is useful for correcting small numbers of errors, but unfortunately, the complexity of the calculations quickly becomes intractable as the number of errors to be corrected increases. An improved version of this algorithm was presented by Gorenstein and Zierler in 1961, but the first *efficient* decoding algorithm was not discovered until 1967 by Berlekamp. Further improvements to this algorithm rather quickly followed.

There are two main decoding strategies for BCH codes that work for all BCH codes. These strategies are based on the algebraic structure of BCH codes and can be implemented at high speeds using shift register circuitry. The first approach is based on an algorithm known as the Berlekamp/Massey algorithm. The second is based on an algorithm known as Euclid's algorithm. The Berlekamp/Massey algorithm is the slightly better of the two, but Euclid's algorithm has widespread usage (partly because Euclid never got a patent[2] and the other fellows did). Both algorithms are very interesting, also involve finite-state machines rather than a combinatorial syndrome table, but an explanation of how and why they work is a bit beyond the scope of this text.

5.6.3 Burst-Correcting Codes

In some applications, channel errors tend to occur in bursts rather than as individual random bit errors. One example of such a situation can occur in magnetic disk or tape systems. A defect in the disk or tape is a spot on the recording medium (disk or tape) where the magnetic material has been damaged and cannot reliably record information. Since the area occupied by a bit in these systems is small, a defect tends to destroy a string of bits rather than an individual bit. A similar defect mechanism exists in compact optical disks. This situation can also arise in telephone channels because of interference from various sources, such as noise from electric motors. Cases also exist in wireless systems such as mobile radio or telephone systems, because of fading of the transmitted electromagnetic wave from reflections.

In the previous section we defined the length of an error burst as the total number of bits from the first error to the last error, inclusive. Burst error-correcting codes can be decoded using the error-trapping decoder of Section 5.5. The minimum possible number of parity-check bits required to correct a burst of length b or less is given by the *Rieger bound*

$$r \geq 2b. \qquad 5.6.1$$

The best understood codes for correcting burst errors are cyclic codes. Unlike the Hamming and BCH codes, there are at this time few known general procedures for

---

[2]As far as I know, Euclid was not interested in error-correcting codes.

algebraically designing binary burst correction codes from a given burst length specification. (One exception to this is the class of codes known as *Fire codes*.) Instead, good burst-correcting codes have been found mainly by computer search. Some known good burst-correcting codes are given in the following table:

TABLE 5.6.3   Good Burst-Correcting Codes

n	k	b	g(x) (octal)
7	3	2	35
15	10	2	65
15	9	3	171
31	25	2	161
63	56	2	355
63	55	3	711
511	499	4	10451
1023	1010	4	22365

Codes for correcting longer bursts can be constructed from these codes by a method known as *interleaving*. We will discuss this method in Section 5.7.

5.6.4 Cyclic Redundancy Check Codes

Cyclic redundancy check, or CRC, codes are error-detecting codes typically used in automatic repeat request (ARQ) systems. CRC codes have no error correction capability but they can be used in combination with an error-correcting code, both to improve the performance of the system, as previously discussed in Chapter 4, and also to detect miscorrects by the error-detecting code. When used in this fashion, the error control system is in the form of a *concatenated* code. This is illustrated in Figure 5.6.1.

CRC codes are an important application of *shortened* cyclic codes. We will discuss shortening a cyclic code in the next section. For now, it is sufficient to say that CRC codes are generated and their syndromes are calculated in precisely the same manner as the cyclic codes we have been looking at previously. Since these codes are error detecting rather than error-correcting, there is no error correction circuit.

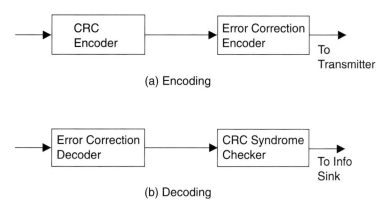

Figure 5.6.1: Concatenated Coding System

TABLE 5.6.4 Some Primitive Polynomials of Degree 3 to 18

Degree	p(x) (octal)	Degree	p(x) (octal)
3	13	11	4003
4	23	12	10123
5	45	13	20033
6	103	14	42103
7	211	15	100003
8	435	16	210013
9	1021	17	400011
10	2011	18	1000201

Instead, the syndrome of the received block is checked to see if it is zero (no error detected) or nonzero (error detected).

CRC codes are often (but not always) constructed of polynomials of the form

$$g(x) = (x + 1)p(x), \qquad 5.6.2$$

where $p(x)$ is an irreducible polynomial of $\deg[p(x)] = r - 1$ that divides $x^{2^{r-1}-1} - 1$ but that does not divide $x^m - 1$ for any integer $m < 2^{r-1} - 1$. Such a polynomial is called a *primitive* binary polynomial. The factor $(x + 1)$ in Equation 5.6.2 is included to ensure that all error patterns with odd Hamming weight are detected. The maximum block length of a CRC code generated by Equation 5.6.2 is $n = 2^{r-1} - 1$ and $k = n - r$. Usually, r is selected to be rather large, and the CRC block length is taken as less than the maximum just given. This is termed *shortening* the code. Shortening a cyclic code does not impact the generation of the code or the calculation of its syndrome (as we will see in Section 5.7).

There are a great many primitive polynomials of large degree. A representative sample is given in Table 5.6.4. Any of these may be used in Equation 5.6.2 to obtain a CRC code. However, the error detection performance achieved by a particular CRC code does depend on the primitive polynomial or polynomials of which it is composed, and two primitive polynomials of the same degree do not necessarily provide equivalent detection capabilities. Usually computer search methods are used to test different candidate CRC polynomials, and codes that offer optimal or near-optimal performance have been found by this method. Table 5.6.5 lists some of the best known CRC code polynomials.

CRC codes are typically evaluated in terms of their *error pattern coverage,* their *burst error detection capability,* and the probability of an undetected error. For an (n, k) CRC code, the coverage is the ratio of the number of invalid blocks of length n to the total number of blocks of length n. This ratio is a measure of the probability that a randomly chosen block is not a valid code block. From the definition, the coverage is

$$\lambda = 1 - 2^{-r}. \qquad 5.6.3$$

For some of the codes in Table 5.6.5, we have the following coverage:

Code	Coverage
CRC-12	0.999756
CRC-ANSI	0.999985
CRC-32A	0.99999999977

TABLE 5.6.5 Some Near-Optimal CRC Codes

Maximum n	Degree	$g(x)$ (octal)	Name
1023	24	140050401	CRC-24
1023	32	50020114342	CRC-32A
2047	12	14017	CRC-12
16383	16	320227	CRC-SDLC
32767	16	300005	CRC-ANSI

The burst error detection capability is a measure of the fraction of all bursts of length b that will be detected by the code. CRC codes of degree r have the following burst detection properties:

1. The code will detect *all* bursts of length $b = r$ or less.
2. The code will detect the fraction $1 - 2^{1-r}$ of all bursts of length $b = r + 1$.
3. The code will detect the fraction $1 - 2^{-r}$ of all bursts of length $b > r + 1$.

The probability of an undetected error for a CRC code usually cannot be determined exactly. However, a lower bound on this probability can be established. If the block length is n (with n less than the maximum block length of the code) and if the code is used on the binary symmetric channel, the undetected error probability approaches 2^{-n} as the BSC crossover probability and the dimension k of the code increase.

5.7 SIMPLE MODIFICATIONS TO CYCLIC CODES

There are a number of simple modifications we can make to a linear code to obtain a new code with slightly different properties. In fact, these modifications are often necessary in practical systems and even provide an extra element of fun in working with these codes. Since the cyclic codes are the most important class of linear block codes, we will take a look at some of these techniques as they apply to cyclic codes.

5.7.1 Expanding a Code

Expanding a code (also called *extending* a code) means increasing the length of the code by adding more parity-check bits. We encountered this in Chapter 4 when we extended the Hamming code. If the minimum Hamming distance of the code generated by $g(x)$ is odd, we can increase the minimum distance by one with the addition of an extra parity bit

$$c_p = \sum_{i=0}^{n-1} c_i.$$

If $c(x) = c_0 + c_1 x + \cdots + c_{n-1} x^{n-1}$ is the code polynomial generated by $g(x)$, the expanded code polynomial created by this operation is equivalent to

$$c'(x) = \sum_{j=0}^{n} c'_j x^j = xc(x) + [xc(x)]/(x+1) = xc(x) + c(x)/(x+1). \quad 5.7.1$$

Equation 5.7.1 is equivalent to first encoding $m(x)$ using a $g(x)$ encoder and then encoding the resulting code word using an $(x+1)$ encoder. The two encoders are cascaded together to give the final systematic code polynomial. This is *not* the equivalent of encoding by $g'(x) = (x+1)g(x)$, and generally, the resulting code is *not* cyclic. For that reason, many people prefer to analyze this coding in matrix form, as we did in Chapter 4.

EXAMPLE 5.7.1

Show that the expanded Hamming code generated by $g(x) = x^3 + x + 1$ is not cyclic.

Solution: Let $m(x) = x \Rightarrow c(x) = x^4 + x^2 + x \Rightarrow c'(x) = x^5 + x^3 + x^2 + 1$. If this code is cyclic, then $c''(x) = x^6 + x^4 + x^3 + x$ must also be a code word. This implies that the original Hamming code must contain the word $x^5 + x^3 + x^2 + 1$. However, $(x^5 + x^3 + x^2 + 1)/(x^3 + x + 1) = 1$. Therefore, the numerator term is not a legal code word in the original Hamming code, and thus, $c''(x)$ is not a legal code word in the expanded code. Since this violates the definition of a cyclic code, the expanded code is not cyclic.

5.7.2 Shortening a Code

The block length of a cyclic error-correcting code is usually fixed at a value $n = 2^j - 1$ for some integer j. There are many practical applications where this is annoying, since many information systems organize the source information into blocks of integer powers of two bits per block. If we were confined to the values of n given earlier, our code rate would barely exceed $1/2$, since we must have $n > k$. Fortunately, this problem can be avoided by shortening the code.

Suppose we wish to have information blocks of l bits, but we use an (n, k) code with $l < k$. Since the code works for all blocks of k information bits, it must work for the subset

$$\overline{m} = \begin{bmatrix} m_0 \, m_1 \cdots m_{l-1} \underbrace{0 \, 0 \cdots 0}_{k-l} \end{bmatrix}.$$

If the code is a cyclic code in systematic form, the code words generated by this subset are

$$\overline{c} = \begin{bmatrix} c_0 \, c_1 \cdots c_{r-1} \, m_0 \, m_1 \cdots m_{l-1} \underbrace{0 \, 0 \cdots 0}_{k-l} \end{bmatrix}.$$

Since the leading $k - l$ bits are all zero, and since the receiver will know this (because we're designing it), there is no need to transmit the first $k - l$ zeroes. The syndrome calculation will be unaffected, because it is governed by Equation 5.4.9. Therefore, we may encode and decode these *shortened* code words using the same encoders and decoders as discussed previously. We must adjust the *timing* of our encoder and decoder to account for the shortened block length $n' = l + r = n - k + l$ and reduce the length of our decoding buffer to n' bits.

The syndrome calculation for this shortened code *will* produce a different final value than the one we had in the previous section. The reason for this is that the high-

Section 5.7 Simple Modifications to Cyclic Codes

est order bit is now in position $l+r-1$, rather than position $n-1=k+r-1$. Our definition of the syndrome in Equation 5.4.13 was selected so that our syndromes would have the convenient form given by Equation 5.5.2. We lose this convenient form (and the considerable simplification of the logic in the syndrome matcher it usually permits) by shortening the code. Fortunately, there is a very simple way to deal with this effect and recover the simplicity of our syndrome matcher.

Let $b = k - l$, and redefine the syndrome as

$$s(x) = [x^{r+b}v(x)]/g(x). \qquad 5.7.2$$

If our error pattern has a "1" in the leading position, so that

$$e(x) = e_{n'-1}x^{n'-1} = e_{n'-1}x^{(n-b)-1},$$

then the resulting syndrome is

$$s(x) = e_{n'-1}[x^{r+b+n-b-1}]/g(x) = e_{n'-1}[x^{n+r-1}]/g(x).$$

This expression is in the same form as Equation 5.5.1. Since

$$x^n/g(x) = [x^n - 1 + 1]/g(x) = ([x^n - 1]/g(x) + 1)/g(x) = 1,$$

we have

$$s(x) = e_{n'-1}x^{r-1},$$

and a single error in the first bit position results in a syndrome that is all zero, except for a "1" in its most significant bit position. This modification recaptures the simplicity we previously had in the logic for implementing our syndrome matcher without a big lookup table.

To implement Equation 5.7.2, we define

$$a(x) \equiv x^{r+b}/g(x). \qquad 5.7.3$$

From Equation 5.3.4, this gives us an equivalent definition for $s(x)$ of

$$s(x) = [a(x)v(x)]/g(x),$$

and therefore our new syndrome is equivalent to premultiplying $v(x)$ by $a(x)$. How can we do this? We see from Equation 5.7.3 that the degree of $a(x)$ must be less than or equal to $r-1$. We can therefore write $a(x)$ in the form

$$a(x) = a_0 + a_1 x + \cdots + a_{r-1}x^{r-1},$$

so that

$$s(x) = [(a_0 + a_1 x + \cdots + a_{r-1}x^{r-1})v(x)]/g(x). \qquad 5.7.4$$

Equation 5.7.4 implies that our syndrome is the result of a "multiply-by-$a(x)/g(x)$" circuit. The circuit of Figure 5.7.1 combines the multiply and divide operations into a single concise operation at the expense of only a few more addition-modulo-two (exclusive-or) circuits. By replacing the topmost "divide by $g(x)$" circuit in Figure 5.4.4 with the circuit of Figure 5.7.1, we accommodate our new syndrome calculation for the shortened code without any further change in the Meggitt decoder. The cyclic syndrome calculation in the syndrome matcher of Figure 5.5.1 is unchanged from the non-shortened form.

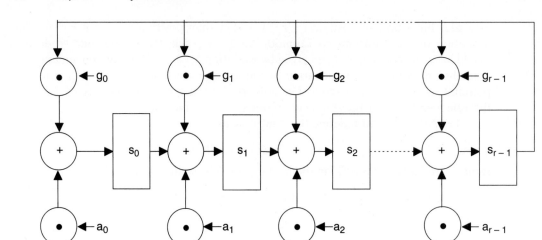

Figure 5.7.1: Multiply-by-$a(x)/g(x)$ Circuit

The coefficients of $a(x)$ may be calculated by factoring x^{r+b} as a product of terms x^j for $j \leq r$. This is illustrated in the following example.

EXAMPLE 5.7.2

Find $a(x)$ for a $(272, 260)$ shortened code constructed from the $(511, 499)$ burst-correcting code in Table 5.6.3.

Solution: We have $b = n - n' = 511 - 272 = 239$, so $r + b = 239 + 12 = 251$. From Table 5.6.3, we have

$$10451 \Rightarrow g(x) = x^{12} + x^8 + x^5 + x^3 + 1.$$

We factor

$$x^{251} = (x^{12})^{20} \cdot x^{11} = (x^{12})^{16} (x^{12})^4 x^{11}.$$

We have factored $x^r = x^{12}$ in such a way that our expression involves squares of $(x^{12})^2$ in order to make use of a very handy property of binary polynomials, namely, that if we have a binary polynomial

$$p(x) = \sum_{j=0}^{J} p_j x^j$$

with $p_i \in \{0, 1\}$, then

$$p^2(x) = \left(\sum_{j=0}^{J} p_j x^j \right)^2 = \sum_{j=0}^{J} p_j x^{2j} = p(x^2). \qquad 5.7.5$$

Using Equation 5.3.4 and
$$x^{12}/g(x) = x^8 + x^5 + x^3 + 1,$$
we obtain
$$(x^{12})^4/g(x) = [(x^{12}/g(x))^4]/g(x) = (x^{32} + x^{20} + x^{12} + 1)/g(x).$$
We factor
$$x^{32} + x^{20} = x^8((x^{12})^2 + x^{12})$$
and note that $(x^{12} + 1)/g(x) = x^8 + x^5 + x^3, x^8/g(x) = x^8$, to get
$$(x^{12})^4/g(x) = [x^8(x^{16} + x^{10} + x^8 + x^6 + x^5 + x^3 + 1) + x^8 + x^5 + x^3]/g(x)$$
Reexpressing these terms as various polynomials of the form
$$x^{12} \cdot f(x) + f'(x), \quad \deg[f'(x)] < 12,$$
we continue to apply Equation 5.3.4, with the eventual result that
$$(x^{12})^4/g(x) = x^{10} + x^8 + x^6 + x^5 + x + 1.$$
We can now use this equation to reduce $(x^{12})^{16}$ to $[(x^{12})^4]^4$, to eventually obtain
$$a(x) = x^{11} + x^9 + x^7 + x^3 + x^2 + 1.$$

The process we have just outlined probably seems a bit tedious to you. However, the tedium pales in comparison with directly computing $x^{251}/g(x)$ by long division. If you are thinking, "Is it worth this just to save a few logic gates?" be reminded that if you do *not* go through this exercise to obtain simple syndromes, you have an even messier job on your hands calculating the syndrome table by exactly the same long-division process (repeated for each syndrome!). With the syndrome defined as we have done here, the syndromes are of the simple form of Equation 5.5.2.

You do, of course, also have the prerogative of simply programming the long-division algorithm into the computer and obtaining $a(x)$ that way. (I'm sure your boss and the shareholders will appreciate the gain in productivity; in this case, think of the method we have just outlined as an efficient way to get results to debug your program with).

5.7.3 Noncyclicity of Shortened Codes

It should be obvious from our previous expression for $\bar{c}$ that a code obtained from shortening a cyclic code is no longer cyclic. CRC codes are a typical example of a shortened cyclic code. This actually is of some advantage in shortened burst-correcting codes, because an error burst can no longer "wrap around" from bit position $n-1$ (which is now one of the virtual "0" bits; we never make an error in these bits) to bit position 0. Consequently, a shortened burst-correcting code can always be decoded with the Meggitt decoder of Figure 5.6.1 after substituting Figure 5.7.1 for the top "divide by $g(x)$" circuit. A second pipeline stage is unnecessary, because the absence of "burst wraparound" means that we cannot "miss" the first error (as we did in Example 5.5.1).

For single-error-correcting codes, such as the Hamming codes, it is usually unnecessary to modify the definition of the syndrome, since these codes only have one specific syndrome that triggers a correction in the Meggitt decoder. The target error pattern is
$$e(x) = x^{n'-1} = x^{n-b-1},$$

for which Equation 5.4.14 gives

$$s(x) = [x^r x^{n-b-1}]/g(x) = [x^{n+r-b-1}]/g(x).$$

This syndrome may be calculated using the same procedure as outlined previously for $a(x)$.

EXAMPLE 5.7.3

Calculate the target syndrome for a shortened Hamming code when the message blocks contain $l = 16$ bits.

Solution: We require a Hamming code with $n = 31$. From Table 5.6.1, we have $r = 31 - 26 = 5$, and

$$g(x) = x^5 + x^2 + 1.$$

With $l = 16$, we have $n' = 16 + 5 = 21$, so $b = n - n' = 31 - 21 = 10$ and $n + r - b - 1 = 25$. We write

$$x^{25} = (x^5)^5 = x^5(x^5)^4; \quad x^5/g(x) = x^2 + 1$$

and use

$$(x^5)^2/g(x) = x^4 + 1, \quad (x^5)^4/g(x) = (x^8 + 1)/g(x) = [x^3(x^2 + 1) + 1]/g(x) = x^3 + x^2,$$

to obtain

$$s(x) = [(x^5/g(x))^4 \cdot x^5/g(x)]/g(x) = (x^5 + x^4 + x^3 + x^2)/g(x) = x^4 + x^3 + 1.$$

The main step in the technique we have outlined is to continue factoring the expressions into powers of x^r and terms of degree $< r$ multiplied by powers of x^r and then apply the reducing equations for x^r modulo $g(x)$ and Equation 5.3.4. If you have not noticed it already, it is always true that

$$x^r/g(x) = g(x) + x^r, \quad \deg[g(x)] = r,$$

for polynomials in $GF(2)[x]$.

5.7.4 Interleaving

Interleaving is one of the most useful and widely used techniques for constructing burst-correcting codes. By combining interleaving with shortening, we can customize a burst-correcting code to almost any circumstance. The basic process for interleaving is illustrated in Figure 5.7.2 for an interleave of 2. The message vector is broken up into two sequences

$$\overline{m} = [m_0 \, m_1 \cdots] \Rightarrow \overline{m}_{even} = [m_0 \, m_2 \cdots], \overline{m}_{odd} = [m_1 \, m_3 \cdots],$$

and each sequence is separately encoded using a burst error-correcting code defined by $g(x)$. The resulting code words are then recombined as shown and transmitted. At the receiver, the received word is again split into separate sequences for decoding, and the resulting message sequences are de-interleaved to recover the original message sequence.

If $g(x)$ is a rate (n, k) encoder, each encoded sequence will be rate k/n, and so will the recombined sequence. Hence, interleaving does not change the basic code rate of

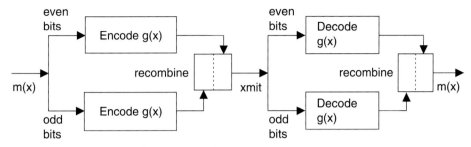

Figure 5.7.2: Two-way Interleaving

the system. Now assume that, during transmission an error burst of length b occurs. Since the two sequences are encoded and decoded separately, each decoder sees a burst of length $b/2$ or less. (One of the decoders will see a burst of less than $b/2$ if b is odd.) If $g(x)$ can correct a burst of length l, the total burst correction capability of the system is $2l$. This basic scheme can be extended to interleave any number of times to achieve a burst correction capability of zl using z encoders and decoders.

In addition to extending the burst correction capability of a basic burst-correcting code, such as one of the codes in Table 5.6.3, interleaving has another advantage. If we use z interleaves, each encoder and decoder needs to shift/process only one bit per z bits transmitted. If the channel transmission rate, in code bits per second, is very high, this can be advantageous in that the logic circuits which implement the encoders and decoders only need to run $1/z$ times the bit clock rate. This can be particularly advantageous in implementing the syndrome matcher in the decoders. The resulting interleaved code remains a cyclic code, unless we also shorten the code for the purpose of obtaining a convenient overall k' in the original message word. If each encoder is an (n, k) encoder, the original message block has $k' = zk$ bits. (If we shorten the code, the k term in the encoder rate occurs *after* shortening when computing k'.) Analysis of the target syndromes for interleaved codes is relatively simple, since each encoder/decoder pair functions as an independent cyclic (or shortened cyclic) code. Therefore, determination of the target syndromes is independent of the number of interleaves and is the same for each decoder.

The circuit that combines the encoder outputs into a single-bit stream is typically called the *interleaver*, while the circuit that "splits up" the coded bit stream at the receiver is typically called the *de-interleaver*. When interleaving is performed as in Figure 5.7.2, the interleaver and the de-interleaver are quite simple and typically consist of a parallel-to-serial converter circuit for the interleaver and a serial-to-parallel converter circuit for the de-interleaver.

In a number of applications, the use of duplicated parallel encoders and decoders is undesirable for reasons of cost. This can especially be true when the speed of the data is slow relative to the speed of the logic circuits that implement the encoders and decoders. In such a case, it is possible—and often desirable—to reuse the *same* encoder circuit and the *same* decoder circuit to calculate all the interleaves. The way this is done is as follows. An entire message block is encoded by the encoder and sent to an interleaver capable of storing z blocks of coded bits. The second block is then encoded and sent to the interleaver, and so on until all z blocks have been encoded. The interleaver then interleaves the code bits of these z blocks. At the receiver, essentially the inverse operation is carried out. The sequence is illustrated in Figure 5.7.3.

This approach basically minimizes the circuitry spent doing the encoding, and decoding functions at the expense of additional buffering in the interleaver and de-interleaver. The two most common types of interleavers for such schemes are the *block* interleaver and the *cross interleaver*. The block interleaver can be viewed as a memory matrix containing z rows (the number of interleaves) and n columns (the number of coded bits per block). The block interleaver and de-interleaver are illustrated in Figure 5.7.4. Coded bits are read into the interleaver row by row, such that each row contains one entire coded block. The data is read out in columns so that the transmitted bit stream consists of interleaved code bits. At the receiver, the bits are again read in by rows and read out by columns, only now each *column* corresponds to a single coded block. The number of rows in the interleaver is selected such that zt is greater than or equal to the maximum burst length, where t is the error correction capability of the code. Thus, a burst of length $b = zt$ causes at most t errors in any given code word.

The block interleaver and de-interleaver are identical circuits, differing only in the order in which the rows are read in and the columns are read out. The block interleaver/de-interleaver is probably the most commonly encountered type of interleaver scheme, because the circuit is implemented with large-capacity memory chips at reasonably low cost.

In addition to the block interleaver, there is another interleaver scheme known as the cross interleaver. (This interleaver is also known as the periodic interleaver or the convolutional interleaver). The cross interleaver is more memory efficient for the same burst length capability by a factor of approximately 2; that is, it requires only about half of the memory of the block interleaver. It is constructed from n parallel delay lines. The first delay line is a zero-delay line, the second has a one-bit delay, the third has a two-bit delay, and so on. The first coded bit is placed on the first delay line, the second on the second delay line, etc. Bits are read out in the same way. The cross interleaver is described in more detail in the paper by J. L. Ramsey listed in the references at the end of the chapter. We will not go into a detailed description of this interleaver here, because, in spite the greater memory efficiency of the cross interleaver, the block interleaver is generally simpler to implement with available integrated circuit memory technology and, for that reason, is more commonly encountered.

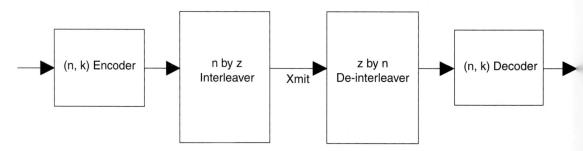

Figure 5.7.3: Interleaving with Time-Shared Encoder/Decoder

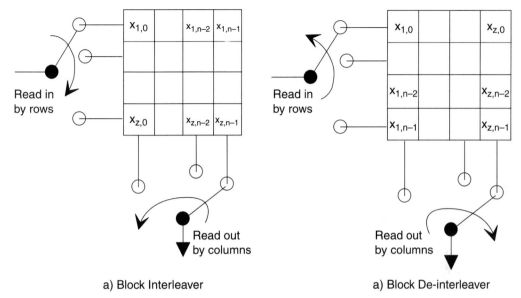

Figure 5.7.4: Block Interleaver and De-Interleaver

SUMMARY

In this chapter, we have looked at the basics of cyclic block codes. After introducing the cyclic shift property, we saw how these codes can be represented algebraically by using polynomials as a kind of "bit position transform" representation of the vector theory of Chapter 4. The basics of polynomial arithmetic in the binary field were described, and Meggitt's important theorem was introduced. Meggitt's theorem is the basis for efficient hardware implementation of cyclic encoders and decoders.

We next looked at the generation of cyclic codes and at the decoding of cyclic codes using simple shift register–based circuits. The Meggitt decoder was discussed in depth. We saw how we could extend this basic idea to the error-trapping decoder for burst error correction.

Some standard block codes were given in Section 5.6, followed by a discussion of simple modifications to cyclic codes in Section 5.7. For those of us whose occupational role lies with choosing and implementing codes (rather than discovering new classes of codes), much of the fun in coding is to be found in the crafting of existing codes to fit particular purposes. To do so, though, requires us to know what we're doing (which, of course, is the reason we study the underlying theory). The specific "tricks of the trade" we examined were expanding a code, shortening a code, and interleaving a simple code to obtain a more powerful one.

REFERENCES

R. E. Blahut, *Theory and Practice of Error Correcting Codes,* Addison-Wesley, Reading, MA, 1983.

S. B. Wicker, *Error Control Systems for Digital Communication and Storage,* Prentice Hall, Upper Saddle River, NJ, 1995.

R. W. Hamming, "Error detecting and error-correcting codes," *Bell Systems Technical Journal*, vol. 29, pp. 147–160, 1950.

J. E. Meggitt, "Error correcting codes for correcting bursts of errors," *IBM Journal*, vol. 4, pp. 329–334, July, 1960.

T. Kasami, "A decoding procedure for multiple-error-correcting cyclic codes," *IEEE Trans. on Information Theory*, vol. IT–10, pp. 134–138, Apr., 1964.

J. L. Ramsey, "Realization of optimum interleavers," *IEEE Transactions on Information Theory*, vol. IT–16, pp. 338–345, May, 1970.

EXERCISES

5.1.1: The code words of one possible (7, 4) Hamming code are as follows:

$$
\begin{array}{cccc}
0000000 & 1101000 & 0110100 & 1011100 \\
1110010 & 0011010 & 1000110 & 0101110 \\
1010001 & 0111001 & 1100101 & 0001101 \\
0100011 & 1001011 & 0010111 & 1111111
\end{array}
$$

Determine whether or not this is a cyclic code.

5.2.1: Find the polynomial transform of the following binary vectors:
  a) [1 1 1 0]   **b)** [1 0 1 1]   **c)** [1 0 0 0 0 1 0 1]   **d)** [0 0 1 0 1 0].

5.2.2: Carry out the indicated arithmetic on the following polynomials in GF(2)[x]:
  a) $v(x) = (1+x)(1+x)$
  b) $v(x) = (1+x)(1+x^2) + x^3$
  c) $v(x) = (1+x^2)^2 + 1 + x$.

5.2.3: Find the degree of each polynomial representation of the vectors in Exercise 5.2.1.

5.3.1: Perform the following calculations in GF(2)[x]:
  a) $x + x^4 \bmod x^2 + 1$
  b) $1 + x^4 \bmod 1 + x$
  c) $1 + x + x^2 + x^3 \bmod 1 + x$
  d) $1 + x + x^2 \bmod 1 + x$.

5.3.2: For polynomials in GF(2)[x], show that
$$(1 + x^n)^2 = 1 + x^{2n}.$$

5.3.3: Factor the polynomial $x^5 + 1$ in GF(2)[x] into a product of irreducible polynomials.

5.3.4: Divide $f(x) = x^7$ by $g(x) = 1 + x^2 + x^3$ in GF(2)[x].

5.3.5: List all of the polynomials in the ring GF(2)[x]/($x^4 - 1$).

5.3.6: Construct the remainder table for division by $g(x) = 1 + x + x^2$ for polynomials in the ring GF(2)[x]/($x^4 - 1$), and find $\rho(x) = f(x)/g(x)$ for
  a) $f(x) = x^3 + x + 1$ **b)** $f(x) = x^3 + x^2$ **c)** $f(x) = x^2 + 1$ **d)** $f(x) = x + 1$.

5.4.1: Show that $1 + x$ is a factor of any $f(x) = x^n + 1$ in GF(2)[x].

5.4.2: Construct a systematic (7, 3) cyclic code.

5.4.3: Use Equation 5.4.9 to verify the code words for the following source messages for the Hamming code of Example 5.4.1.
  a) $m(x) = x^2$ **b)** $m(x) = 1 + x$ **c)** $m(x) = x^3$ **d)** $m(x) = 1 + x^3$.

5.4.4: Find the Γ matrix and sketch the encoder for a $(7, 3)$ code generated by
$$g(x) = 1 + x + x^2 + x^4.$$

5.4.5: Repeat Exercise 5.4.4 for a $(7,4)$ code having generator
$$g(x) = 1 + x^2 + x^3.$$

5.4.6: An (n,k) cyclic code defined by generator polynomial $g(x)$ uses the decoder of Figure 5.4.3. Show that the syndrome associated with error pattern $e(x) = x^j$, $0 \leq j \leq n-1$, is given in vector form by

$$S = \Gamma^j \begin{bmatrix} g_{r-1} \\ g_{r-2} \\ \vdots \\ g_0 \end{bmatrix}.$$

5.4.7: Use the result of Exercise 5.4.6 to construct the syndrome table in Figure 5.4.3 for a $(7, 4)$ Hamming code defined by $g(x) = 1 + x + x^3$. Then implement an efficient decoder for this code by applying Meggitt's theorem.

5.4.8: Find the Meggitt set E_{meg} and the syndromes required for the error correction circuit of Figure 5.4.4 for a $(7, 3)$ code defined by $g(x) = 1 + x + x^2 + x^4$.

5.4.9: Design a Meggitt decoder for the code of Exercise 5.4.8.

5.4.10: Using the decoder of Exercise 5.4.8, decode the following received polynomials:
  a) $v(x) = x + x^4 + x^6$
  b) $v(x) = 1 + x + x^4 + x^5 + x^6$
  c) $v(x) = 1 + x^4 + x^5 + x^6$.

5.4.11: The polynomial $g(x) = x^8 + x^7 + x^6 + x^4 + 1$ defines a $(15, 7)$ code that can correct up to $t = 2$ errors. What is the Meggitt set for the decoder syndrome table of Figure 5.4.4? (You do not have to calculate the syndromes in this problem.)

5.5.1: The generator polynomial $g(x) = x^4 + x + 1$ defines a $(15, 11)$ Hamming code. Find the required syndrome table for the syndrome matcher of Figure 5.5.1 for this code.

5.5.2: Given the $(15, 7)$ code of exercise 5.4.11, suppose we wished to implement an error correction circuit using only one syndrome matcher and one buffer. In this case, our syndrome table must contain syndromes for all elements of the Meggitt set. Find the complete set of syndromes required for the syndrome table.

5.5.3: For each of the following codes, determine whether the code can be decoded using a pipelined Meggitt decoder in an error-trapping configuration:
  a) a $(63, 57)$ code with $t = 1$
  b) a $(15, 7)$ code with $t = 2$
  c) a $(31, 21)$ code with $t = 2$
  d) a $(31, 16)$ code with $t = 3$.

5.5.4: Find the burst length for the following error patterns:
  a) $e3(x) = x^2 + x^8$ **b)** $e(x) = 1 + x + x^{10}$
  c) $e(x) = x^5 + x^6 + x^7$ **d)** $e(x) = x^7 + x^6 + 1$.

5.6.1: Express the following polynomials in octal format:
  a) $x^4 + x^3 + 1$ b) $x^5 + x^2 + 1$ c) $x^6 + x^4 + x^3 + x + 1$
  d) $x^9 + x^5 + x^3 + x^2 + 1$.

5.6.2: Give the polynomial specified by the following octal representations:
  a) 23   b) 45   c) 103   d) 721.

5.6.3: For a BSC with 8-dB signal-to-noise ratio, select a Hamming code to achieve a lower bound on the error rate of less than 10^{-6}. (*Hint:* See Exercise 2.3.1.)

5.6.4: Design an error-trapping decoder for the three-error-correcting (15, 5) BCH code.

5.6.5: Find the Meggitt set and the associated error syndromes for all correctable errors for the (15, 9), $b = 3$ burst-correcting code in Table 5.6.3. Design the decoder.

5.7.1: Why do we express an expanded code in the form of Equation 5.7.1 instead of in the form of some generator polynomial $g'(x)$?

CHAPTER 6

Convolutional Codes

6.1 DEFINITION OF CONVOLUTIONAL CODES

In the previous two chapters we have examined linear block codes. Convolutional codes are the second major form[1] of error-correcting channel codes. Convolutional codes differ significantly from block codes in both structural form and error correction properties. In block coding, the data stream is divided into a number of blocks of length k, and each block is encoded into an n-bit code word. A convolutional code converts the entire data stream, regardless of its length, into a single code word. On the average, the ratio of source bits to code bits in this code word is k/n. This ratio is called the *rate* of the convolutional code.

Block codes with rates above 0.95 are common. We saw some examples in Chapter 5. However, high-rate block codes of the type we have studied typically have limited error correction capabilities, on the order of one or two symbol errors per block[2]. These codes are well-suited for use in channels having low raw error probabilities. We examined the performance characteristics of block codes in Chapter 4. In contrast, most common convolutional codes have code rates below 0.90. The relatively low rates of convolutional codes are compensated for by the fact these codes generally have much more powerful error-correcting capability. The convolutional codes, therefore, are well suited for use in very noisy channels with high raw error probabilities.

Convolutional codes are encoded using shift register encoders. Figure 6.1.1 illustrates the basic idea. Source data is broken into *frames* of k_0 bits per frame. $M + 1$ frames of source data are encoded into an n_0-bit code frame, where M is the memory depth of the shift register. As each new data frame is read, the old data is shifted one frame to the right, and a new code word is calculated. The code is characterized by its code rate $R = k_0/n_0$ and its *constraint length* $v = M + 1$. For the most common convolutional codes, the source frame length is $k_0 = 1$, and we will refer to such a code

---

[1]Strictly speaking, convolutional codes are a subset of the class of codes known as *tree* codes. The distinguishing feature of convolutional codes is that they are linear codes, i.e., the sum of any two codewords is also a codeword. In general, tree codes need not be linear. It is accurate to say that convolutional codes are the most widely used type of tree code.

[2]BCH codes (and their nonbinary relatives, the Reed–Solomon codes) can be constructed for correcting many more errors per block. However, the code rates of BCH codes with multiple-error-correcting capability drops off rather steeply as the number of correctable errors increases.

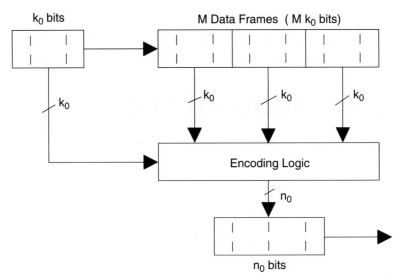

Figure 6.1.1: Shift Register Encoder

as a "binary" convolutional code. The code rate and the constraint length are the key parameters of good convolutional codes.

The encoder of Figure 6.1.1 is a *finite impulse response* or FIR encoder because the effect of any one data frame lasts only over $v = M + 1$ code frames. This is the most common configuration for convolutional encoders. It is also possible, however, to use a feedback shift register configuration (similar to our "divide by $g(x)$" circuit in Chapter 5) to encode a convolutional code. When this is done, the encoder is an *infinite impulse response* or IIR encoder. In practice, most convolutional encoders are of the FIR variety[3].

EXAMPLE 6.1.1

Figure 6.1.2 illustrates a rate $1/2$, $v = 3$ binary convolutional encoder. The adders shown in the figure perform addition in $GF(2)$, i.e., they are exclusive-or circuits, and the polynomial representation for binary polynomials in the ring $GF(2)[x]$ for the message input and the output code bits is used. The input frames are one bit long, while the code frames are two bits long. For each message bit in $m(x)$ the encoder produces two coded output bits with one bit in $c_0(x)$ and one bit in $c_1(x)$. The two code bits are interleaved and sent as a two-bit symbol sequence. The two interleaved code-bit polynomials are defined by code polynomials

$$g_0(x) = 1 + x + x^2,$$
$$g_1(x) = 1 + x^2.$$

6.1.1

---

[3]There are two important classes of codes that use IIR encoders and that are very similar to convolutional codes in many ways (but, of course, very different in some other important ways). These are the codes for "trellis-coded modulation" or TCM and the newly developed "turbo codes." Both of these types of codes are very important. However, TCM is not a convolutional code (although it shares many properties of convolutional codes), and turbo codes are rather complex and we will not be dealing with them in this introductory text.

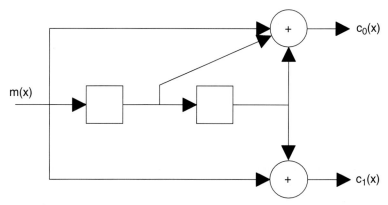

Figure 6.1.2: Rate 1/2, $M = 2$ Convolutional Encoder

The resulting code polynomials are then

$$c_0(x) = m(x)g_0(x),$$
$$c_1(x) = m(x)g_1(x). \qquad 6.1.2$$

Each code polynomial in Equation 6.1.2 is in the same form as the nonsystematic cyclic codes of Chapter 5. The code frames can be looked at in vector form, and the encoded sequences can be expressed as a vector of polynomials. For Example 6.1.1, we can write

$$C(x) = [c_0(x)c_1(x)] = m(x)[g_0(x)g_1(x)] = m(x)G(x).$$

This convenient notation can be extended to rate $1/n$ convolutional codes by defining the generator vector

$$G(x) = [g_0(x) \cdots g_{n-1}(x)] \qquad 6.1.3$$

and writing the code frame polynomials as

$$C(x) = [c_0(x) \cdots c_{n-1}(x)] = m(x)G(x). \qquad 6.1.4$$

In this text, we will be concerned only with binary codes. However, the ideas and notation we have set forth can be readily extended to nonbinary, i.e., $k_0 > 1$, codes in a straightforward fashion. It might seem at this point like we are limiting ourselves to rate $1/n$ codes by restricting ourselves to binary convolutional codes. The discerning reader may be a bit indignant about the way this restriction seems to confine us to *very low-rate* codes. Fear not! Before we are finished, we will see how we can use binary codes to achieve higher code rates through a method known as *puncturing* a code. The majority of high-rate convolutional codes used in practice are punctured codes. This is because punctured codes are known that have performance properties which match their nonbinary counterparts. These punctured codes are much more cost effective to implement and, consequently, are very popular compared with nonbinary convolutional codes.

EXAMPLE 6.1.2

Find the code sequence given by the encoder in Example 6.1.1 for the source sequence

$$m(x) = 1 + x + x^3.$$

Solution: Our code polynomials for this sequence are

$$c_0(x) = m(x)g_0(x) = 1 + x^4 + x^5,$$
$$c_1(x) = m(x)g_1(x) = 1 + x + x^2 + x^5.$$

Under our polynomial notation convention, the highest power of x is the first symbol to be transmitted. Assuming that the interleave sends the c_0 bit first and the c_1 bit second for each code frame, the transmitted bit sequence, written with "time" running from left to right, is

$$11, 10, 00, 01, 01, 11.$$

We have sent only four message bits, $[1, 0, 1, 1]$, yet the transmitted code sequence contains 12 transmitted bits. This is not rate 1/2! What is going on? The answer to this puzzle is simple. The encoder has $M = 2$ memory elements. The encoder must "flush" its buffer to complete the code sequence. The last two code symbols, $[01, 11]$, in the transmitted sequence correspond to this emptying of the encoder's shift register. The first eight bits correspond directly to the four message bits at rate 1/2. Our *effective code rate* in this example was $4/12 = 1/3$, even though the code itself is a "rate 1/2" code. This reduction in the code rate is known as the *fractional rate loss*. If we send a message containing K information bits using a convolutional code with rate R and a memory depth of M, our effective code rate is

$$R_{\text{eff}} = \frac{K}{R^{-1}K + R^{-1}M} = \frac{RK}{K + M} = \frac{R}{1 + M/K}. \qquad 6.1.5$$

Convolutional codes are therefore most effective when $K \gg M$. Under such conditions, the effective code rate asymptotically approaches "the" code rate (as defined earlier).

Convolutional codes are linear codes, since the sum of any two code words is also a code word

$$m_1(x)G(x) + m_2(x)G(x) = [m_1(x) + m_2(x)]G(x).$$

(As in Chapters 4 and 5, we are using addition for binary vector spaces and polynomials.) There is a strong similarity between Equation 6.1.4 and our expressions for cyclic codes in Chapter 5 and, indeed, convolutional codes have many of the same properties of cyclic codes. However, there are a number of important differences as well. First, the code length of a cyclic code is constrained to a particular block length, while there is no length restriction whatsoever for a convolutional code. Second, while systematic convolutional codes exist, there is a marked difference between their performance and the performance of nonsystematic convolutional codes. (There was no performance difference in cyclic codes between the systematic and nonsystematic form). In general, the typical nonsystematic convolutional code is superior in performance to a systematic convolutional code of the same constraint length and code rate. For this reason, nonsystematic convolutional codes are typically used.

For a given code rate, the performance of a convolutional code is a function of the code's constraint length, with error rate performance improving with increased constraint length. The memory depth M of a binary convolutional code is given by

$$M = \max \deg[g_0(x), \cdots, g_{n-1}(x)], \qquad 6.1.6$$

and the constraint length is[4]

$$\nu = M + 1. \qquad 6.1.7$$

Among the known good convolutional codes, larger constraint lengths generally lead to better error rate performance compared with good codes having shorter constraint lengths. This improvement comes at the expense of increased decoder complexity, which, for binary convolutional codes, increases as $2^{\nu-1}$.

6.2 STRUCTURAL PROPERTIES OF CONVOLUTIONAL CODES

6.2.1 The State Diagram and Trellis Representations

A convolutional encoder is a finite-state automaton or "state machine." As such, it is convenient to represent its operation using a state diagram. If there are M memory elements in the encoder, the state diagram will have 2^M states, which we may designate as $S_0, S_1, \cdots, S_{2^M-1}$. It is convenient to associate the state labels with the contents of the shift register. For example, the encoder in Figure 6.1.2 has four states, which we will define as

$$S_0 \leftrightarrow (00), S_1 \leftrightarrow (10), S_2 \leftrightarrow (01), S_3 \leftrightarrow (11).$$

We will use this naming convention throughout the chapter.

The state diagram is found from direct analysis of the encoder diagram. For simple encoders, it may be possible to draw the state diagram directly from the circuit. Figure 6.2.1 shows the state diagram of the encoder in Figure 6.1.2. For more complex codes, it may be helpful to first construct a state table and then draw the state diagram from the table.

There is a strong similarity between analysis of convolutional encoders and the analysis of channels with memory in Chapter 2. As was the case in Chapter 2, it is often useful to represent the operation of the encoder by means of a trellis diagram, since, as you recall, a trellis diagram is a state diagram that explicitly shows the time dependence of the state transitions. Figure 6.2.2 illustrates the trellis diagram for the encoder of Example 6.1.1. The state labeling convention used here is the same as the one we introduced in Chapter 2. The input/output designations, read left to right, are associated with the arrows in the diagram, read in the clockwise direction.

The performance characteristics of a convolutional code can be analyzed from the code's trellis diagram. Indeed, we will later utilize the trellis diagram to construct the *decoder* for a convolutional code. The error-correcting properties of a convolutional code are determined by the *adversary paths* through the trellis. Two paths are said to be adversaries if they both begin in the same state, S_i, end in the same state, S_f, and have no state in common at any step between the initial and final states. Figure 6.2.3 illustrates several adversary paths for our example code of Figure 6.2.2. Any

---

[4]The definition of the constraint length is not consistent in the literature. Some authors define the constraint length to be equal to M. One must check which definition is being used in a particular paper or book.

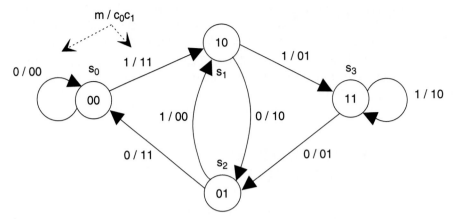

Figure 6.2.1: State Diagram of Encoder in Figure 6.1.2

particular path can be specified by listing its state transitions. For convenience, let us assign the following path names:

$$P_0 = S_0 \to S_0 \to S_0 \to S_0 \to S_0,$$
$$P_1 = S_0 \to S_1 \to S_2 \to S_0 \to S_0,$$
$$P_2 = S_0 \to S_1 \to S_3 \to S_2 \to S_0,$$
$$P_3 = S_0 \to S_0 \to S_1 \to S_2 \to S_0.$$

Paths P_0 and P_1 are adversaries from time index t to time index $t+3$. By comparing the codebit sequences between these two paths, we see that these adversaries are at a Hamming distance of 5. Likewise, P_3 and P_2 are adversaries from t to $t+3$ and also have a relative Hamming distance of 5. Similarly, P_0 and P_3 are adversaries from $t+1$ to $t+4$ with Hamming distance 5. Paths P_0 and P_2 are adversaries from t to $t+4$ with a Hamming distance of 6.

We will see a little later that the performance of a convolutional code is determined by the Hamming distance between the adversary paths in its trellis. Because of

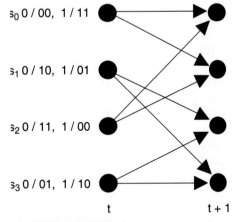

Figure 6.2.2: Trellis diagram for encoder of example 6.1.1

this, analysis of a convolutional code is based on the Hamming distance properties of its adversaries. The simple example illustrated in Figure 6.2.3 shows how very quickly the numbers of possible adversary paths can grow, and we must ask ourselves how we can handle the combinatorics involved. Fortunately, the fact that we are dealing with *linear* codes brings a great simplification to our trellis path analysis.

Consider two code-bit sequences $\bar{c}_1$ and $\bar{c}_2$. The Hamming distance between these sequences is

$$d_H(\bar{c}_1, \bar{c}_2) = w_H(\bar{c}_1 + \bar{c}_2)$$

(and we must remember that $1+1=0$). Since the code is linear, the sum of any two code sequences is itself a code sequence $\bar{c}_3 = \bar{c}_1 + \bar{c}_2$. Therefore,

$$d_H(\bar{c}_1, \bar{c}_2) = w_H(\bar{c}_3) = d_H(\bar{c}_3, \bar{0}), \qquad 6.2.1$$

where $\bar{0}$ is the all-zeroes code sequence. For a linear binary code, $\bar{0}$ must be a valid code sequence (since any code sequence is its own additive inverse). Therefore, *the Hamming distance properties of any two code sequences in the trellis are equivalent to the Hamming distance properties between some code sequence and the all-zeroes code sequence*. To understand the performance of the convolutional code, we do not need to study all possible adversary path pairs. We only need to study the Hamming weights of paths that are adversary to the all-zeroes path.

6.2.2 Transfer Functions of Convolutional Codes

We can obtain the information we need for this analysis from a simple modification of the state diagram and the introduction of the idea of the *transfer function* (also known as a generating function or a weight enumerator) of a convolutional code. We will first look at the state diagram. Since we are interested in finding the Hamming weights of

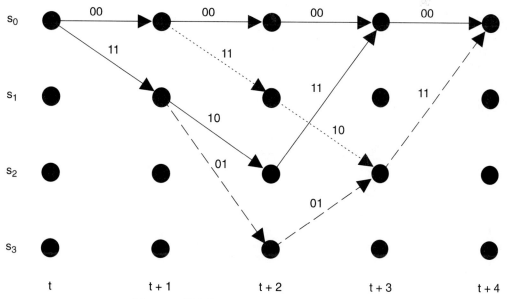

Figure 6.2.3: Adversary Path Examples

the nonzero adversary paths, we modify the state diagram by (1) removing the self-loop at the all-zero state S_0 and (2) adding a new S_0 node, representing the termination of the nonzero adversary path. This is pictured for our code of Example 6.1.1 in Figure 6.2.4. By making these modifications to the state diagram, we have reconfigured the state diagram to show *only* the nonzero adversary paths that begin and end in state S_0.

We now introduce some *transfer function operators*. Referring to Figure 6.2.4 consider the transition $S_0 \to S_1$. This transition involves a code symbol of Hamming weight 2 and a source symbol of Hamming weight 1. What we want is a convenient "bookkeeping" system that lets us keep track of the accumulated Hamming weight of the code symbols and the accumulated Hamming weight of the source symbols for any given path starting at S_0 and terminating later in S_0 after some number of state transitions. We will also find it useful to keep track of how many steps through the trellis are made by the adversary path. We can accomplish all the necessary bookkeeping by introducing the code-word weight operator D, the source-symbol weight operator N, and the time index operator J. Each of these operators is an indeterminate. We will not "solve" to find a value for them any more than we "solve" to get a value for the digit position operator x in our generator polynomials. However, we will let D, N, and J have exponents that count the number of "1" bits in a symbol (for D and N) or the number of trellis steps (for J). For instance, the transition $S_0 \to S_1$ has a code symbol of weight 2 and so is assigned the operator D^2. Likewise, its source symbol is weight 1 and the transition involves one step, so the *transfer function operator* for the transition is $S_0 \to S_1$ is $J N D^2$. Transfer function operators can be assigned in a similar manner to the other branches in the state diagram. An operator raised to the zeroth power is set to 1. Figure 6.2.5 shows these various operators for the state diagram of Figure 6.2.4.

To see how the bookkeeping works, consider the series of transitions

$$S_0 \to S_1 \to S_3 \to S_3.$$

In making these transitions, we pass through transfer function operators

$$(J N D^2)(J N D)(J N D) = J^3 N^3 D^4.$$

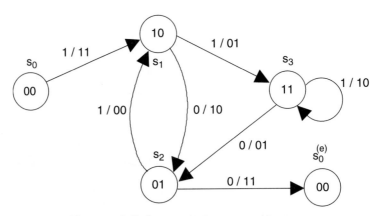

Figure 6.2.4: Non-zero Adversary Paths

Suppose the accumulated transfer function at state S_0 at the beginning of this set of transitions was some transfer function $X_0 = J^a \, N^b \, D^c$. Then, after the three transitions shown previously, the transfer function upon reaching the final S_3 transition is

$$X_3 = J^3 \, N^3 \, D^4 \, X_0.$$

We can solve for the transfer function for all possible paths starting at S_0 and ending at S_0 by writing a set of state equations for the transfer function diagram. There will be one state equation for each node. For instance, node S_1 would have the associated state equation

$$X_1 = J N D^2 X_0 + J N X_2.$$

The general form of each of these equations is X_i equals the X of each node that can transition into it, multiplied by the transfer function of the arrow connecting X to X_i. For Figure 6.2.5, the set of state equations is

$$X_0^{(e)} = J D^2 X_2,$$
$$X_1 = J N D^2 X_0 + J N X_2,$$
$$X_2 = J D X_1 + J D X_3,$$
$$X_3 = J N D X_1 + J N D X_3,$$

where $X_0^{(e)}$ represents the *ending* state S_0, while X_0 represents the *beginning* state S_0. The transfer function $T(J, N, D)$ is found by solving this set of equations for $X_0^{(e)}$, assuming that $X_0 = 1$. We have four equations in four unknowns and can solve for $X_0^{(e)}$ using the usual rules of linear algebra (with $1 + 1 = 2$). Carrying this out for the code of Example 6.1.1, we get

$$T(J, N, D) = \frac{D^5 \, N \, J^3}{1 - D N J (1 + J)}.$$

To see the individual adversary path terms, we apply long division to this and obtain

$$T(J, N, D) = D^5 J^3 N + D^6 J^4 (1 + J) N^2 + \cdots + D^{j+5} J^{j+3} (1 + J)^j N^{j+1} + \cdots,$$

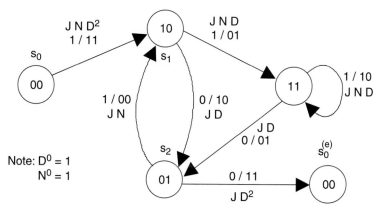

Figure 6.2.5: Transfer Function Diagram

or

$$T(J, N, D) = \sum_{j=0}^{\infty} D^{j+5} J^{j+3} (1 + J)^j N^{j+1}.$$

The transfer function supplies us with all the information we will need to completely characterize the structure and performance of the code. For instance, consider the term

$$D^6 J^4 (1 + J) N^2 = D^6 J^4 N^2 + D^6 J^5 N^2.$$

This term tells us that there are exactly two adversary paths of Hamming weight 6 and both paths involve source sequences with Hamming weight 2. One of the paths takes four steps through the trellis and the other takes five steps. By direct examination of either the trellis diagram or the transfer function diagram, we find that these two path sequences are

$$S_0 \to S_1 \to S_3 \to S_2 \to S_0,$$
$$S_0 \to S_1 \to S_2 \to S_1 \to S_2 \to S_0.$$

The method for finding the transfer function of a given convolutional code is straightforward if somewhat tedious. Unfortunately, it rapidly becomes even more tedious as the constraint length of the code increases, since the number of states in the state diagram doubles every time we increase the constraint length by one. On the other hand, there are only a relatively few known good convolutional codes (at least at present) and oftentimes the discoverer of a code is kind enough to carry out the derivation of the transfer function for us. The homework problems include a few simple codes (some of them not particularly good) which will give us some experience with transfer function analysis. The *important* idea to take away from here is this: Convolutional codes can be represented by transfer functions, and these transfer functions tell us everything we need to know to evaluate the performance of the code. We will discuss the evaluation of a code's performance in Section 6.5. First, however, we turn our attention to decoding convolutional codes.

6.3 THE VITERBI ALGORITHM

For very simple convolutional codes with very limited error correction capability, it is possible to carry out decoding using syndrome methods and shift register circuitry somewhat similar to those used in Chapter 5 for cyclic block codes. In practice, though, convolutional codes are employed when significant error correction capability is required. In such cases, a more powerful method is needed. Such a method was published in 1965 by Andrew Viterbi and rather quickly became known as the Viterbi algorithm. The Viterbi algorithm is of major practical importance, and we shall introduce it in this section primarily by means of an example.

Under a particular set of conditions, the Viterbi algorithm is a maximum-likelihood decoding algorithm. Why this is and what these conditions are will be discussed in the next section. For now, we will take an in-depth look at what the Viterbi algorithm is and how it works.

We have seen how to represent a convolutional code using a trellis diagram. A binary convolutional code with constraint length v has 2^{v-1} states in its trellis. One way to view a Viterbi decoder is to construct it as a network of simple, identical processors with one processor for each state in the trellis. The processors are interconnected and which processors are connected to each other is determined by the branches of the trellis diagram. As an example, consider the code having the trellis diagram in Figure 6.2.2. The Viterbi decoder will have four "node processors." The processor assigned to state S_0 receives inputs from itself and from the node processor assigned to state S_2. This processor supplies outputs to itself and to the node processor for state S_1. Similarly, the node processor assigned to state S_1 receives inputs from the node processors for states S_0 and S_2 and supplies outputs to the node processors for states S_2 and S_3. In addition, each processor monitors the received code sequence $Y(x)$. We assume that this sequence is equal to the transmitted sequence $C(x)$ plus an error sequence $e(x)$, so that

$$Y(x) = C(x) + e(x).$$

At each time step, each processor always assumes it is processing the received sequence arising from the true transmitted code sequence. Each processor is responsible for the following tasks:

1. Each processor must calculate a number, called a likelihood metric, that is related to the probability that the received sequence arises from the true transmitted sequence; there are several popular types of likelihood metrics; for now, in this section, we will use as a likelihood metric the accumulated Hamming distance between the received sequence and the expected transmitted sequence; the larger the Hamming distance becomes, the *less* likely it is that this processor is decoding the true transmitted message (the Hamming distance metric is therefor sort of an "unlikelihood metric");

2. Each processor must supply, as an output, its likelihood metric to each node processor connected to it on its output side;

3. For each of its input paths, the node processor must calculate the Hamming distance between the n-bit code symbol Y it has just received and the n-bit code symbol it *should* have received if the path of the transmitted message had just made a transition from the input-side node processor; this is called the *likelihood update;* the node processor adds the likelihood update to the likelihood supplied to it by the source node processor; it then compares the two updated likelihood metrics and selects the path associated with the input-side node processor having the smallest accumulated Hamming distance (i.e., it selects the *most likely* path); the node processor will replace its own likelihood metric with the newly selected likelihood metric (this is task 1 and occurs at the end of the processing cycle);

4. Finally, based on which input path it selects, the node processor must decode the message bit associated with the selected path and update a record (called a survivor path register) of all of the decoded message bits associated with the selected path.

An example will make this process easier to understand.

EXAMPLE 6.3.1

Assume that we have the convolutional code of Example 6.1.1 and the trellis of Figure 6.2.2. At time $t-1$, assume that the processors have the following initial conditions:

Node Processor	Likelihood Metric	Survivor Path Register
S_0	3	0 0 0 1 0 0 x x x x x
S_1	3	1 1 1 0 0 1 x x x x x
S_2	1	1 0 1 1 1 0 x x x x x
S_3	2	1 1 1 0 1 1 x x x x x

Assume that the received code-word symbol at time t is $Y = 1\ 1$. Find the resulting likelihoods and survivor path registers for each of the node processors at time t.

Solution: Refer to Figure 6.2.2. Processor S_0 receives likelihood inputs of $\mu_0 = 3$ (from itself) and $\mu_2 = 1$ from node processor S_2. The expected code-word for the transition $S_0 \rightarrow S_0$ is 0 0, so the likelihood update for this path is $\Delta\mu_{00} = 2$, giving a total likelihood metric for this path of $3 + 2 = 5$. The expected code-word for the transition $S_2 \rightarrow S_0$ gives $\Delta\mu_{20} = 0$, and the total likelihood is $1 + 0 = 1$. Processor S_0 therefore selects transition $S_2 \rightarrow S_0$ as the most likely transition. The decoded bit associated with a transition into S_0 is a 0. Processor S_0 therefore copies the contents of the survivor path register for node S_2 into its own survivor path register and appends a new "0" in the next bit position. The resulting survivor path register for S_0 becomes 1 0 1 1 1 0 0 x x x x. The new likelihood metric for S_0 becomes $\mu_0 = 1$.

Similar calculations for the other node processors are as follows:

S_1: $\mu_0 + \Delta\mu_{01} = 3 + 0 = 3;\quad \mu_2 + \Delta\mu_{21} = 1 + 2 = 3;$

The likelihoods are *tied*; in this case, the node processor has no valid statistical way to choose between the paths; it resolves this dilemma by "tossing a coin" (fetching a random number) to break the tie; let's say node processor S_2 "wins the toss" and is selected as the winning path; the survivor path is updated to become 1 0 1 1 1 0 1 x x x x, and the new likelihood metric becomes $\mu_1 = 3$.

S_2: $\mu_1 + \Delta\mu_{12} = 3 + 1 = 4;\quad \mu_3 + \Delta\mu_{32} = 2 + 1 = 3;$

The path from processor S_3 is selected, and its survivor path register is copied by S_2, with a "0" padded on the end to give 1 1 1 0 1 1 0 x x x x. The new likelihood becomes $\mu_2 = 3$.

S_3: $\mu_1 + \Delta\mu_{13} = 3 + 1 = 4;\quad \mu_3 + \Delta\mu_{33} = 2 + 1 = 3;$

The path from processor S_3 is selected, and its survivor path register is updated by padding a "1" on the end to give 1 1 1 0 1 1 1 x x x x. The new likelihood becomes $\mu_3 = 3$.

These simple steps constitute the Viterbi algorithm. It really is quite simple. In our example above, the method of keeping track of the survivor path registers is called the "register transfer method." This method is the easiest to understand, but is not the most hardware-efficient method of updating path registers. There is another method of keeping track of the decoded message sequences, called the "traceback method," which is more commonly used in practice because it has a significantly less expensive method of hardware implementation. We will discuss the traceback method a bit later after we've had a little more practice with the Viterbi algorithm.

EXAMPLE 6.3.2

For the convolutional code of Example 6.1.1, assume that it is known the encoder's initial state is s_0. Decode the received sequence 10 10 00 01 10 01.

Solution: Since we know the initial state, we initialize the likelihoods to

$$\mu_0 = 0; \quad \mu_1 = \mu_2 = \mu_3 = \infty \text{ (actually, any \emph{large} number will do).}$$

The results of applying the Viterbi algorithm are illustrated in the following figure:

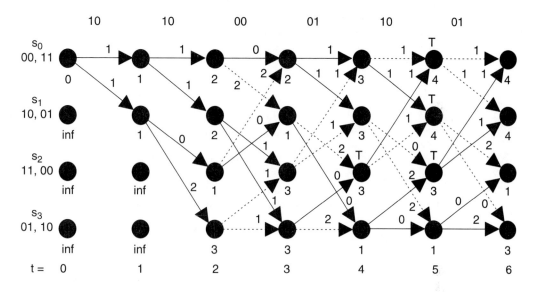

The paths selected by the algorithm at each step are shown as solid lines. Paths rejected by the algorithm are shown as dashed lines. The accumulated likelihoods (Hamming distances) are shown below each state node at each time step. $\Delta\mu$ for each path is shown above the branch for that path. A "T" above a node indicates a tie which was resolved by a random coin toss. This occurs four times during the algorithm. Let us look rather closely at the behavior of the algorithm step by step.

The first two time steps are relatively uninteresting. Because we knew that S_0 was the initial state, paths leading back to S_0 at $t=0$ "win." The first two steps correspond to the initialization of the two shift register bits in the encoder of Figure 6.1.2. The process becomes interesting at $t=3$, where the first adversary paths rejoin. It is worth noticing that this always occurs at the vth time step. The survivor path registers at $t=3$ are

$$S_0: \quad 0\,0\,0\,x\,x\,x,$$
$$S_1: \quad 1\,0\,1\,x\,x\,x,$$
$$S_2: \quad 0\,1\,0\,x\,x\,x,$$
$$S_3: \quad 0\,1\,1\,x\,x\,x.$$

The last two bits of each path register match the associated state number. With the state numbering convention we employed in Section 6.1, this always must occur. (Do you see why?)

After three steps, we are unable to decide on the correct decoding of even the first bit (since the four path registers disagree on what this bit should be). At $t = 4$, we have path registers

$$S_0: \quad 0\ 0\ 0\ 0\ x\ x,$$
$$S_1: \quad 0\ 0\ 0\ 1\ x\ x,$$
$$S_2: \quad 0\ 1\ 1\ 0\ x\ x,$$
$$S_3: \quad 1\ 0\ 1\ 1\ x\ x,$$

and there is still no agreement about the decoding of even the first bit. It is *not* correct to assume at this point that the state with the lowest Hamming distance (S_3) is the "correct" decoding. That's not the way the Viterbi algorithm works. We must carry out the calculations for all of the states at each time step of the algorithm.

At $t = 5$, the path selections result in

$$S_0: \quad 0\ 1\ 1\ 0\ 0\ x,$$
$$S_1: \quad 0\ 0\ 0\ 0\ 1\ x,$$
$$S_2: \quad 1\ 0\ 1\ 1\ 0\ x,$$
$$S_3: \quad 1\ 0\ 1\ 1\ 1\ x,$$

and at $t = 6$,

$$S_0: \quad 1\ 0\ 1\ 1\ 0\ 0,$$
$$S_1: \quad 1\ 0\ 1\ 1\ 0\ 1,$$
$$S_2: \quad 1\ 0\ 1\ 1\ 1\ 0,$$
$$S_3: \quad 1\ 0\ 1\ 1\ 1\ 1.$$

Suddenly, all four survivor path registers agree on the first four decoded bits! What happened? If you trace backwards from each state at $t = 6$, you will observe all the surviving paths joining together at state S_3 at $t = 4$. This is the key behavior of the Viterbi algorithm that makes powerful error correction take place. After the algorithm has had a chance to observe a sufficient number of received code symbols, it is able to use this *sequence* of information to pick the globally most likely transmitted sequence, and hence, the most likely original data sequence. Notice how these path selections for the first four steps through the trellis *cannot be changed* by any future decisions the node processors may make. This is because all the node processors now *agree* on the first four steps. We still do not know what the fifth and sixth bits are, but we could safely decode and output the first four at this point.

In this particular example, it turns out that the most likely node at $t = 4$ ended up being a node selected by all four node processors. Why couldn't we have saved ourselves some work and just picked this one at $t = 4$? The answer to this question is: it only worked out this way in this example because the error pattern in the received code sequence was a single-bit error. Comparing the received and most likely code sequences, we have

received: 10 10 00 01 10 01,

most likely: 11 10 00 01 ?? ??.

This convolutional code, however, is capable of correcting much more severe error patterns than this. We would lose this greater correction capability if we simply assumed that the current most likely state is "the" correct state. (The problems at the end of the chapter contain some examples that illustrate exactly this point.) It is also worth noting that we were helped out in this example by the random coin toss which arbitrated the tie at node S_2 at $t = 5$. If this decision had gone the other way, the node processors would still be in disagreement over which path was most likely at $t = 4$, and the decoding decision would not be possible until at least the $t = 7$ code symbol was received.

In any practical implementation of the Viterbi algorithm, we must use a finite number of bits for our survivor path registers. This is called the *decoding depth* of the decoder. If we use too few bits, the performance of the algorithm will be hurt by having to *force* decoding decisions when we run out of path register bits. The common strategy when we are forced to make such a decision is to select as "most likely" those decoded bits belonging to the node processor that has the best likelihood metric. Most of the time, this will result in correct decoding, but *sometimes it will not*. When such a forced decision is erroneous, we call this a *truncation error*. How many bits of decoding depth are required to make the probability of truncation error negligible? This question was studied by Forney in the early 1970s. He found that a decoding depth greater than or equal to 5.8 times the number of bits in the encoder's shift register was sufficient to ensure minimal performance loss due to truncation error. For the code of Example 6.1.1, this gives us a decoding depth greater than or equal to 12 bits in the survivor path decoding registers.

There is another practical consideration when implementing Viterbi decoders. Any hardware implementation will have a finite number of bits in the arithmetic circuits it uses[5]. Notice in the example how the likelihood metrics for the not-most-likely nodes tends to become larger as time marches on. This means that the potential exists for the arithmetic circuits to run out of bits for representing their likelihoods. In very long code sequences with large numbers of errors, this can be a real problem. (After all, we use convolutional codes to handle just this kind of situation!)

The solution to this problem stems from the fact that all node decisions are *relative* decisions. It does not matter what the actual numerical values of the likelihoods for two competing paths are. What matters is, "Which one is smaller?" Consequently, one popular strategy for dealing with arithmetic overflow is to occasionally subtract the value of the smallest likelihood from each node processor's likelihood. This leaves the *relative* likelihoods unchanged, while limiting the range of the likelihood number each node processor must be able to express.

6.4 WHY THE VITERBI ALGORITHM WORKS I: HARD-DECISION DECODING

6.4.1 Maximum Likelihood under Hard-Decision Decoding

We previously stated that the Viterbi algorithm, under appropriate conditions, is a maximum-likelihood decoder. The time has come to look at when this is so and why the Hamming distance serves as a valid measure of the relative statistical likelihoods of each path.

---

[5]Viterbi decoders can also be implemented with analog electronics. In this case, the finite-bit limit is replaced by a maximum voltage (or current) limit in the amplifier output swings.

So far, we have been looking at what is called "hard-decision" decoding by the Viterbi algorithm. "Hard decision" is a term that means the demodulator/bit detector preceding the Viterbi decoder has already "decided" if the received bits in the coded sequence are "0"s or "1"s. The Viterbi decoder, like all error-correcting decoders, bases its decisions on probabilities and selects that decoded sequences which maximizes the joint probability of the received bit sequence and the original information sequence. We now look at the probabilities involved in this decision-making process.

To keep things simple, we will assume transmission over the binary symmetric channel. (The results we obtain for this case are also true for other memoryless channels; we pick the BSC to keep the notation simple.) Each decision made by a node processor in the Viterbi decoder leads to a selection of one code-word sequence, $\bar{x}$, and the rejection of its adversary sequence, $\bar{y}$. We will represent these sequences as $(t+1)$-bit binary vectors

$$\bar{x} = [x_0 \, x_1 \cdots x_t],$$
$$\bar{y} = [y_0 \, y_1 \cdots y_t].$$

The decision is based on the observed received sequence

$$\bar{z} = [z_0 \, z_1 \cdots z_t],$$

and we will select sequence $\bar{x}$ if the joint probabilities are $\Pr[\bar{z}, \bar{x}] > \Pr[\bar{z}, \bar{y}]$.

We can express this condition as

$$\Pr(\bar{z}|\bar{x}) \cdot \Pr(\bar{x}) \stackrel{?}{>} \Pr(\bar{z}|\bar{y}) \cdot \Pr(\bar{y}), \qquad 6.4.1$$

where $\Pr(\bar{x})$ is the probability of transmitting code-word $\bar{x}$ and $\Pr(\bar{y})$ is the probability of transmitting code-word $\bar{y}$. *The first assumption in the Viterbi algorithm is that all possible code-words are equally probable.* This is equivalent to assuming that all message sequences are equally probable (which we saw is a reasonable and common situation when we looked at channel capacity in Chapter 2). Under this condition, the *hypothesis test* of inequality (6.4.1) simplifies to

$$\Pr(\bar{z}|\bar{x}) \stackrel{?}{>} \Pr(\bar{z}|\bar{y}). \qquad 6.4.2$$

If $\bar{x}$ really is the original transmitted code-word, then it must be true that is the sum of $\bar{x}$ plus some error vector $\bar{e}_x$. Under hard-decision decoding, $\bar{e}_x$ is a binary vector

$$\bar{e}_x = [e_{x0} \, e_{x1} \cdots e_{xt}],$$

and

$$\bar{z} = \bar{x} + \bar{e}_x,$$

where, under the rules of binary vector addition, $1 + 1 = 0$.

In a similar fashion, if the actual transmitted code-word was $\bar{y}$, then it must be true that

$$\bar{z} = \bar{y} + \bar{e}_y,$$

and the hypothesis test becomes

$$\Pr(\bar{z}|\bar{x}) = \Pr(\bar{x} + \bar{e}_x|\bar{x}) \stackrel{?}{>} \Pr(\bar{y} + \bar{e}_y|\bar{y}) \Rightarrow \Pr(\bar{e}_x) \stackrel{?}{>} \Pr(\bar{e}_y), \qquad 6.4.3$$

since the error probability in the BSC is independent of the transmitted sequence. (This is the second assumption built into the Viterbi algorithm; the error probabilities must be independent of the transmitted sequence.)

In a memoryless channel such as the BSC, the errors are statistically independent. Suppose the channel error probability is p and the error vectors have Hamming weights

$$w_H(\bar{e}_x) \equiv w_x,$$
$$w_H(\bar{e}_y) \equiv w_y.$$

Then

$$\Pr(\bar{e}_x) = p^{w_x}(1-p)^{t+1-w_x},$$

and

$$\Pr(\bar{e}_y) = p^{w_y}(1-p)^{t+1-w_y}.$$

(There is no binomial coefficient in either of these expressions because the two error sequences are *specific,* i.e., there is one and only one $\bar{e}_x$ for which $\bar{z} = \bar{x} + \bar{e}_x$, etc.) Inserting these two expressions into Equation 6.4.3 and taking the natural logarithm of each side, we see that

$$w_x \ln(p) + (t+1-w_x)\ln(1-p) \stackrel{?}{>} w_y \ln(p) + (t+1-w_y)\ln(1-p).$$

Rearranging terms, we find that

$$(w_x - w_y)[\ln(p) - \ln(1-p)] \stackrel{?}{>} 0 \Rightarrow (w_x - w_y)\ln\left(\frac{1}{p} - 1\right) \stackrel{?}{<} 0.$$

If $p < 0.5$, then

$$\frac{1}{p} - 1 > 1 \Rightarrow \ln\left(\frac{1}{p} - 1\right) > 0,$$

and the maximum-likelihood test becomes simply

$$w_x \stackrel{?}{<} w_y. \qquad 6.4.4$$

Inequality 6.4.4 is the result we are looking for. Since

$$w_x = w_H(\bar{e}_x) = d_H(\bar{z}, \bar{x}), \qquad w_y = d_H(\bar{z}, \bar{y}),$$

the maximum-likelihood sequence is that sequence having the minimum Hamming distance between the received code-word and the assumed path through the trellis, under the four assumptions we made during our derivation of Inequality 6.4.4. Under these conditions, the Viterbi algorithm is the optimum decoding algorithm.

6.4.2 Error Event Probability

We are now ready to look at the error rate performance of convolutional codes using hard-decision Viterbi decoding. Suppose

$$d_H(\bar{x}, \bar{y}) = d,$$

and let $\bar{x}$ be the true transmitted sequence. What is the probability of the Viterbi decoding selecting $\bar{y}$ instead of $\bar{x}$? (This sort of error is called an *error event of distance* d.) We have

$$\bar{z} = \bar{x} + \bar{e}_x = \bar{y} + \bar{e}_y.$$

In Section 4.3, we introduced the mathematical idea of a *metric function* and defined its properties. One of these properties was the triangle inequality (see Sec. 4.3 to review this). The Hamming distance is a metric function and obeys the triangle inequality. Consequently,

$$d_H(\bar{z}, \bar{x}) \leq d_H(\bar{x}, \bar{y}) + d_H(\bar{z}, \bar{y}) \Rightarrow w_H(\bar{e}_x) - w_H(\bar{e}_y) \leq d.$$

The Viterbi decoder will make an error if

$$w_H(\bar{e}_x) - w_H(\bar{e}_y) > 0.$$

Since $\bar{x}$ is the transmitted sequence, we can represent the Hamming weight of $\bar{e}_y$ as

$$w_H(\bar{e}_y) = K w_H(\bar{e}_x),$$

where K is some positive constant. From the two preceding inequalities, we have

$$0 < (1 - K) w_H(\bar{e}_x) \leq d.$$

The left-hand inequality in this expression tells us that $0 < K \leq 1$, so the error sequence $\bar{e}_x$ is constrained by

$$w_H(\bar{e}_x) \leq \frac{d}{1 - K}.$$

This must hold true for all permissible values of K. The smallest upper bound on the Hamming weight occurs when $K = 0$, so

$$w_H(\bar{e}_x) \leq d.$$

If $w_H(\bar{e}_x) > d$, this means that there will be some error bits in $\bar{z}$ at bit positions where $x_i = y_i$. When we compare Hamming distances $d_H(\bar{z}, \bar{x})$, and $d_H(\bar{z}, \bar{y})$, these "excess" bit errors appear in both terms and will cancel out during the comparison. This is why we only need to concern ourselves with bit errors in positions where $x_i \neq y_i$.

To obtain the lower limit, we look at *only* those bit positions where $x_i \neq y_i$. There are precisely d such bit positions (by the definition of the Hamming distance between the adversary sequences). Clearly, an error will be made if $w_H(\bar{e}_x) > d/2$ in these bit positions. If d is an odd number, the probability of a trellis error is

$$\Pr(E|d) = \Pr\left(\frac{d+1}{2} \leq w_H(\bar{e}_x) \leq d \mid e_{xi} = 0 \text{ if } x_i = y_i\right).$$

Since there are d bit positions where a "harmful" bit error can occur, and since the errors are statistically independent, this probability is given by the binomial distribution as

$$\Pr(E|d) = \sum_{j=\frac{d+1}{2}}^{d} \binom{d}{j} p^j (1-p)^{d-j}, \quad d \text{ odd}. \qquad 6.4.5$$

Section 6.4 Why the Viterbi Algorithm Works I: Hard-Decision Decoding

When d is an even number, there is a slight complication, since it is possible for $w_H(\bar{e}_x) = d/2$. In this case, we would have a *tie* between the adversary paths. As we saw in the previous section, a tie implies that we have no statistically valid way to pick one sequence over the other. The Viterbi decoder must, however, pick one or the other of the two paths. Since this is a pure guess, the decoder has, at best, only a 50% chance of picking the correct path. Therefore, if d is an even number,

$$\Pr(E|d) = \sum_{j=\frac{d}{2}+1}^{d} \binom{d}{j} p^j (1-p)^{d-j} + \frac{1}{2}\binom{d}{d/2} p^{d/2}(1-p)^{d/2}, \; d \text{ even.} \quad 6.4.6$$

The second term on the right in Equation 6.4.6 gives us the probability of error due to ties.

EXAMPLE 6.4.1

Compare the error event probabilities for adversaries with $d=5$ and $d=6$ over a BSC having $p = 0.01$.

Solution: For $d=5$, we have

$$\Pr(E|d=5) = \sum_{j=3}^{5}\binom{5}{j}(10^{-2})^j(.99)^{5-j} = 10^{-5}\cdot(.99)^2 + 49.5\cdot 10^{-8} + 10^{-10} = 9.8506\cdot 10^{-6}.$$

For $d=6$, we have

$$\Pr(E|d=6) = \frac{1}{2}\binom{6}{3}\cdot(10^{-2})^3(.99)^3 + \sum_{j=4}^{6}\binom{6}{j}(10^{-2})^j(.99)^{6-j} = 9.8506\cdot 10^{-6}.$$

This rather surprising result comes about because of the error probability of the *tie* event. This example illustrates how detrimental ties are for Viterbi decoding. It is a consequence of the fact that, for odd d,

$$\binom{d}{(d+1)/2} \equiv \frac{1}{2}\binom{d+1}{(d+1)/2}.$$

If $p \ll 1$, both $\Pr(E|d_{odd})$ and $\Pr(E|d_{odd}+1)$ are dominated by the $j=(d+1)/2$ term, and these terms are approximately equal in the two expressions.

Equations 6.4.5 and 6.4.6 give us the probabilities of a trellis error during Viterbi decoding. Usually, however, we are more interested in finding the bit error rate. Unfortunately, an exact expression for the bit error rate is generally very difficult to find. We must usually be content to know only upper and lower *bounds* on the error rate.

6.4.3 Bounds on Bit Error Rate

Any given convolutional code contains adversary paths of differing Hamming distances. We saw this in Section 6.2 when we looked at the transfer functions of convolutional codes. Suppose

$$\Delta = \{d_0, d_1, d_2, \cdots\} \quad 6.4.7$$

220 Chapter 6 Convolutional Codes

is the set of all possible Hamming distances between adversaries in the transfer function of the convolutional code. By convention, we let $d_0 < d_1 < d_2 < \ldots$, and, in most common literature on convolutional codes, the minimum distance d_0 is called the *minimum free distance* and is given the symbol $d_f \equiv d_0$. Errors can be made for adversaries at *any* of these distances. The transfer function of the code gives us the information we need to obtain bounds on the bit error rate.

EXAMPLE 6.4.2

Find the expected number of bit errors associated with a decoding error for each possible Hamming distance for the code of Example 6.1.1.

Solution: We found the transfer function for this code in Section 6.2. It is

$$T(J, N, D) = \sum_{j=0}^{\infty} D^{j+5} J^{j+3}(1 + J)^j N^{j+1}.$$

The exponent of N gives us the Hamming weight of the message sequence associated with the various paths of distance $j + 5$. Since the transfer function is referenced to the all-zeroes path, this exponent also tells us that the number of decoded bits which will be in error if the decoder makes a path selection error at this distance. If we set $J = N = 1$, the number of different adversary paths at each given Hamming distance is found from

$$T(1, 1, D) = \sum_{j=0}^{\infty} 2^j D^{j+5}.$$

To find the bit error rate, we need to know the number of error bits which result from decoding paths of various Hamming distances. Since this is given by the exponent of N, we find it using

$$\left.\frac{\partial T(J, N, D)}{\partial N}\right|_{J=1, N=1} = \sum_{j=0}^{\infty} (j + 1)2^j D^{j+5} = \sum_{j=0}^{\infty} n_j D^{j+5},$$

where

$$n_j \equiv (j + 1)2^j.$$

Since each adversary path is equally likely, the expected number of bit errors for each distance $d_j = j + 5$ is found by summing the expected number of errors for each path over the number of paths. The number of paths is 2^j and the expected number of bit errors in the path is $j + 1$, so

$$p_b(d = d_j) = (j + 1)2^j \Pr(E|d = d_j) = n_j \Pr(E|d = d_j). \qquad 6.4.8$$

The expected number of bit errors associated with the adversaries at the different possible distances is therefore given by the coefficients in the series expansion of the code's transfer function

$$\left.\frac{\partial T(J, N, D)}{\partial N}\right|_{J=1, N=1}.$$

EXAMPLE 6.4.3

Find a lower bound for the bit error rate of a convolutional code over the binary symmetric channel.

Solution: Since the error rate must be greater than the error rate for error events at any one given distance,

$$p_b > n_0 \Pr(E|d_0). \qquad 6.4.9$$

EXAMPLE 6.4.4

Find an upper bound for the bit error rate of a convolutional code.

Solution: Recall from elementary probability theory that if A and B are not mutually exclusive events, then

$$\Pr(A \cup B) < \Pr(A) + \Pr(B).$$

This inequality is known as the *union bound*. Applying this to our problem,

$$p_b < \sum_{j=0}^{\infty} n_j \Pr(E|d = d_j). \qquad 6.4.10$$

Equations 6.4.9 and 6.4.10 give us our expressions for lower and upper bounds on the bit error rate of a convolutional code. The error rate coefficients n_j are found from the transfer function as described earlier. (This fulfills the earlier promise that the transfer function would tell us everything we need to know about the performance of the convolutional code.)

EXAMPLE 6.4.5

Estimate upper and lower bounds for the error rate of the convolutional code from Example 6.1.1 using a BSC with $p = 0.01$. Use the first 12 terms in Equation 6.4.10 to estimate the upper bound.

Solution: From Inequality 6.4.9, $p_b > 9.8506 \cdot 10^{-6}$. By direct calculation from Equation 6.4.10,

$$p_b < 6.8395 \cdot 10^{-5}.$$

The convolutional code improves the error rate by over two orders of magnitude, compared with the raw BSC error rate.

6.5 SOME KNOWN GOOD CONVOLUTIONAL CODES

There are a number of families of linear block codes for which an elegant algebraic procedure is known for finding members of the family through an algebraic construction. Examples include the Hamming codes, the BCH codes, the Fire codes, the Reed–Muller codes, and the Reed–Solomon codes. In sharp contrast, there is at present no known procedure for finding the members of families of convolutional codes other than by computer search. Convolutional codes are classified by their constraint length v and code rate R. They are specified by listing their generator polynomials. It is common practice, as it was with cyclic codes, to abbreviate the generator polynomials using octal notation. In the following tables, the octal representations are read from left to right (dropping leading zeroes on the left) to get generator polynomials in the form

$$g(x) = g_0 + g_1 x + g_2 x^2 + \cdots + g_{v-1} x^{v-1}.$$

For instance,

$$(15) \Rightarrow g(x) = 1 + x + x^3.$$

The following codes are taken from the work of Odenwalder (1970), Larsen (1973), and Daut et al. (1982):

Codes with rate 1/2. The following codes were discovered by Odenwalder (1970) and Larsen (1972):

TABLE 6.5.1 Known-Good Codes with Rate 1/2

Constraint Length	Generator Polynomials	d_f
3	(5, 7)	5
4	(15, 17)	6
5	(23, 35)	7
6	(53, 75)	8
7	(133, 171)	10
8	(247, 371)	10
9	(561, 753)	12
10	(1167, 1545)	12

The transfer function coefficients for most of the codes in Table 6.5.1 have been published. The first eight values for n_j (corresponding to distances $d_f + j$) are tabulated as follows:

TABLE 6.5.2 Coefficients for Bit Error Rate Calculation for Rate 1/2 Codes

Constraint Length	$j=0$	1	2	3	4	5	6	7
3	1	4	12	32	80	192	448	1024
4	2	7	18	49	130	333	836	2069
5	4	12	20	72	225	500	1324	3680
6	2	36	32	62	332	701	2342	5503
7	36	0	211	0	1404	0	11,633	0
8	2	22	60	148	340	1008	2642	6748
9	33	0	281	0	2179	0	15,035	0

Codes with rate 1/3. The following codes were discovered by Odenwalder (1970) and Larsen (1972):

TABLE 6.5.3 Known-Good Codes with Rate 1/3

Constraint Length	Generator Polynomials	d_f
3	(5, 7, 7)	8
4	(13, 15, 17)	10
5	(25, 33, 37)	12
6	(47, 53, 75)	13
7	(133, 145, 175)	15
8	(225, 331, 367)	16
9	(557, 663, 711)	18
10	(1117, 1365, 1633)	20

Values for n_j for some of the rate 1/3 codes are given in Table 6.5.4.

TABLE 6.5.4 Coefficients for Bit Error Rate Calculation for Rate 1/3 Codes

Constraint Length	$j=0$	1	2	3	4	5	6	7
3	3	0	15	0	58	0	201	0
4	6	0	6	0	58	0	118	0
5	12	0	12	0	56	0	320	0
6	1	8	26	20	19	62	86	204
7	1	0	20	0	53	0	184	0
8	1	0	24	0	113	0	287	0

It is worthwhile to compare Tables 6.5.1 and 6.5.2 with Tables 6.5.3 and 6.5.4. The rate 1/3 codes are generally much more powerful codes than their rate 1/2 counterparts of the same constraint length.

Codes with Rate 1/4. The following codes were discovered by Larsen (1973):

TABLE 6.5.5 Known-Good Codes with Rate 1/4

Constraint Length	Generator Polynomials	d_f
3	(5, 7, 7, 7)	10
4	(13, 15, 15, 17)	13
5	(25, 27, 33, 37)	16
6	(53, 67, 71, 75)	18
7	(133, 135, 147, 163)	20
8	(235, 275, 313, 357)	22
9	(463, 535, 733, 745)	24
10	(1117, 1365, 1633, 1653)	27

Codes with Rate 1/5. The following codes were discovered by Daut et. al. (1982):

TABLE 6.5.6 Known-Good Codes with Rate 1/5

Constraint Length	Generator Polynomials	d_f
3	(5, 5, 7, 7, 7)	13
4	(13, 15, 15, 17, 17)	16
5	(25, 27, 33, 35, 37)	20
6	(57, 65, 71, 73, 75)	22
7	(131, 135, 135, 147, 175)	25
8	(233, 257, 271, 323, 357)	28

6.6 WHY THE VITERBI ALGORITHM WORKS II: SOFT-DECISION DECODING

6.6.1 Euclidean Distance and Maximum Likelihood

As we discussed in Section 6.4, ties during Viterbi decoding are detrimental to code performance. These ties originate as a consequence of hard-decision decoding. The hard-decision decoder is illustrated in Figure 6.6.1. The bit quantization carried out by the binary

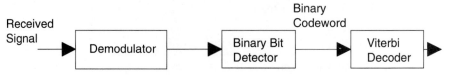

Figure 6.6.1: Hard-Decision Decoding

bit detector maps the continuous-valued demodulated signal into a discrete two-level signal. This operation is information lossy, as we discussed long ago in Chapters 1 and 2.

One question we might ask ourselves is, "Is the bit detector really necessary?" As it happens, the answer to this question is "No!" Let's think about the operation of the Viterbi algorithm some more. The Viterbi algorithm makes its path decisions based on a "likelihood metric." So far, we have used Hamming distance as our likelihood metric. Now, what does the Hamming distance measure? From our discussion of the Hamming cube in Chapter 4 (see Figure 4.3.1), we saw that the Hamming distance had a geometrical interpretation. It was related to "how far apart" two code symbols are in a vector space. In the case of the Hamming distance, this was the number of bit positions in which two code-words differed.

In the hard-decision decoder, the binary bit detector takes the real-valued signal at its input and makes a decision as to whether this signal corresponds to a "0" or "1". Suppose, in the absence of noise, the sampled input signal to the bit detector can take on values of either $+A$ or $-A$, and let us further suppose the binary bit detector will decide the binary output bit is a "1" for $+A$ and a "0" for $-A$. In that case, if the samples of the demodulator's output is the sequence

$$+A \quad -A \quad -A \quad +A \quad +A \quad +A \quad -A,$$

the bit detector's output sequence would be

$$1 \quad 0 \quad 0 \quad 1 \quad 1 \quad 1 \quad 0.$$

Instead of going through the step of converting the $+/-A$ values into "1" or "0", why couldn't we just work directly with the continuous-valued sequence of numbers from the demodulator output? The answer to this question is: "We can; however, if we do so, we can no longer use Hamming distance as our likelihood metric because Hamming distance has no meaningful definition for sequences of real numbers."

Suppose we have two vectors having as elements real-valued numbers

$$\bar{x} = [x_0 \, x_1 \cdots x_L],$$
$$\bar{y} = [y_0 \, y_1 \cdots y_L].$$

What is the "distance" between these two vectors? As it happens, there are many ways to define the "distance" between them because "distance" is measured in terms of a "metric function" (which we defined in Section 4.3). The most widely used metric function for real-valued vectors is the *Euclidean distance* metric

$$d_E(\bar{x}, \bar{y}) \equiv \sqrt{(x_0 - y_0)^2 + (x_1 - y_1)^2 + \cdots + (x_L - y_L)^2}. \qquad 6.6.1$$

Can we simply use Equation 6.6.1 to replace the Hamming distance as our likelihood metric? The answer to this question is: "Almost."

Section 6.6 Why the Viterbi Algorithm Works II: Soft-Decision Decoding

Huh? What does that mean? Here is a very important fact: *We can use as our likelihood metric any metric function that preserves the maximum-likelihood property of the Viterbi algorithm.* Let's see what consequences this has. Suppose the output of the demodulator is a sequence $\bar{z}$ of real-valued samples and let us further suppose $\bar{z}$ consists of the original transmitted sequence of $+/-A$ values *plus* zero-mean additive white Gaussian noise. The node processors in the Viterbi decoder must choose either sequence $\bar{x}$ or sequence $\bar{y}$, based on the measurement $\bar{z}$, and must do so in a manner that maximizes the joint probability of the selected sequence and $\bar{z}$. In other words, the Viterbi algorithm must choose between hypotheses

$$\bar{z} = \bar{x} + \bar{\eta}_x \quad \text{or}$$
$$\bar{z} = \bar{y} + \bar{\eta}_y,$$

where $\bar{\eta}_x$ and $\bar{\eta}_y$ are the samples of an additive white Gaussian noise (AWGN) process that would be necessary if $\bar{z}$ was to be observed given $\bar{x}$ or $\bar{y}$, respectively.

By following exactly the same probability argument we used in Section 6.4, $\bar{x}$ is the most likely sequence if

$$\Pr(\bar{z}|\bar{x}) = \Pr(\bar{x} + \bar{\eta}_x|\bar{x}) \stackrel{?}{>} \Pr(\bar{z}|\bar{y}) = \Pr(\bar{y} + \bar{\eta}_y|\bar{y}) \Rightarrow \Pr(\bar{\eta}_x) \stackrel{?}{>} \Pr(\bar{\eta}_y). \quad 6.6.2$$

Now, suppose $\bar{\eta}$ is some vector of zero-mean Gaussian random variables

$$\bar{\eta} = [\eta_0 \; \eta_1 \cdots \eta_L].$$

Since each term η_i is a Gaussian random variable with a mean value of zero, its probability is

$$\Pr(\eta_i) = \frac{1}{\sqrt{2\pi}\sigma} \exp\left(\frac{-\eta_i^2}{2\sigma^2}\right), \quad 6.6.3$$

where $\sigma^2 = E[\eta_i^2 - E^2(\eta_i)] = E[\eta_i^2]$ is the variance of the zero-mean Gaussian random process.

Since we are assuming that our AWGN process is uncorrelated (that's what "white" means), the individual noise samples are statistically independent. Therefore, the probability of the vector $\bar{\eta}$ is found by multiplying the probabilities of the individual elements of $\bar{\eta}$. This gives us

$$\Pr(\bar{\eta}) = \frac{1}{[2\pi]^{(L+1)/2}\sigma^{L+1}} \exp\left[\frac{-1}{2\sigma^2} \sum_{i=0}^{L} \eta_i^2\right]. \quad 6.6.4$$

Let's save ourselves some writing by defining the notation

$$\|\bar{\eta}\|^2 \equiv \sum_{i=0}^{L} \eta_i^2. \quad 6.6.5$$

Applying Equations 6.6.4 and 6.6.5 to Inequality 6.6.2, and canceling common terms, we find that

$$\exp\left[\frac{-1}{2\sigma^2}\|\bar{\eta}_x\|^2\right] \stackrel{?}{>} \exp\left[\frac{-1}{2\sigma^2}\|\bar{\eta}_y\|^2\right].$$

If we take the natural logarithm of each side and cancel common terms, our maximum-likelihood condition becomes

$$\|\bar{\eta}_x\|^2 \stackrel{?}{<} \|\bar{\eta}_y\|^2.$$

However,

$$\bar{\eta}_x \equiv \bar{z} - \bar{x} \Rightarrow \|\bar{\eta}_x\|^2 \equiv d_E^2(\bar{z}, \bar{x}),$$

and, similarly,

$$\bar{\eta}_y \equiv \bar{z} - \bar{y} \Rightarrow \|\bar{\eta}_y\|^2 \equiv d_E^2(\bar{z}, \bar{y}),$$

so our likelihood test is

$$d_E^2(\bar{z}, \bar{x}) \stackrel{?}{<} d_E^2(\bar{z}, \bar{y}). \qquad 6.6.6$$

Our maximum-likelihood condition depends on the *square* of the Euclidean distance, rather than the Euclidean distance itself. (That's what I meant by "almost" earlier). Now, obviously, if inequality 6.6.6 is satisfied, then it will also be true that $d_E(\bar{z}, \bar{x}) \stackrel{?}{<} d_E(\bar{z}, \bar{y})$. Consequently, you could truthfully argue that we could substitute Euclidean distance for Hamming distance in the Viterbi algorithm. But let's think about that for a second. Look at Equation 6.6.1 again. In order to calculate the Euclidean distance, we must first calculate the *squared* Euclidean distance and take the square root. Why would we want to do this? It's extra work and more expense, and the squared Euclidean distance already satisfies our maximum-likelihood condition.

Summarizing all this, we see that we *do not have to* use the binary bit detector in Figure 6.6.1. Instead, we simply replace the Hamming distance with the squared Euclidean distance as our Viterbi decoder's likelihood function. Nothing else changes in the Viterbi algorithm. When we do this, our decoder is called a *soft-decision* Viterbi decoder.

6.6.2 Elimination of Ties and Information Loss

Why is this advantageous? It certainly looks like more work. We will see shortly that it isn't really as much extra work as you might be thinking right now. First, though, let's look at why we would want to use soft-decision decoding.

The first reason is: We eliminate ties! The probability of having two real-valued squared Euclidean distances that are *exactly* equal is zero for all practical purposes. (There is a rather elegant mathematical theory that establishes this, but we'll not go into that; for the insatiably curious, I recommend a mathematics course on measure theory.) The detrimental effect of ties was demonstrated rather clearly in our previous examples. Eliminating them improves the error rate of the Viterbi decoder.

It does not stop there, however. The error rate of the decoder improves by much more than is accounted for simply by eliminating ties. Let's see why this is. Suppose we have a binary bit detector and a hard-decision decoder. The binary bit detector gives us a binary symmetric channel as seen by the Viterbi decoder. The crossover probability p

Section 6.6 Why the Viterbi Algorithm Works II: Soft-Decision Decoding

of this channel is equal to the probability of getting a noise sample $\eta_i > A$ (thus turning a "0" into a "1", or vice versa. The probability of this happening is

$$p = \frac{1}{\sqrt{2\pi}\sigma} \int_A^\infty e^{-u^2/2\sigma^2}\, du \equiv Q\left(\frac{A}{\sigma}\right). \qquad 6.6.7$$

Therefore, the BSC crossover probability is a function of the rms noise value σ.

Now let's look at our Euclidean distances between real-valued code vectors. We have

$$d \equiv d_E(\overline{x}, \overline{y}) = \left[\sum_{i=0}^{L} (x_i - y_i)^2\right]^{1/2}.$$

Each term $(x_i - y_i)^2$ in this expression is either 0 or $4A^2$. Suppose the demodulator has been constructed so that its output delivers sample values normalized to $A = 1/2$. Then each term in the summation is either zero (if both terms are equal) or one (if the terms are not equal). Since the demodulator will apply the same gain to both the signal and the noise, we can do this without changing the ratio A/σ. If we do this, then Equation 6.6.7 becomes

$$p = Q\left(\frac{1}{2\sigma}\right), \qquad 6.6.8$$

and

$$d_E^2(\overline{x}, \overline{y}) = d_H(\overline{x}, \overline{y}). \qquad 6.6.9$$

This normalization convention makes the *squared* Euclidean distance of the convolutional code and the minimum free Hamming distance of the code *equal*. (This can be a handy time-saver.)

Now let's find the probability of a trellis decoding error. If $\overline{z} = \overline{x} + \overline{\eta}$, an error will occur only if $\overline{\eta}$ is large enough to cause $d_E(\overline{z}, \overline{y}) < d_E(\overline{z}, \overline{x})$. Since the η_i are statistically independent zero-mean Gaussian random variables, it is rather easy to prove that the probability of this is

$$\Pr(E|d) = \frac{1}{\sqrt{2\pi}\sigma} \int_{d/2}^\infty e^{-u^2/2\sigma^2}\, du \equiv Q\left(\frac{d}{2\sigma}\right). \qquad 6.6.10$$

This expression replaces Equations 6.4.5 and 6.4.6 in our error rate analysis of Section 6.4. The overall error rate bounds are still given by Equations 6.4.9 and 6.4.10 as before, only now Equation 6.6.10 is used as $\Pr(E|d)$. Let's look at some numerical values.

EXAMPLE 6.6.1

Repeat Example 6.4.5 using soft-decision decoding, and compare the results with the earlier example.

Solution: To make a fair comparison, we must use the same noise in each case. We are given the crossover probability $p = 0.01$. Therefore,

$$p = 0.01 = Q\left(\frac{1}{2\sigma}\right) \Rightarrow \frac{1}{2\sigma} \approx 2.33.$$

(This expression is solved either numerically or by using the table of values of the Q function.) For soft-decision decoding,

$$\Pr(E|d = 5) = Q(2.33\sqrt{5}) \approx 10^{-7},$$
$$\Pr(E|d = 6) = Q(2.33\sqrt{6}) \approx 6 \cdot 10^{-9},$$
$$\Pr(E|d = 7) = Q(2.33\sqrt{7}) \approx 0.353 \cdot 10^{-9},$$

etc. Notice how each term above drops by slightly more than one order of magnitude for each increment of d. The distance terms for 8, 9, and so on rapidly cease to contribute to the error rate. Applying the n_j coefficients, we get (approximately)

$$10^{-7} < p_b < 1.24 \cdot 10^{-7}.$$

This is *more than two orders of magnitude* better than the previous example.

The error rate improvement in Example 6.6.1 is much more than can be accounted for by eliminating the ties. What's going on? In hard-decision decoding, the bit detector *quantizes* the output of the demodulator. This is an *information-lossy* process. The information being lost in this case is information carried by the *noise*. Huh? Why is the loss of *that* information bad? One way to think about it is like this: To the binary bit detector, an input of -0.000001 is just as much a logic "0" as an input of -0.5 would be. I can't speak for you, but if *I* were to observe an input to the bit detector that was almost exactly halfway between $+0.5$ and -0.5, I wouldn't be so sure if it should be called a "0" or a "1". In soft-decision decoding, the Viterbi algorithm is *informed* of such borderline cases (that which is *informative* is *information*) and the decoder can use this information in making its decoding decisions.

6.6.3 Calculation of the Likelihood Metric

The error rate improvement from soft-decision decoding is often enough to justify using Euclidean distances rather than Hamming distances in our Viterbi decoder. However, as I pointed out earlier, calculation of the squared Euclidean distance likelihood metric is not as much work as we might be thinking. Consider this. The likelihood updates $\Delta\mu$ are made up of terms of the form

$$(z_i - x_i)^2 = z_i^2 - 2z_i x_i + x_i^2,$$
$$(z_i - y_i)^2 = z_i^2 - 2z_i y_i + y_i^2.$$

These terms are summed up to get likelihoods μ_x and μ_y. The path decision involves comparing these likelihoods, and this comparison is basically a subtraction of one from the other (followed by checking the sign of the result). Now, the term z_i^2 is common to *both* likelihoods. Therefore, it makes absolutely no contribution to the final decision, and it can be *dropped* without changing the outcome of the Viterbi algorithm's path selection. The expected path elements x_i, y_i are constants (for each node processor), and if we apply our normalization convention, then $2x_i \in \{-1, 0, +1\}$. If we define our likelihood update to be

$$\Delta\mu_x = -2x_i z_i + x_i^2, \qquad 6.6.11$$

(with a similar definition for $\Delta\mu_y$), our likelihood metric can be calculated without even using any multiplications (other than the trivial cases of multiplying by zero or making a sign change to z_i). Using this redefined likelihood is completely equivalent to using squared Euclidean distance for our binary codes. (The situation is only slightly more complicated for *nonbinary* convolutional codes; we won't be covering those, but it's still nice to know a similar kind of trick can be applied to the nonbinary case.)

There is one last point to make before ending this section on soft-decision decoding. In many practical applications, one wishes to use digital rather than analog circuits to implement the Viterbi decoder. This means that the signal z_i must be processed through an analog-to-digital converter, which quantizes z_i into a finite number of bits of representation. Isn't this process also information lossy? Of course. However, as the number of bits of resolution of the analog-to-digital converter is increased, the *amount* of information loss decreases. It has been observed by a large number of different people that using five or six bits in the analog-to-digital converter usually gives performance results extremely close to those of an analog soft-decision decoder. An analysis of the effect of quantization on z_i can be found in the paper by Bartles and Wells (see the references at the end of the chapter). This paper also discusses some other creative ways to compute likelihood metrics for the Viterbi algorithm.

6.7 THE TRACEBACK METHOD OF VITERBI DECODING

The Viterbi algorithm is a decoding algorithm, and so it is required to output the decoded message sequence. Until now, we have been using the *register transfer* method of keeping track of the decoded message. Each node processor manages a *path survivor register* in which is stored that node processor's best estimate of the correct decoded sequence. Example 6.3.2 illustrated the register transfer method.

The register transfer method is the easiest method to understand, but it is not the most cost effective method of keeping track of the decoded message when a high-speed decoder is required. This is because the register transfers in such a decoder must take place in parallel at each step through the trellis. Consequently, the survivor path registers must be interconnected to permit parallel transfers, and this interconnection can be very costly to implement in an integrated circuit.

The *traceback* method is an alternative way of keeping track of the decoded message sequence. It is a little more difficult to understand (at first), but it is very popular because its implementation in integrated circuit form is more cost effective than the register transfer method for high-speed decoders. We begin our explanation of the traceback method with the reminder of how we have been assigning our state designations. Referring to Figure 6.1.2 as an example, our state designations have been simply the contents of the encoder's shift register. This was illustrated in Section 6.2 for our example encoder.

Because of this convention, the least significant bit (lsb) of the state designator at time t gives us the input message bit at time $t - v + 1$. For instance, if the state at time t is "01", the message bit at time $t - 3 + 1 = t - 2$ must have been a "1". Therefore, the lsb of the state designator can be used to identify our decoded path sequence if we are willing to accept an additional output lag of $v - 1$ clock cycles. Since we already have a decoding delay of at least $\Gamma = 5.8(v - 1)$ clock cycles (in order to avoid truncation errors in the Viterbi algorithm), this additional decoding delay is a small price to pay for obtaining a lower cost hardware solution.

The basic idea behind traceback decoding is illustrated in Figure 6.7.1. Instead of transferring the contents of the survivor path registers each time, each node processor is assigned a unique register in which it stores the lsb of the state picked by that node processor as the survivor path. For the survivor path selections made in Figure 6.7.1, the node processor for state "00" stores a "1" (because it selected the path from state "01"), while the other processors each store a "0" because of their path selections.

Since we are dealing with binary convolutional codes, each node processor only has two path possibilities at each decision event. Notice how the lsb of each source state is different for the two path choices (see Figure 6.2.1). This will always be true with the state-naming convention we are using. (In fact, this is one important reason why we picked this convention in the first place.) Therefore, in addition to providing us with the decoded message sequence (as seen by each node processor), the contents of the survivor path registers also provide us with the information we need to trace the selected state histories backwards in time.

An example will help illustrate this point. Figure 6.7.2 shows the Viterbi algorithm decoding process for eight time steps through the trellis (using the code from Example 6.1.1). For clarity, only the surviving path decisions are shown at each time step. The solid line in the figure is the survivor path agreed on by all four node processors at the last time step shown in the figure. This path traces back to state "0 0". Whatever else may have happened during the time prior to the start of this figure, we know that the last two bits leading into state "0 0" must have been "0, 0" so the decoded message sequence corresponding to the solid line in the figure must be

$$0\ 0\ 1\ 1\ 1\ 0\ 0.$$

The entries into each node processor's traceback (i.e., survivor path) register at each trellis step are also shown in the figure. The traceback process is also illustrated. The traceback begins at the far right side of the figure and proceeds backwards in time

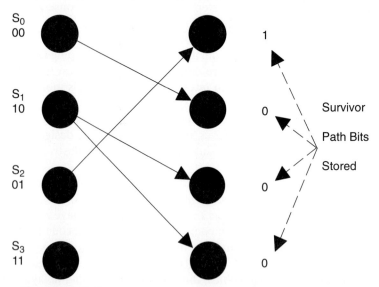

Figure 6.7.1: Survivor Path Storage by Traceback Method

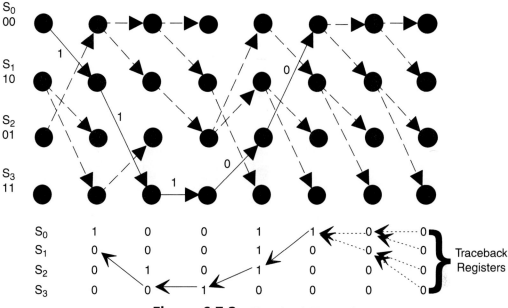

Figure 6.7.2: Traceback Example

as shown by the arrows. Notice that it does not matter which state we assume for the initial state of the traceback. This will always be the case provided the Viterbi algorithm does not have a truncation error. Once the traceback is accomplished, the decoded bit sequence is read from left to right. The decoded sequence is 0 0 1 1 1 0 0, which is the correct sequence. The last two message bits, corresponding to the final two steps through the trellis, have not been decoded yet. (This is due to the extra decoding lag mentioned above.)

The traceback method can be efficiently implemented using two-port random-access memory chips[6]. Each word in the RAM consists of 2^{v-1} bits. At each trellis step, we must write one new survivor path word into the traceback RAM and, at the same time, trace back through the previous Γ entries (where $\Gamma \geq 5.8(v-1)$ is the traceback depth) before we can begin outputting the decoded message bit. The memory size M must be large enough so that every decoded bit can be read before the location in which it is stored gets overwritten by a new survivor path entry. For example, our code from Example 6.1.1 has a constraint length of 3, and so the minimum traceback depth is $\Gamma = 12$. If we read one decoded output bit each time we step through the trellis, we must do $\Gamma + 1 = 13$ read accesses (traceback depth plus one more read to get the current decoded bit) for every write access to the RAM.

This simple scheme will require us to use a very high-speed RAM, since it must be accessed 13 times every trellis step. For high-speed systems, this can require very fast (and relatively expensive) RAM circuits. This expense can be avoided by using a

---

[6]A two-port memory is memory that can be written into, in one location, from the first port and simultaneously read out of, in a second location, from the second port.

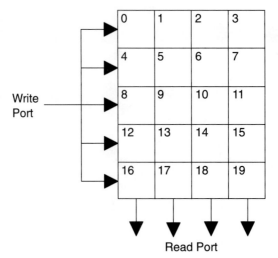

Figure 6.7.3: Survivor Path Memory Row/Column Configuration

"row-column" scheme, where the traceback process and the survivor path writes are pipelined. We will now look at such a method.

Suppose we organize the RAM into R rows and C columns as shown in Figure 6.7.3. Each column is a separate RAM chip, so we can read C words of RAM per read access. At each trellis step, we write one word into the RAM (proceeding from left to right and top to bottom) and read C words (from right to left, bottom to top). This lets us trace back C steps for each trellis step. We require $C-1$ rows for traceback, plus one row for the decoded message readout, plus one row for the newly written survivor path data, so $R = C + 1$. If our minimum traceback depth is Γ, the number of columns must satisfy

$$C(C-1) \geq \Gamma,$$

or

$$C \geq \frac{1 + \sqrt{1 + 4\Gamma}}{2}. \qquad 6.7.1$$

EXAMPLE 6.7.1

Determine the memory configuration required for a minimum traceback depth of $\Gamma = 12$.

Solution: Applying inequality 6.7.1, we have $C = 4$. Therefore, we have $R = 5$. Figure 6.7.4 illustrates the traceback process for this example. The first survivor path write goes into memory word 8 of the figure. During this process, words 7, 6, 5, and 4 are read, and the traceback is performed. At the next trellis step, the survivor path write goes into memory word 9, and the traceback read is performed for locations 3, 2, 1, and 0. This brings the traceback to the "top of memory" so the next read access will wrap around to the bottom.

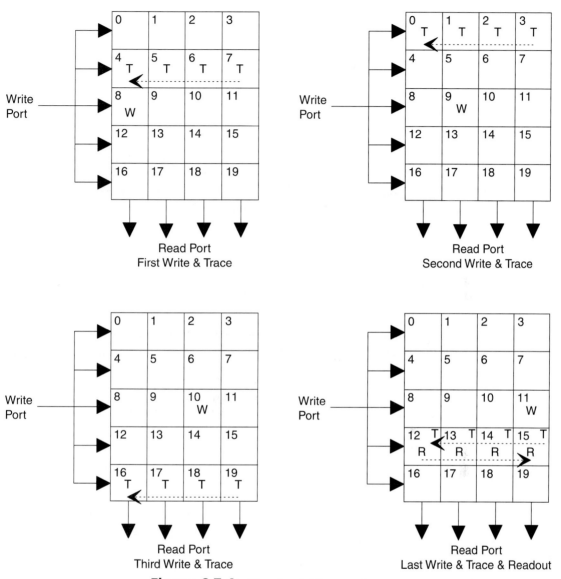

Figure 6.7.4: Traceback Memory Accesses

The third trellis step writes the survivor path information into location 10 and the traceback accesses locations 19, 18, 17, and 16. We now have a total of 12 traceback reads, which equals the minimum traceback depth. The fourth trellis step writes the survivor path information into location 11. Locations 15, 14, 13, and 12 are traced to obtain the decoded message bits. The decoded message is read from locations 12, 13, 14, and 15 to obtain the proper time ordering of the output message string.

Observe how locations 12 through 15 are decoded "just in time," since the next trellis step will be written into location 12, and the process begins again. The traceback accompanying the write into location 12 will trace memory locations 11, 10, 9, and 8. Try tracing through the process for this next row and confirm that the message bits in locations 16, 17, 18, and 19 will be decoded "just in time."

6.8 PUNCTURED CONVOLUTIONAL CODES

6.8.1 Puncturing

So far, we have been looking at convolutional codes with rate $1/n$. Sections 6.4 and 6.6 discussed the error rate performance of these convolutional codes. We saw that the error rate performance was determined by the minimum free distance, d_f, and the tables of known good codes in Section 6.5 shows that relatively simple convolutional codes can obtain large free distances. Consequently, convolutional codes provide very powerful error correction capability, compared with the block codes of Chapter 5.

This is obtained at a price, however. The convolutional codes we have looked at so far have relatively low code rates. From our study of channel capacity in Chapter 2, we are led to ask ourselves how we might achieve higher code rates for convolutional codes without sacrificing the attractive error performance properties of these codes.

One possible approach is to use nonbinary convolutional codes. Referring to Figure 6.1.1, it is clear that we could enlarge the source frame to some $k > 1$ to achieve a rate k/n convolutional code. An example of a rate 2/3 code and its trellis is shown in Figure 6.8.1. This code has $d_f = 3$. It is a 4-ary code, since each trellis node has four exit paths and must deal with four input paths. The number of inputs each node processor must deal with is a disadvantage for nonbinary convolutional codes, since the add/compare/select (ACS) operation of the Viterbi algorithm must now compare four input paths rather than only two. More generally, a rate k/n convolutional code requires the ACS operation to deal with 2^k input paths, and so the complexity of the Viterbi decoder increases geometrically with k.

The complexity growth associated with 2^k-ary convolutional codes is a severe enough problem that codes with $k > 1$ have never been particularly popular in any

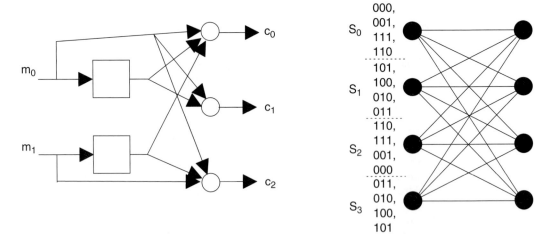

Figure 6.8.1: Rate 2/3 Code

broad areas of application. In 1979, Cain, Clark, and Geist published a paper (see references) dealing with an alternative way of achieving the higher code rates of nonbinary convolutional codes without incurring the dramatic cost disadvantages. These codes are called *punctured* convolutional codes.

The basic idea behind punctured convolutional codes is a simple one. Suppose we constructed a rate $1/n$ convolutional encoder. Such an encoder produces n code bits per message bit. For instance, a rate $1/2$ code might produce code bits $(c_0\, c_1)$. Now suppose we delete one of the code bits every two code symbols to produce code sequences of the form

$$(c_0\, c_1)_t,\ (c_0\, -)_{t+1},\ (c_0\, c_1)_{t+2},\ (c_0\, -)_{t+3}, \cdots.$$

Since the deleted code bits are not transmitted, this scheme has an average of three code bits for every two message bits which gives us a rate $2/3$ code.

Deleting code bits is called *puncturing* a code. Puncturing increases the rate of a code at the expense of reducing the minimum free distance of the original (unpunctured) base code. However, it is not fair to compare the d_f of a punctured code with the d_f of its base code because the punctured code has a higher code rate. Instead, d_f should be compared to the d_f one obtains from a nonbinary convolutional code having the same code rate as the punctured code and the same number of memory elements in its encoder. Cain et al. published a number of rate $2/3$ and $3/4$ codes in their paper and demonstrated that, in almost all cases, punctured codes exist that have the *same* minimum free distance as the best known nonbinary convolutional codes of the same code rate and constraint length. Following the original work of Cain et al., Hagenauer (1988), Lee (1988), and a number of researchers have investigated punctured convolutional codes. Punctured codes with rates up to $9/10$ are now known.

EXAMPLE 6.8.1

Figure 6.8.2 illustrates the encoder for a rate $2/3$ punctured code. The trellis for this encoder is shown in Figure 6.8.3. The encoder is the same as the rate $1/2$ encoder from Example 6.1.1. Code bit c_1 is punctured in every second code-word as shown by the trellis in Figure 6.8.3. Decoding of

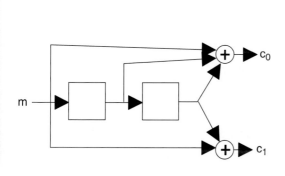

Figure 6.8.2: Rate 2/3 Encoder

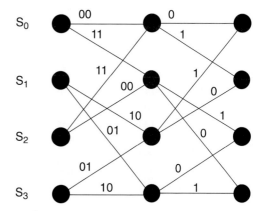

Figure 6.8.3: Trellis Rate 2/3 Code

the punctured code is accomplished using the Viterbi algorithm with the obvious modification for the different number of bits per code symbol for even and odd code-words. The punctured code is still a *linear* code, but is no longer shift invariant, since the resulting sequence of code bits has the time dependence shown in the trellis. This code can still be described by a transfer function, except the state diagram now requires eight states rather than four (four states for "time-even" trellis states plus four states for "time-odd" trellis states). The Viterbi decoder, however, requires only four node processors. Each node processor must be state dependent, i.e., must keep track of whether two code bits are expected at the present time step or if only one code bit is expected.

The minimum free distance of this code is $= 3$, which is the same as the rate 2/3 nonbinary code given earlier. Adversaries having this minimum distance begin in the "odd" time columns, where only one code bit is transmitted, and rejoin after three trellis steps.

6.8.2 Good punctured Convolutional Codes

Punctured codes are specified in a manner similar to the octal generator notation used in Section 6.5. The notation is modified by the need to specify the puncturing sequence of the code. This is usually done by writing the base code in parenthesis followed by the octal notation of the generator polynomial used at successive time slots. For example, the code of Example 6.8.1 would be denoted by

$$(7, 5), 7.$$

This notation tells us that the first message bit is encoded using generator polynomials 7 $(1 + x + x^2)$ and 5 $(1 + x^2)$ and the second message bit is encoded using only generator polynomial 7. The *puncturing period* is the number of message bits that are encoded before the returning to the base code. In this example, the puncturing period is 2.

Let us look at another example. There is a rate 3/4 punctured code (with constraint length 4) defined by the code polynomials

$$(15, 17), 15, 17.$$

For this code, the first message bit is encoded into two code bits using generator polynomials

$$15 \Rightarrow 1 + x + x^3$$

and

$$17 \Rightarrow 1 + x + x^2 + x^3.$$

The second message bit is encoded by polynomial 15, and the third message bit is encoded by polynomial 17. Thus, three message bits are encoded into four code bits, resulting in a rate 3/4 code. The puncturing period is 3. The base code is the same as the rate 1/2, constraint-length-4 code given in Table 6.5.1. This code has $d_f = 4$ (compared with $d_f = 6$ for the base code). The Viterbi decoder for this code requires eight node processors, while the state diagram for this code contains 24 states (8 times the puncturing period). The Viterbi node processors must be state dependent, since the expected code symbol depends on where we are in the puncturing sequence.

Section 6.8 Punctured Convolutional Codes

In the two examples we have looked at so far, the punctured codes were merely punctured versions of known-good rate 1/2 codes. However, it is not always true that puncturing a known good rate $1/n$ code yields a good punctured code. For example, the rate 1/2, constraint-length-4 code (15, 17) does *not* yield a good rate 2/3 punctured code. Instead, the best rate 2/3 constraint-length-4 punctured code is (15, 13), 15. There is no known systematic procedure for generating good punctured convolutional codes. Good codes are discovered by computer search. Some of the best known punctured codes are tabulated as follows:

Rate 2/3 Punctured Codes: (Discovered by Cain et al.)

TABLE 6.8.1 Best Rate 2/3 Punctured Codes

Constraint Length	Generators (Octal Representation)	d_f
3	(7, 5), 7	3
4	(15, 13), 15	4
5	(31, 33), 31	5
6	(73, 41), 73	6
7	(163, 135), 163	6
8	(337, 251), 337	8
9	(661, 473), 661	8

Rate 3/4 Punctured Codes: (Discovered by Cain et al.)

TABLE 6.8.2 Best Rate 3/4 Punctured Codes

Constraint Length	Generators (Octal Representation)	d_f
3	(5, 7), 5, 7	3
4	(15, 17), 15, 17	4
5	(35, 37), 37, 37	4
6	(61, 53), 53, 53	5
7	(135, 163), 163, 163	6
8	(205, 307), 307, 307	6
9	(515, 737), 737, 737	6

Rate 4/5 Punctured Codes: (Published by Lee, 1988)

TABLE 6.8.3 Best Rate 4/5 Punctured Codes

Constraint Length	Generators (Octal Representation)	d_f
4	(17, 11), 11, 11, 13	3
5	(37, 35), 25, 37, 23	4
6	(61, 53), 47, 47, 53	4
7	(151, 123), 153, 151, 123	5
8	(337, 251), 237, 237, 235	5
9	(765, 463), 765, 765, 473	5

Rate 5/6 Punctured Codes: (Published by Lee, 1988)

TABLE 6.8.4	Best Rate 5/6 Punctured Codes	
Constraint Length	Generators (Octal Representation)	d_f
4	(17, 15), 13, 15, 15, 13	3
5	(37, 23), 23, 23, 25, 25	4
6	(75, 53), 75, 75, 75, 75	4
7	(145, 127), 133, 127, 145, 133	4
8	(251, 237), 235, 235, 251, 251	5
9	(765, 473), 765, 473, 463, 457	5

Additional codes, up to rate 9/10, can be found in Lee's 1988 paper.

SUMMARY

We have covered much ground in this chapter. We began by introducing convolutional codes. The strategy employed by these codes is very different from that of block-coding methods. Convolutional codes are very powerful and, for that reason, are often used in environments with low signal-to-noise ratios.

Convolutional codes can be analyzed using state diagrams and trellis diagrams (which are nothing other than state diagrams "stretched out in time"). The notion of the transfer function of a convolutional code was introduced, and we saw how code properties can be described in terms of "operators" that perform the bookkeeping needed to keep track of the Hamming weights of adversary paths in the trellis.

The Viterbi algorithm was discussed in depth. This important algorithm is the key to the efficient implementation of decoders for convolutional codes and, as we shall see in the next chapter, is fundamental for the decoding of trellis modulation codes as well. We looked first at *how* the Viterbi algorithm works and then turned to the question of *why* it works. We examined the conditions under which Viterbi decoding is maximum-likelihood decoding and presented the theory behind the error rate performance of convolutional codes.

After introducing a number of known-good convoultional codes, we explored the difference between the hard-decision strategy of Section 6.4 and the more powerful soft-decision decoding strategy in Section 6.6. We examined the importance of likelihood ties between adversary paths and how soft-decision decoding alleviates the problem presented by ties. We also saw that the performance benefits of soft-decision decoding involves more than merely reducing the negative impact of ties. Soft-decision decoding is less information lossy than hard-decision decoding, since the soft-decision decoder does not quantize the received information.

We then looked at the practical decoding strategy known as traceback decoding and contrasted it with the simpler-to-understand but harder-to-implement method of register transfer decoding. We saw that the "trick" to traceback decoding is to exploit the a priori information the decoder has about the trellis structure of the code. We examined the memory depth requirements needed to implement traceback decoding and saw, by means of an example, how the memory can be efficiently used in a "just-in-time" fashion to achieve a minimum traceback memory size.

The simple binary convolutional codes suffer from the drawback of low code rates. In Section 6.8, we introduced the important idea of *puncturing* a convolutional code to obtain a higher average code rate. Some practical punctured codes were given, and we saw that reasonably high code rates are achievable using this technique. The resulting punctured codes are still *linear* codes, but their decoding structure is now time varying. Decoding of punctured convolutional codes can be carried out using the Viterbi algorithm, provided that the node processors in that algorithm are given a state dependency determined by where we are in the trellis within the puncturing period of the code.

REFERENCES

G. D. Forney, Jr., "The Viterbi algorithm," *Proc. IEEE,* vol. 61, no. 3, pp. 268–278, March, 1973.

G. D. Forney, Jr., "Convolutional codes I: Algebraic structure," *IEEE Trans. Inform. Th.,* vol. IT-16, no. 6, pp. 720–738, Nov., 1970.

G. D. Forney, Jr., "Convolutional codes II: Maximum likelihood decoding," *Information and Control,* vol. 25, pp. 222–266, July, 1974.

J. P. Odenwalder, *Optimal Decoding of Convolutional Codes,* Ph.D. Dissertation, University of California at Los Angeles, 1970.

K. Larsen, "Short convolutional codes with maximal free distances for rates 1/2, 1/3, and 1/4," *IEEE Trans. Inform. Th.,* vol. IT-19, pp. 371–372, 1973.

D. Daut, J. Modestino, and L. Wismer, "New short constraint length convolutional code construction for selected rational rates," *IEEE Trans. Inform. Th.,* vol. IT-28, no. 5, pp. 793–799, Sept., 1982.

R. Wells and G. Bartles, "Simplified calculation of likelihood metrics for Viterbi decoding in partial response systems," *IEEE Trans. Magnetics,* vol. 32, no. 5, Pt. III, pp. 5226–5237, Sept., 1996.

J. Cain, G. Clark, Jr., and J. Geist, "Punctured convolutional codes of rate (n-1)/n and simplified maximum likelihood decoding," *IEEE Trans. Inform. Th.,* vol. IT–25, no. 1, pp. 97–100, Jan., 1979.

P. Lee, "Construction of rate (n-1)/n punctured convolutional codes with minimum required SNR criterion," *IEEE Trans. Commun.,* vol. 36, no. 10, pp. 1171–1174, Oct., 1988.

J. Hagenauer, "Rate compatible punctured convolutional codes and their applications," *IEEE Trans. Commun.,* vol. 36, no. 4, pp. 389–400, April, 1988.

EXERCISES

6.1.1: The source sequence $M(x) = 1 + x + x^4 + x^6 + x^7$ is input to the convolutional encoder of Figure 6.1.2. Find the coded output sequence. Express the output bit sequence in both polynomial form and as a binary vector. What is the Hamming weight of the code sequence?

6.1.2: How many source bits must be encoded in order for the effective code rate of the convolutional encoder of Figure 6.1.2 to come within 1% of its asymptotic rate?

6.1.3: A rate 1/3 convolutional code is defined by the generator polynomials

$$g_0(x) = 1 + x^2 + x^3,$$
$$g_1(x) = 1 + x + x^3,$$
$$g_2(x) = 1 + x + x^2 + x^3.$$

Draw the encoder for this code.

6.1.4: A systematic convolutional encoder with an FIR response always has $g_0(x) = 1$. Draw the encoder for a rate 1/2 systematic convolutional encoder defined by $g_1(x) = 1 + x + x^3$.

6.2.1: A convolutional encoder is defined by the generator polynomials

$$g_0(x) = 1 + x + x^2 + x^3 + x^4,$$
$$g_1(x) = 1 + x + x^3 + x^4,$$
$$g_2(x) = 1 + x^2 + x^4.$$

a) What is the constraint length of this code?
b) How many states are in the trellis diagram for this code?
c) What is the code rate of this code?

6.2.2: Draw the state diagram and the trellis diagram for the code of Exercise 6.1.3.

6.2.3: A convolutional encoder has M memory elements. What is the fewest number of steps through its trellis required to produce a pair of adversary paths?

6.2.4: Two source sequences, $m_1(x) = 0$ and $m_2(x) = 1$, define a pair of adversary paths in the convolutional code of Exercise 6.2.2. Sketch these paths through the trellis, and find the Hamming distance between the code-words produced by these source messages.

6.2.5: Find the transfer function for the convolutional code defined by

$$g_0(x) = 1 + x,$$
$$g_1(x) = 1 + x^2.$$

6.3.1: Suppose you are given the convolutional code generated by

$$g_0(x) = 1 + x + x^2,$$
$$g_1(x) = 1 + x + x^2,$$
$$g_2(x) = 1 + x.$$

a) Draw the trellis for this code.
b) A six-bit source message with an additional two "0" bits appended to the end is sent using this code. Find the original source sequence and the most likely number of code-bit errors in the received sequence if the received sequence (read left to right) is

$$\{y\} = 101\ 100\ 001\ 011\ 111\ 101\ 111\ 110.$$

6.3.2: What decoding depth is required to avoid truncation-error performance loss for the convolutional codes given in Table 6.5.1?

6.4.1: If d is an odd integer, prove that

$$\binom{d}{(d+1)/2} = \frac{1}{2}\binom{d+1}{(d+1)/2}.$$

6.4.2: Given a BSC with $p = 0.10$, what is the probability of an error event for adversary paths at a Hamming distance of 5?

6.4.3: Write an expression in terms of the set of conditional probabilities $\Pr(E|d_j)$ for the union bound on the error rate for the rate 1/2 code with a constraint length of 7 given in Table 6.5.1.

6.4.4: Repeat Exercise 6.3.1 (b) for a received code-word sequence of

$$\{y\} = 101\ 100\ 001\ 011\ 110\ 110\ 111\ 110.$$

If the source message is the same as for Exercise 6.3.1, explain the difference in the results for these two exercises.

6.5.1: Comparing the codes given in Section 6.5, what conclusions can you draw about convolutional code performance as a function of constraint length and code rate?

6.6.1: In your own words, explain the difference between soft-decision decoding and hard-decision decoding. Based on your knowledge of information theory, explain which method you would expect to have superior error rate performance and justify your explanation.

6.7.1: Using the results of Exercise 6.3.1, find the contents of each node processor's survivor path register using the traceback method.

6.7.2: Using the results from Exercise 6.7.1, decode the source message sequence.

6.7.3: Determine the number of rows, number of columns, and total number of bits required for a two-port traceback RAM for a convolutional code of constraint length 6.

CHAPTER 7

Trellis-Coded Modulation

7.1 MULTIAMPLITUDE/MULTIPHASE DISCRETE MEMORYLESS CHANNELS

7.1.1 I–Q Modulation

Thus far, we have dealt primarily with the binary symmetric channel. The BSC is important in its own right, but it is not the only commonly used communication channel. In this chapter, we will take a look at channel coding for several classes of discrete memoryless channels (DMC) that are common in radio link communication systems, wireless telephony, and many other popular communication systems. Our information sources will be *m-ary* discrete memoryless sources, and our channels will also have *m-ary* output alphabets[1]. The object of our study in this chapter is a class of *nonlinear* error-correcting codes known as trellis-coded-modulation (TCM) codes.

There are a great many different kinds of DMCs used in carrier-modulated wireless communication systems. However, the principal differences among these channels primarily involve modulation methods and signal-processing methods, which are outside the scope of this text. Fortunately, we need not be concerned with the particular details and differences here, since, as far as channel coding goes, these signal-processing details are largely irrelevant to our coding, and those few details that are important can be understood easily without going in to the channel-dependent details. We will use two different channels, the *m-ary* phase shift keying (*m*-PSK) and the quadrature amplitude-modulation (QAM) channels, to illustrate the design and application of TCM codes. Both of these channels are widely used and are important in their own right. The procedures we will develop here for these channels are easily extended to the horde of other modulation channels one encounters in practice.

Our basic transmitter block diagram is shown in Figure 7.1.1. We will assume that the information source is memoryless and has entropy $H(A) = \log_2(m)$. The source symbols *a* are drawn from the *m-ary* source alphabet and are applied to the channel encoder and a digital modulator. The digital modulator produces two continuous-time analog pulse signals, $p_I(t)$ and $p_Q(t)$, called the *in-phase* and the *quadrature* signals,

---

[1]An *m-ary* source is a source having a symbol alphabet containing *m* distinct symbols.

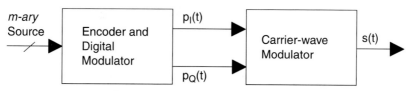

Figure 7.1.1: *m-ary* Transmitter

respectively. These signals are sent to the carrier-wave modulator which produces an *information-bearing waveform signal*

$$s(t) = p_I(t)\cos(2\pi ft) - p_Q(t)\sin(2\pi ft), \qquad 7.1.1$$

where f is called the carrier frequency. The information-bearing waveform $s(t)$ is transmitted.

Figure 7.1.2 shows the receiver for the system. It is essentially the inverse of the transmitter function. The noisy demodulated pulse signals $p_I(t)$ and $p_Q(t)$ are sampled and measured by the bit detector and decoded to reproduce the original *m-ary* symbol. In the absence of noise, the pulse signals $p_I(t)$ and $p_Q(t)$ have discrete amplitudes, and there are $n \geq m$ possible combinations of their two amplitudes. Without TCM encoding, $n = m$. As we will see later in this chapter, TCM encoding will call for $n = 2m$ combinations of amplitudes for $p_I(t)$ and $p_Q(t)$.

7.1.2 The *n*-ary PSK Signal Constellation

Since $p_I(t)$ and $p_Q(t)$ are each multiplied by a sinusoid to produce $s(t)$, and since these two sinusoids have a relative phase angle of ninety degrees, it is convenient and useful to represent the pulses $p_I(t)$ and $p_Q(t)$ as points in a two-dimensional vector space. The two axes of this space are called the in-phase, or *I* axis (corresponding to $p_I(t)$), and the quadrature, or *Q* axis (corresponding to $p_Q(t)$). The two components of the vector are given by the measured amplitudes of the $p_I(t)$ and $p_Q(t)$ signals. The set of n possible vectors resulting from this is called the *signal constellation*. Different types of modulation differ only in the configuration of their signal constellations. Figure 7.1.3 below shows the signal constellation for 4-*ary* phase shift keying (commonly known in the communications business as QPSK).

Each solid dot in the signal constellation of Figure 7.1.3 represents a point in the two-dimensional *I-Q* vector space. Each signal point is assigned one of the symbols in the *n-ary* signal alphabet, typically represented using $\log_2(n)$ bits. In *n-ary* PSK, the n signal points are located on a circle with an angular spacing of $2\pi/n$ radians between

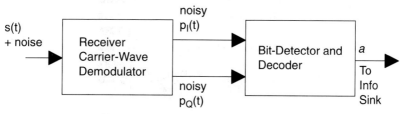

Figure 7.1.2: *m-ary* Receiver

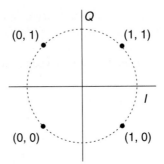

Figure 7.1.3: QPSK Signal Constellation

adjacent signal points. To reduce the impact of noise on the system, it is customary that the binary vectors assigned to nearest neighbor signal points are unit Hamming distance apart. This has been done in Figure 7.1.3.

It is important to notice that the binary vectors assigned to each signal point do *not* represent coordinates in *I–Q* space. The *I–Q* coordinate vector is $[p_I\, p_Q]$ and is an element of a real-valued, two-dimensional vector space. Distance in *I–Q* space is measured in terms of Euclidean distance and not Hamming distance. The binary vector representation of the symbol corresponding to a point in the signal constellation can be arbitrarily assigned without regard to the *I–Q* coordinate. It is the digital modulator that produces the $[p_I\, p_Q]$ vector corresponding to a given *n-ary* symbol.

In the presence of zero-mean additive white Gaussian noise (AWGN), the vectors in *I–Q* space may be written

$$\overline{v} = [p_I + \eta_I \quad p_Q + \eta_Q] \equiv \overline{p} + \overline{\eta} \qquad 7.1.2$$

where η_I and η_Q are statistically independent Gaussian random variables with variance

$$E[\eta_I^2] = E[\eta_Q^2] = \sigma^2. \qquad 7.1.3$$

With the addition of noise, the discrete points in the signal constellation become "fuzzy," due to the addition of the random noise components in Equation 7.1.2. This is illustrated in Figure 7.1.4. Because the *I* and *Q* noise components have identical variances, the contours of constant probability $\Pr(\overline{v}|p_I, p_Q)$ are circles centered on the original (noiseless) signal constellation point. The probability of having noisy received signal point $\overline{v}$ a distance *r* away from its desired location $\overline{p}$ in the signal constellation is simply

$$\Pr[d_E(\overline{v}, \overline{p}) = r] = \frac{r}{\sigma^2} \exp(-r^2/2\sigma^2), \qquad 7.1.4$$

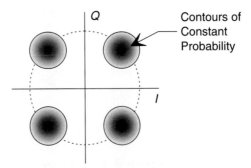

Figure 7.1.4: Noisy QPSK Signal Constellation

Section 7.1 Multiamplitude/Multiphase Discrete Memoryless Channels

where $d_E(\cdot)$ is the Euclidean distance metric. Equation 7.1.4 is known as the Rayleigh probability distribution function. It is a standard probability distribution in probability theory and describes the probability distribution of the square root of the sum of the squares of two independent zero-mean Gaussian random variables, each with variance σ^2.

7.1.3 PSK Error Rate

The relative magnitude of the signal vector $\bar{p}$ to the noise vector $\bar{\eta}$ is generally expressed in terms of a *signal to noise ratio* (SNR)

$$SNR \equiv \frac{E[|\bar{p}|^2]}{2\sigma^2}. \qquad 7.1.5$$

It is standard practice to express Equation 7.1.5 in *decibels*, which is defined as

$$SNR_{dB} \equiv 10 \log_{10}(SNR). \qquad 7.1.6$$

The error rate performance of the channel is determined by the SNR. For PSK channels, it is clear that $|\bar{p}|^2$ is a constant. In the absence of coding, an error will occur if signal $\bar{p}_i$ is transmitted, but because of noise, the received signal $\bar{v}$ is such that

$$d_E(\bar{v}, \bar{p}_i) > d_E(\bar{v}, \bar{p}_j), \qquad j \neq i.$$

For an optimum *n-ary* PSK receiver, in the absence of coding, it can be shown that the bit error rate for the *n-ary* PSK channel is approximately given by

$$p_e \approx \frac{2}{\log_2(n)} Q(\sin(\pi/n)\sqrt{2 \cdot SNR}), \qquad 7.1.7$$

where

$$Q(x) \equiv \frac{1}{\sqrt{2\pi}} \int_x^\infty e^{-u^2/2} \, du \qquad 7.1.8$$

is the error rate function we encountered in Chapter 6, Equation 6.6.10 and where we have assumed nearest neighbor signal constellation points are assigned symbols that are a unit Hamming distance from each other.

There is an alternative way of expressing Approximation 7.1.7 that will prove useful when we look at coding for these channels. From Equation 7.1.5, we know that the signal to noise ratio depends on $|\bar{p}|^2$. It is also clear that the minimum Euclidean distance between points in the signal constellation must also depend on $|\bar{p}|^2$. Let us denote the minimum Euclidean distance between signal constellation points using the notation

$$\Delta_0 \equiv \min_{\bar{p}_i \neq \bar{p}_j} \{d_E(\bar{p}_i, \bar{p}_j)\}, \qquad i,j = 0, \cdots, n-1. \qquad 7.1.9$$

Further, let us assume that the transmitter transmits one *n-ary* signal vector every T seconds and define the *noise power spectral density* as

$$N_0 \equiv \sigma^2 T. \qquad 7.1.10$$

Using these definitions, it can be shown from modulation theory that for *n-ary* PSK

$$\sin(\pi/n)\sqrt{2 \cdot SNR} = \frac{\Delta_0}{\sqrt{2N_0}},$$

and therefore

$$p_e = \frac{2}{\log_2(n)} Q\left(\frac{\Delta_0}{\sqrt{2N_0}}\right). \qquad 7.1.11$$

The bit error rate is a function of the minimum Euclidean distance between points in the signal constellation. Equation 7.1.11 will prove useful later when we look at error rate performance of TCM codes.

Remark: Equation 7.1.10 might give us the impression that we could improve the error rate simply by signaling at a faster rate. Unfortunately, this happy idea is not true, because faster signaling requires more bandwidth and increasing the bandwidth increases σ^2. The decrease in T is exactly canceled out by the increase in σ^2. (We have not shown this here, but it is true nevertheless.) The noise power spectral density N_0 is a more fundamental quantity than σ^2, because it does not depend on any of the details of the receiver, whereas σ^2 does.

EXAMPLE 7.1.1

Calculate the bit error rate for 4-PSK, 8-PSK, and 16-PSK at a signal to noise ratio of 8 dB. Also calculate the minimum Euclidean distance for 8-PSK and 16-PSK as a function of the Euclidean distance of 4-PSK.

Solution: $SNR_{dB} = 8 \Rightarrow SNR = 10^{8/10} \approx 6.31$. Applying approximation 7.1.7 we have

4-PSK: $\quad p_e = \dfrac{2}{\log_2(4)} Q[\sin(\pi/4)\sqrt{12.62}] = Q(2.512) \approx 6 \cdot 10^{-3}$,

8-PSK: $\quad p_e = \dfrac{2}{\log_2(8)} Q[\sin(\pi/8)\sqrt{12.62}] = \dfrac{2}{3} Q(1.36) \approx 57.9 \cdot 10^{-3}$,

16-PSK: $\quad p_e = \dfrac{2}{\log_2(16)} Q[\sin(\pi/16)\sqrt{12.62}] = \dfrac{1}{2} Q(0.693) \approx 121 \cdot 10^{-3}$.

For the minimum Euclidean distance, we have

$$\Delta_0 = \sqrt{2 \cdot N_0} \sqrt{2 \cdot SNR} \cdot \sin(\pi/n),$$

so the *relative* distances depend only on n. From this,

$$\Delta_{0(8-PSK)} = \frac{\sin(\pi/8)}{\sin(\pi/4)} \Delta_{0(4-PSK)} = 0.541 \Delta_{0(4-PSK)},$$

$$\Delta_{0(16-PSK)} = \frac{\sin(\pi/16)}{\sin(\pi/4)} \Delta_{0(4-PSK)} = 0.2759 \Delta_{0(4-PSK)}.$$

At a fixed signal-to-noise ratio, we see that the error rate performance becomes poorer rapidly as we add points to the signal constellation. The reason is due to the decrease in the minimum Euclidean distance between points in the signal constellation for a fixed noise variance. To obtain equivalent bit error rate performance, the signal-to-noise ratio would need to improve as

Section 7.1   Multiamplitude/Multiphase Discrete Memoryless Channels   247

the number of bits per point in the signal constellation increases. The amount by which the SNR must improve can be approximated (in a "back of the envelope" fashion) as

$$\Delta SNR_{dB} \approx 20 \log_{10}(\Delta_{0(4-PSK)}/\Delta_{0(n-PSK)}), \qquad 7.1.12$$

which, for this example, gives us about 5 dB for the 8-PSK case and about 11 dB for the 16-PSK case.

7.1.4 Quadrature Amplitude Modulation

We now turn to quadrature amplitude modulation (QAM) channels. Like PSK, QAM channels are represented in terms of an I–Q diagram. For PSK, the n points in the signal constellation are arranged in a circle. For QAM, the points in the signal constellation are laid out on a rectangular grid. Figure 7.1.5 illustrates two QAM constellations for 8-ary and 16-ary QAM.

Consider the 16-QAM constellation in Figure 7.1.5. Let the spacing between nearest-neighbor points in the QAM diagram be d. The individual points themselves are located at (I, Q) coordinates

$$(i \cdot d/2, j \cdot d/2)$$

for integer values of i and j. As was the case for PSK, the error rate of the channel can be expressed as a function of the spacing d and the noise power spectral density. The spacing d has the same role as Δ_0 in the PSK expressions.

Suppose we have $M = 2^b$ points in our constellation, where b is the number of bits represented by a signal point. For reasons of power efficiency, if b is an even number, the signal constellation always has the same number of rows and columns (such as 2 by 2, 4 by 4, and so on). Each row and each column have the same number of points.

If b is an odd number, it is not possible for every row and every column to have the same number of points. However, the signal constellation is often constructed so that the points in it come as close as possible to forming a square array of points. This is illustrated in Figure 7.1.6 for a 32-ary ($b = 5$) constellation. Notice how the "corners"

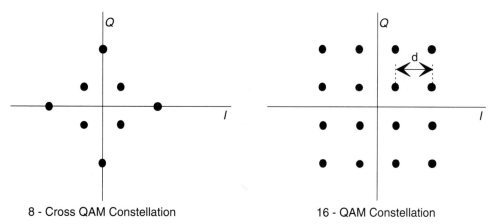

Figure 7.1.5: QAM Signal Constellations

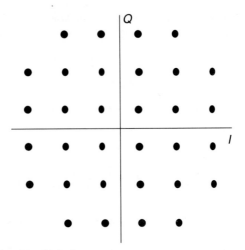

Figure 7.1.6: 32-Cross QAM Constellation

of the square are eliminated (because we don't have enough signal points to "fill them in"). Such a constellation is called a *cross*.

Setting $d = \Delta_0$, and assuming that the Hamming distance between nearest neighbor points in the signal constellation is unity, the bit error rate of M-QAM can be upper bounded by

$$p_e \cong \frac{4}{b} Q\left(\frac{\Delta_0}{2\sigma}\right). \qquad 7.1.13$$

In practice, communication systems often are designed under a constraint of maximum transmitted power, and so it is common to express Equation 7.1.13 in a different form

$$p_e \cong \frac{4}{b} Q\left(\sqrt{\frac{3}{M-1} \cdot SNR}\right), \qquad 7.1.14$$

which expresses the error rate in terms of the signal to noise ratio.

7.2 SYSTEMATIC RECURSIVE CONVOLUTIONAL ENCODERS

In the previous chapter, we dealt exclusively with convolutional codes having encoders in nonsystematic form. It is also possible to do convolutional encoding in systematic form, where the source bits appear unaltered in the code symbols. This is not normally done in "standard" convolutional coding because systematic convolutional codes using finite-impulse-response type encoders typically have inferior error rate performance compared to nonsystematic FIR encoders. However, if we are willing to use recursive encoders, it is possible to find systematic convolutional codes with good performance.

Standard TCM encoders employed in practice generally use what is known as an Ungerboeck encoder. An Ungerboeck encoder may be viewed as a convolutional encoder followed by a signal mapper that translates b-bit code symbols into a 2^b-*ary* signal constellation. The known good TCM codes are typically expressed as a set of

Section 7.2 Systematic Recursive Convolutional Encoders

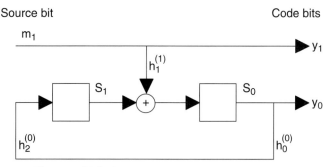

Figure 7.2.1: Rate 1/2 Systematic Encoder

polynomials, called *parity-check polynomials*, which are implemented using a systematic recursive convolutional encoder. Let us look at how such an encoder works.

Figure 7.2.1 illustrates the encoder for a rate 1/2 systematic convolutional code. Its trellis, shown in Figure 7.2.2, is easily found by constructing the next-state table for the encoder. Examination of the trellis will show that the minimum Hamming distance between adversary paths for this encoder is three. If we compare this with the 4-state convolutional code from Chapter 6, we see that the encoder shown below would have an inferior Hamming distance and, consequently, would not make as good a convolutional encoder as the nonsystematic rate 1/2, constraint-length-3, convolutional code of the previous chapter. However, the encoder shown below *is* a suitable encoder for a TCM code, as we shall see later.

This encoder is defined by a pair of parity check polynomials

$$H^{(0)} = h_0^{(0)} + h_1^{(0)}x + h_2^{(0)}x^2 = 1 + x^2,$$
$$H^{(1)} = h_0^{(1)} + h_1^{(1)}x + h_2^{(1)}x^2 = x,$$

and has $M = 2$ memory elements. The coefficients of the parity check polynomials tell us the connections that are to be made to the encoder. This is illustrated in Figure 7.2.1.

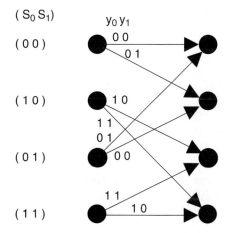

Figure 7.2.2: Trellis for Rate 1/2 Encoder

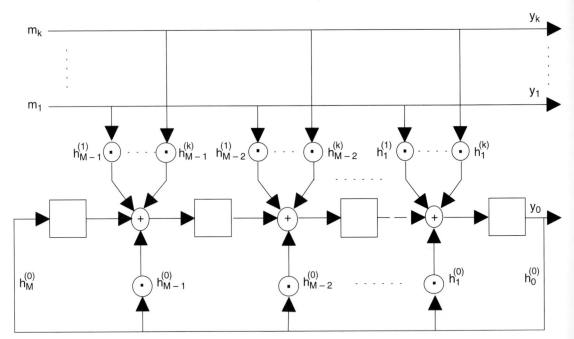

Figure 7.2.3: Canonical Systematic $k/(k+1)$ Encoder

A coefficient of "0" implies "no connection," while a "1" calls for a connection. If our state variables, S_i, are enumerated from *right to left*, a coefficient $h_i^{(\cdot)} = 1$ calls for a connection to be made at the output of the storage element for state S_i. Figure 7.2.3 illustrates this convention.

More generally, for a rate $k/(k+1)$ encoder, we label the source bits as $m_1, \cdots, m_k$ and associate parity-check polynomial $H^{(j)}$ with message bit m_j. The coefficients of $H^{(j)}$ specify the connections made by bit m_j to the canonical encoder shown in Figure 7.2.3. As implied by the figure, for TCM codes the parity-check polynomial $H^{(0)}$, which is associated with the feedback connections, is always of the form

$$H^{(0)} = 1 + h_1^{(0)} x + \cdots + x^M,$$

while $h_0^{(j)} = h_M^{(j)} = 0$ for $j \neq 0$. Also, it is typical to specify the $H^{(j)}$ polynomials in octal (as was done in Chapter 6). When this is done, the octal designation is read left to right, which corresponds to $h_0^{(\cdot)}, h_1^{(\cdot)}, \cdots, h_M^{(\cdot)}$. As an example, the polynomial "23" for an $M = 4$ encoder is

$$H^{(0)} = 1 + x^3 + x^4.$$

The systematic convolutional codes used within a TCM encoder generally are poorer convolutional codes than those of Chapter 6. However, the convolutional codes of that chapter rarely make good TCM codes. As was the case in Chapter 6, most TCM codes are found by computer search.

7.3 SIGNAL MAPPING AND SET PARTITIONING

We now look at the process of mapping the $k+1$ code bits from the systematic convolutional encoder into the signal points in the I–Q constellation of the transmitted signal. This is accomplished through a procedure known as *set partitioning*. The fundamental idea of set partitioning is to group the points of the signal constellation into sets that have the maximum possible Euclidean distance between points. For example, suppose we have an 8-PSK signal constellation. The eight points lie on a circle with nearest neighbor points separated by an angle of 45 degrees. Let us call this constellation set $A0$. We can group these eight points into two sets of four points each with nearest neighbors separated by angles of 90 degrees. Call these sets $B0$ and $B1$. Each of these two sets can, in turn, be separated into two sets of two points each separated by 180 degrees. We'll call the four total sets resulting from this partitioning $C0$, $C1$, $C2$, and $C3$. This process is illustrated in Figure 7.3.1.

This process can clearly be continued until each set contains only one signal point. Each time a set is partitioned, each partition is assigned a binary number. When we reach the final stage (where each set has only one point), the code word associated with that point is obtained by reading the binary numbers associated with each set

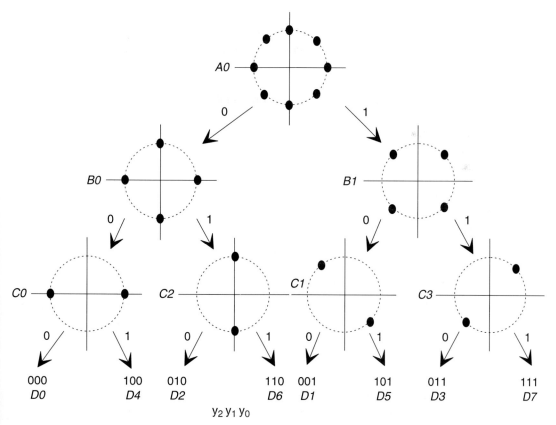

Figure 7.3.1: Set Partitioning of 8-PSK

partition beginning from the bottom of the graph and tracing back to the original *A0* signal constellation. Note that code bit y_0 (which corresponds to the partitioning from *A0* into *B0* and *B1*) will be the bit output by the shift register encoder of Figure 7.2.3.

The specific assignments of which signal points go into each partition is arbitrary, except for the requirement that the points within a partition must have the maximum Euclidean distance possible.

The fundamental idea behind TCM is to use the convolutional encoder to control the permitted sequence of constellation signals in such a way that the Euclidean distance between adversary received signal sequences is greater than it would be without coding. To look at a specific example, suppose we wanted to send two information bits per signal. We could do this using a 4-PSK signal constellation, such as that in Figure 7.1.3. Suppose the signal points in this figure lie on a circle with a radius of one. The minimum distance between signal points is then $d = \sqrt{2}$, and the error rate is given by Equation 7.1.11 with $n = 4$ and $\Delta_0 = d$.

Ungerboeck's great idea was to add redundancy to the system by adding more points to the signal constellation and using convolutional encoding to control the sequence of transmitted signal points. For our present example, an Ungerboeck encoder would be designed for an 8-PSK signal constellation using set partitioning as in Figure 7.3.1. The encoder and its trellis for a code of $H^{(2)} = 0$, $H^{(1)} = 2$, $H^{(0)} = 5$ is shown in Figure 7.3.2. The "signal mapper" selects which point in the signal constellation is to be transmitted from the outputs y_2, y_1, y_0 of the encoder according to the code assignments given in Figure 7.3.1.

The labeling of the trellis diagram is worth commenting on. Each node in the trellis shows two exit paths. The labels C0, C2, and so forth specify which subset in Figure 7.3.1 is associated with each trellis path. The labels are read according to our usual convention where the labels (read left to right) correspond to the arrows of the trellis (read clockwise from the top). Thus, in the top trellis, node C0 corresponds to the transitions from that node to itself, C2 corresponds to the lower transition from that node, and so on for the rest of the trellis.

From Figure 7.3.1, we know that each subset C0, C2, etc. contains *two* signal points. In the encoder of Figure 7.3.2, the information bit m_2 is *not used* by the convolutional encoder. However, $y_2 = m_2$ *is* used by the signal mapper. Each of the transitions shown in the trellis really represents *two* parallel transitions. Which of the two transitions is actually used is determined by the code bit y_2. For example, the output for C0 is actually either D0 or D4, depending on whether $y_2 = 0$ or $y_2 = 1$, respectively. From Figure 7.3.1, the Euclidean distance between points in the same C subset is $d_{par} = 2$ (which is greater than the minimum Euclidean distance in the 4-PSK constellation).

In addition to parallel transitions, we must also consider the minimum Euclidean distance between adversary trellis paths. For example, compare the output sequence C0, C0, C0 with the sequence C2, C1, C2. The minimum Euclidean distance between these two paths can be calculated from the set-partitioning diagram of Figure 7.3.1 as

$$d_{trellis} = \sqrt{(\sqrt{2})^2 + (.76537)^2 + (\sqrt{2})^2} = 2.1414.$$

Analysis of the trellis will show that this is the smallest distance for nonparallel paths. Since $d_{trellis} > d_{par}$, the performance of this code will be determined by the parallel transitions. In general, when the trellis contains parallel paths, the *free distance* of the code is

$$d_{free} = \min(d_{par}, d_{trellis}). \qquad 7.3.1$$

Section 7.3 Signal Mapping and Set Partitioning

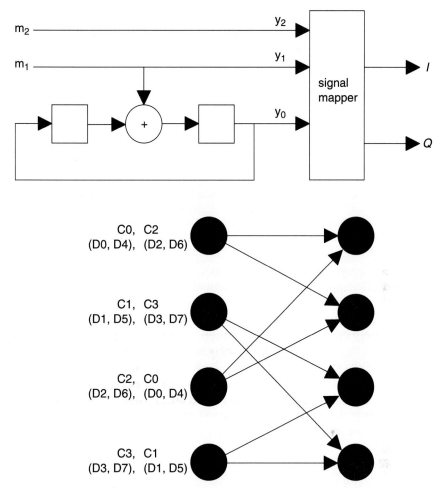

Figure 7.3.2: TCM Encoder and Trellis

In general, for a rate $k/(k+1)$ code, if $H^{(k)} = 0$, the trellis will contain parallel paths, while if $H^{(k)} \neq 0$, the trellis will not have any parallel paths.

The signal mapper is typically a lookup table that translates the output of the encoder into the appropriate I–Q signals. It is important, therefore, for us to be clear about which I–Q signal outputs correspond to which trellis transitions. To analyze a particular code, we construct a next-state truth table. For the example above, this table is given in Table 7.3.1.

In this table, we do not show information bit input m_2, since that bit is not involved in determining the code's trellis diagram. Bit m_2 determines which signal point within a given C subset is actually sent by the transmitter. The C subset is determined from the set-partitioning diagram (Figure 7.3.1 for this example). Published trellis codes are designed assuming the set partition numbering convention $(y_k, y_{k-1}, \cdots, y_0)$ we used in Figure 7.3.1. It is important to bear this in mind when using published known codes, since otherwise, the transmitted I–Q signal from the signal mapper

TABLE 7.3.1: Truth Table for signal mapping in TCM

m_1	S_0	S_1	next S_0	S_1	y_1	y_0	Output Subset
0	0	0	0	0	0	0	C0
1	0	0	1	0	1	0	C2
0	1	0	0	1	0	1	C1
1	1	0	1	1	1	1	C3
0	0	1	1	0	0	0	C0
1	0	1	0	0	1	0	C2
0	1	1	1	1	0	1	C1
1	1	1	0	1	1	1	C3

may not correctly match the convolutional encoder, and the resulting free distance of the code may therefore be less than the designed distance achievable from the encoder.

Although the convolutional encoder part of a TCM encoder is a linear encoder, the signal mapper is not a linear element, and the resulting TCM code is therefore *not* a linear code. This can easily be seen from the signal constellation. Observe that signals in the signal constellation add as vectors. Therefore, the sum of two I–Q signals is *not* a valid point in the signal constellation. This is why TCM codes are nonlinear codes. However, they may still be analyzed from the trellis diagram, and the minimum distance of most published trellis codes may be found by analyzing distances with respect to the trellis path taken by the all-zeroes information input sequence path. This is because of the symmetry in most practical I–Q signal constellations, however, and not because of the linearity theorem we developed in Chapter 6 (TCM codes are not linear!).

7.4 KNOWN GOOD TRELLIS CODES FOR PSK AND QAM

An exact error rate performance expression for most TCM codes is very difficult to obtain. Fortunately, for practical systems operating under moderate to high signal to noise ratio conditions (which is frequently the case in commercial applications), the error rate performance of a TCM code is well approximated as

$$p_e = N_{\text{free}} Q\left(\frac{d_{\text{free}}}{2\sigma}\right), \qquad 7.4.1$$

where p_e is the probability of a path select error in the trellis, d_{free} is given by Equation 7.3.1 and N_{free} is called the error coefficient and is equal to the average number of trellis paths having distance d_{free} in the trellis.

Since analysis of the trellis to find N_{free} is usually somewhat involved, it is often more convenient to compare the performance of the TCM system with the performance of an uncoded system. Recall that the main idea in TCM was to achieve redundancy by doubling the number of points in the signal constellation. For example, a 4-PSK uncoded system becomes an 8-PSK TCM system. The *coding gain* of a TCM code is the amount by which the signal to noise ratio of the uncoded system would have to improve in order to obtain the same performance as the coded TCM system. Therefore, if the coding gain of a TCM code is known, the bit error rate can be estimated from the expression for the uncoded system by simply improving the SNR by the amount of the coding gain (using the expressions in Section 7.1). Published TCM codes usually include the coding gain of the code, so this approach to estimating performance is a very practical one.

If doubling the number of points in the signal constellation for TCM encoding is a good idea, why not keep going and quadruple the number of points? As it happens, Ungerboeck looked at this idea in his 1982 paper. He analyzed the channel capacity gains for this strategy and found that there is very little "leftover" performance to be gained by quadrupling the number of points in the signal constellation. Most of the coding gain that *can* be achieved *is* achieved simply by doubling the number of points (resulting in rate $k/(k+1)$ codes).

Simple trellis codes can be found "by hand." However, more complex TCM codes must be found from computer search. Like convolutional codes, there is no known analytical procedure for designing TCM codes. A great many TCM codes have been discovered to date and are used extensively in wireless communication, compressed video systems, voice-frequency high-speed modems, and in deep-space communication. The following tables list some of the best known-good TCM codes. In these tables, Δ_0 is the minimum Euclidean distance between signal points in the signal constellation of the coded system and the coding gains are calculated assuming the same signal power in the coded system as in the uncoded system. The parity check polynomials are specified in octal format.

TCM Codes for PSK: (Ungerboeck, 1987)

Table 7.4.1: Codes for 8PSK TCM (4-PSK base)

No. of trellis states	k	$H^{(2)}$	$H^{(1)}$	$H^{(0)}$	$d_{\text{free}}^2/\Delta_0^2$	Coding Gain (dB)	N_{free}
4	2	0	2	5	4.000	3.01	1
8	2	04	02	11	4.586	3.60	2
16	2	16	04	23	5.172	4.13	2.3
32	2	34	16	45	5.758	4.59	4
64	2	066	030	103	6.343	5.01	5.3
128	2	122	054	277	6.586	5.17	0.5
256	2	130	072	435	7.515	5.75	1.5

TABLE 7.4.2: Codes for 16PSK TCM (8-PSK base)

No. of trellis states	k	$H^{(2)}$	$H^{(1)}$	$H^{(0)}$	$d_{\text{free}}^2/\Delta_0^2$	Coding Gain (dB)	N_{free}
4	3	0	2	5	1.324	3.54	4
8	3	0	04	13	1.476	4.01	4
16	3	0	04	23	1.628	4.44	8
32	3	0	10	45	1.910	5.13	8
64	3	0	024	103	2.000	5.33	2
128	3	0	024	203	2.000	5.33	2
256	3	374	176	427	2.085	5.51	8

Codes for QAM: (Ungerboeck, 1987)

TABLE 7.4.3: Codes for 16-QAM TCM (8-PSK base)

No. of trellis states	k	$H^{(2)}$	$H^{(1)}$	$H^{(0)}$	$d_{\text{free}}^2/\Delta_0^2$	Coding Gain (dB)	N_{free}
4	3	–	2	5	4.000	4.36	4
8	3	04	02	11	5.0	5.33	16
16	3	16	04	23	6.0	6.12	56
32	3	10	06	41	6.0	6.12	16
64	3	064	016	101	7.0	6.79	56
128	3	042	014	203	8.0	7.37	344
256	3	304	056	401	8.0	7.37	44
512	3	0510	0346	1001	8.0	7.37	4

TABLE 7.4.4: Codes for 32CR-QAM TCM (16-QAM base)

No. of trellis states	k	$H^{(2)}$	$H^{(1)}$	$H^{(0)}$	$d_{\text{free}}^2/\Delta_0^2$	Coding Gain (dB)	N_{free}
4	4	–	2	5	4.0	3.01	4
8	4	04	02	11	5.0	3.98	16
16	4	16	04	23	6.0	4.77	56
32	4	10	06	41	6.0	4.77	16
64	4	064	016	101	7.0	5.44	56
128	4	042	014	203	8.0	6.02	344
256	4	304	056	401	8.0	6.02	44
512	4	0510	0346	1001	8.0	6.02	4

TABLE 7.4.5: Codes for 64-QAM TCM (32CR-QAM base)

No. of trellis states	k	$H^{(2)}$	$H^{(1)}$	$H^{(0)}$	$d_{\text{free}}^2/\Delta_0^2$	Coding Gain (dB)	N_{free}
4	5	0	2	5	4.0	2.80	4
8	5	04	02	11	5.0	3.77	16
16	5	16	04	23	6.0	4.56	56
32	5	10	06	41	6.0	4.56	16
64	5	064	016	101	7.0	5.23	56
128	5	042	014	203	8.0	5.81	344
256	5	304	056	401	8.0	5.81	44
512	5	0510	0346	1001	8.0	5.81	4

It is worth noting that the codes for the QAM systems in Tables 7.4.3 through 7.4.5 are all identical. Also, it is worth noting for the codes in Tables 7.4.4 and 7.4.5 that $H^{(3)}$ through $H^{(5)}$ are zero. These codes contain many parallel paths in the trellis.

In this chapter, we have given a brief introduction to trellis-coded modulation. There is much more that could be said about these codes. For example, in some communication systems there is a need for coding the transmitted signal sequence in such a way as to make the receiver less sensitive to phase errors in its carrier recovery circuit. The TCM codes we have discussed here are "two-dimensional" codes, since the *I–Q* signal constellation is two dimensional. TCM codes also exist that incorporate *time redundancy* as well as signal space redundancy. This means that the code words are sequences of *I–Q* signals rather than a single *I–Q* signal. Such codes are known as "higher dimension" TCM codes. For the motivated and interested student, the references at the end of this chapter provide additional details of TCM coding. We will say no more about these codes here because the need for and usefulness of many of the more advanced codes involves signal-processing and modulation theory considerations, which are beyond the scope of this text.

SUMMARY

This chapter has given us a brief introduction to trellis-coded modulation (TCM). We started by looking at two important nonbinary channels, namely, the multilevel phase-shift keying channel and the quadrature amplitude modulation channel. We saw that these channels are described in modulation theory with the use of an *I–Q* diagram. This diagram describes the "in-phase" and the "quadrature" components of a carrier-wave-modulated signal. We looked at the "raw" channel error rate expressions for these channels in preparation for the theory of how this error rate performance can be improved with coding.

We next introduced the systematic *recursive* convolutional encoder. Unlike the convolutional codes in Chapter 6, TCM systems employ feedback in their convolutional encoders. The canonical Ungerboeck encoder was examined. These encoders or modified versions of them are widely used in important commercial systems.

The Ungerboeck encoder must be followed by a signal mapper to transform the binary code words into the form of an *I–Q* signal. The technique of set partitioning was used to show how optimum assignments of code words to *I–Q* signal vectors may be made. This mapping process is nonlinear, so the resulting TCM code is a nonlinear code. However, it has the very important property that it may still be *decoded* by using the notion of a trellis diagram and employing the Viterbi algorithm. We finished up the chapter by introducing a number of known-good TCM codes for PSK and QAM systems.

REFERENCES

G. Ungerboeck, "Channel coding with multilevel/phase signals," *IEEE Trans. Information Theory,* vol. IT–28, no. 1, pp. 55–67, Jan., 1982.

G. Ungerboeck, "Trellis-coded modulation with redundant signal sets part I: Introduction," *IEEE Communications Magazine,* vol. 25, no. 2, pp. 5–11, Feb., 1987.

G. Ungerboeck, "Trellis-coded modulation with redundant signal sets part II: State of the art," *IEEE Communications Magazine,* vol. 25, no. 2, pp. 12–21, Feb., 1987.

A. Calderbank and N. Sloane, "Four-dimensional modulation with an eight-state trellis," *AT&T Technical Journal,* vol. 64, pp. 1005–1017, May-June, 1985.

A. Calderbank and N. Sloane, "An eight-dimensional trellis code," *Proceedings of the IEEE,* vol. 74, no. 5, pp. 757–759, May, 1986.

M. Simon and D. Divsalar, "Combined trellis coding with asymmetric MPSK modulation," *JPL Publication* 85–24, May 1, 1985.

EXERCISES

7.1.1: Draw the signal constellation for an $m=8$ PSK channel. Calculate the minimum Euclidean distance between signal points, assuming that the points are located on a circle with unit radius.

7.1.2: What is the probability of a noisy received signal exceeding a distance r_x away from its desired location in the signal constellation?

7.1.3: The error rate function $Q(x)$ is approximated to within 0.27% for $x > 0$ by

$$Q(x) = \frac{1}{\sqrt{2\pi}} \frac{1}{0.661x + \sqrt{x^2 + 5.510}} \exp(-0.5x^2).$$

Plot the error rate vs. signal to noise ratio for SNRs in the range from 0 dB to 12 dB for 4PSK, 8PSK, and 16PSK.

7.2.1: Construct the next-state table for the encoder of Figure 7.2.1, and verify the trellis of Figure 7.2.2.

7.2.2: Draw the Ungerboeck encoder for the parity-check polynomials

$$H^{(0)} = 1 + x^3,$$
$$H^{(1)} = x^2,$$
$$H^{(2)} = x.$$

7.2.3: Give the code rate and draw the Ungerboeck encoder defined by the parity-check polynomials

$$H^{(0)} = 1 + x^3 + x^4,$$
$$H^{(1)} = x^2,$$
$$H^{(2)} = 0.$$

7.2.4: Give the parity-check polynomials for the following octal representations:
   a) $H^{(0)} = 23$, $H^{(1)} = 04$, $H^{(2)} = 16$
   b) $H^{(0)} = 103$, $H^{(1)} = 030$, $H^{(2)} = 066$.

7.3.1: Perform set partitioning on the 16-QAM signal constellation of Figure 7.1.5.

7.3.2: Perform set partitioning on the 8-CROSS QAM signal constellation of Figure 7.1.5.

7.3.3: For the encoder of Figure 7.3.2 operating with an 8PSK signal constellation,
   a) Find the code bit sequence $Y = (y_2\, y_1\, y_0)$ for a source sequence $(m_2\, m_1)$

   $$\{(1\,1),(0\,0),(0\,1)\},$$

   assuming an initial state of $(0,0)$.
   b) For the code-bit sequence of (a), draw the corresponding I–Q sequence.
   c) Find the I–Q sequence corresponding to an all-zeroes source input, and calculate the Euclidean distance between this sequence and the result of (b) assuming a unit circle for the signal constellation.

7.4.1: Estimate the error rate of the 8-state TCM code for 8PSK in Table 7.4.1 assuming a signal to noise ratio of 8 dB. Compare this result with the result of Example 7.1.1.

CHAPTER 8

Information Theory and Cryptography

8.1 CRYPTOSYSTEMS

8.1.1 Basic Elements of Ciphersystems

Up to this point, we have been occupied with the problems of efficient transmission of information (data compression), reliable use of the communication channel (data translation codes), and reliability of the received information (error-correcting/detecting codes). In this chapter, we will turn our attention to another important aspect of communications, namely, privacy and secrecy of the transmitted information. This topic is the subject of cryptography.

The encrypting of information is an activity far older than information theory and, in fact, constitutes a separate and specialized field of study. However, there are some interesting things that information theory has to say about this subject, and with the rise of the internet as a major vehicle of information exchange in the past few years, cryptography has begun to move out of the shadows of military and intelligence applications and into the mainstream of communications theory in general. Although we can not hope to cover all aspects of cryptography in this little book (or even every *important* aspect of it), we can and should look at its basic elements.

We will be dealing with the fundamentals of cryptosystems and, in particular, with that class of cryptosystems known as *ciphersystems*. The basic problem we are concerned with is easy to describe. Suppose that a sender, whom we shall call Alice, wishes to transmit a message to a receiver, whom we shall call Bob. Let us further suppose that Alice and Bob wish the contents of this message to remain secret even if an eavesdropper (Eve) were to somehow obtain a copy of the transmitted message. How can this be accomplished?

The problem facing Alice and Bob has two aspects. First, Bob must be able to read the message. Second, both Alice and Bob must be reasonably certain that Eve cannot read the message. We call the original message that Alice sends the *plaintext* of the message. Let us suppose the plaintext has n characters drawn from an alphabet X. We represent the plaintext message as a vector of n symbols

$$\bar{x} = (x_1 \, x_2 \cdots x_n), \qquad x_i \in X. \qquad 8.1.1$$

For convenience, we will assume that the x_i are letters taken from the English alphabet.

Before transmitting the message, Alice *encrypts* the message. This is done by using a rule which maps $\bar{x}$ into a new string of characters called a *ciphertext*

$$\bar{y} = (y_1 \, y_2 \cdots y_m), \qquad y \in Y. \qquad 8.1.2$$

Often (but not always), the ciphertext alphabet Y and the plaintext alphabet X are the same. The number of characters in the ciphertext may or may not be the same as in the plaintext. The mapping from $\bar{x}$ to $\bar{y}$ is based on an *encryption rule,* which we may write as

$$\bar{y} = E_k(\bar{x}). \qquad 8.1.3$$

The subscript k denotes a particular rule, called a *key,* employed by the ciphersystem to determine the final ciphertext. We shall say more about this key in a little while.

The ciphertext message is transmitted to Bob. In order to read the original plaintext, Bob must employ another rule, called a *decryption rule,* to recover the original message. This rule can be looked at as an inverse mapping

$$\bar{x} = D_k(\bar{y}). \qquad 8.1.4$$

This decryption rule is clearly the inverse of the mapping defined by the encryption rule. In order for this to work, Alice and Bob must agree in advance as to what E_k and D_k are to be and, as we shall soon see, must also agree on the selection of the key k.

Now, the whole reason for this is to prevent Eve from being able to read the plaintext message, given the ciphertext message. Therefore, it is in the best interests of Alice and Bob to make it as hard as possible for Eve to figure out the decryption rule. However, the task of decrypting $\bar{y}$ must not be made too difficult for Bob to be able to carry out economically. What is needed is something that is easy for Bob to do, but very difficult for Eve to do. The most common approach to solving this problem is to make E_k a rule that is capable of producing any one of a great many possible ciphertexts. The key k is a parameter, chosen in secret by Alice and Bob, that determines which one of the many possible ciphertexts will actually be produced. In this way, even if Eve knows the general structure of the encryption rule, she must still be able to figure out which key has been used before she will be able to "break" the ciphertext and recover the original plaintext. In other words, the encryption and decryption rules are *fixed,* and the key supplies the randomness needed so that the ciphertext contains the "spurious" information needed to thwart Eve's efforts.

Good encryption rules (i.e., ciphersystems that are hard to "break") oftentimes require quite a bit of work to find. Once you have one, you'd probably like to be able to use it for a long time. Common sense, then, tells us that we should like to have a great many possible keys, each of which specifies one of K possible encryptions. If the size of the keyset, K, is too small, Eve could decipher the message simply by trying all of the possible keys. This does, of course, assume that Eve already knows the general encryption algorithm. However, this is usually an assumption we must make. After all, if good encryption methods are hard to find, that also implies that there are not a great many of them. If Eve is in the business of reading your mail, you should assume that she knows most or even all of the encryption rules that have been invented.

This raises some fundamental questions we must consider. First, is it possible to come up with a ciphersystem that is absolutely unbreakable? Second, how many possible keys are required to prevent Eve from being able, in a practical amount of time, to find the correct key? Third, is a large number of keys a sufficient condition to ensure that Eve cannot practically accomplish breaking the ciphersystem? Fourth, how many times can a given key be reused before the secrecy of the system is compromised? These are the basic issues we will deal with in this chapter.

8.1.2 Some Simple Ciphersystems

To appreciate the nature of the encryption problem, we will find it useful to have some example ciphersystems to look at. In this section, we will look at four simple systems, each of which has been around for a long time. These examples do not exhaust all of the different ciphersystems that have been invented. However, they are interesting not only as examples, but also, because some much more sophisticated ciphersystems employ one or more of them in creative ways to obtain much more formidable ciphersystems.

EXAMPLE 8.1.1 The Shift Cipher

The shift cipher is one of the oldest and simplest cryptosystems. It was the system used by Julius Caesar more than 2000 years ago. Suppose we wish to send a plaintext in English. For simplicity, we will assume that we shall ignore the difference between upper- and lowercase letters and that we will omit all spaces and punctuation marks from the text. (As it happens, spaces and punctuation marks actually make it easier for Eve to break the code.) We assign each of the 26 letters of the alphabet to a unique integer from 0 to 25 as shown in Table 8.1.

In this and other examples, we will follow the common practice of using lowercase letters to represent plaintext and uppercase letters to represent the equivalent ciphertext. (This is done merely for explanation purposes here in the text and is not part of the actual ciphersystem itself). The encryption rule for the shift cipher is to simply replace each plaintext character in the message by a ciphertext character given by

$$y_i = (x_i + k) \bmod 26 = E_k(x_i).$$

Decryption is carried out as

$$x_i = (y_i - k) \bmod 26 = D_k(y_i).$$

TABLE 8.1: Alpha Numeric Assignments

letter	number	letter	number	letter	number
a	0	b	1	c	2
d	3	e	4	f	5
g	6	h	7	i	8
j	9	k	10	l	11
m	12	n	13	o	14
p	15	q	16	r	17
s	18	t	19	u	20
v	21	w	22	x	23
y	24	z	25		

The key, k, is an integer chosen one time at random, by Alice and Bob, in the range from 0 to 25. Since it is obvious that $k = 0$ is not an excellent choice, we have, for all practical purposes, 25 possible keys.

Suppose Alice wishes to send the message "my dog has fleas", and let's assume that she and Bob have chosen $k = 16$ as the key. We represent the plaintext message

$$\bar{x} = \text{mydoghasfleas}$$

using the equivalent sequence of numbers

$$12 \quad 24 \quad 3 \quad 14 \quad 6 \quad 7 \quad 0 \quad 18 \quad 5 \quad 11 \quad 4 \quad 0 \quad 18.$$

Applying the encryption rule, we add 16 to this sequence, modulo 26, and get

$$2 \quad 14 \quad 19 \quad 4 \quad 22 \quad 23 \quad 16 \quad 8 \quad 21 \quad 1 \quad 20 \quad 16 \quad 8.$$

Again using Table 8.1, we get the ciphertext

$$\bar{y} = \text{COTEWXQIVBUQI}.$$

One obvious disadvantage of the shift cipher is the small keyset. With only 25 keys, this cipher is easily broken simply by an exhaustive search. Eve merely needs to try all of the possible keys, one after another, until plaintext messages that "make sense" are found. In this example, there is only one such "sensible" plaintext message, and it is, of course, "my dog has fleas".

The analysis we have just described Eve carrying out is called *cryptoanalysis*. In particular, when Eve analyzes the ciphersystem using only the intercepted ciphertext, it is called a "ciphertext-only attack." The simple shift cipher is pathetically vulnerable to such an attack.

EXAMPLE 8.1.2: The Vigenère cipher

The Vigenère cipher is a slightly more powerful ciphersystem. It was named after Blaise Vigenère, who invented it in the 16th century. The Vigenère cipher uses a *keyword* to shift different letters in the plaintext by different amounts. The algorithm is otherwise identical to the shift cipher. Suppose the keyword has m letters

$$k = k_1 k_2 \cdots k_m.$$

Let us again assume the 26-letter alphabet in Table 8.1. The encryption rule is

$$E_k(\bar{x}) = (x_1 + k_1, x_2 + k_2 \cdots x_m + k_m) \bmod 26,$$

while decryption is given by

$$D_k(\bar{y}) = (y_1 - k_1, y_2 - k_2 \cdots y_m - k_m) \bmod 26.$$

If the keyword is shorter than the length of the plaintext message, it is simply repeated. For instance, suppose we have the same plaintext message as in Example 8.1.1 and the keyword is TRICKEM. This corresponds to the sequence 19 17 8 2 10 4 12. The encryption is carried out as

$$\text{TRICKEM} + \text{mydogha} \,, \quad \text{TRICKE} + \text{sfleas}$$

to yield

$$5 \quad 15 \quad 11 \quad 16 \quad 16 \quad 11 \quad 12 \quad 11 \quad 22 \quad 19 \quad 6 \quad 10 \quad 22,$$

or
$$\bar{y} = \text{FPLQQLMLWTGKW}.$$

With a 26-letter alphabet and an m-letter keyword, the number of possible keys is 26^m. For the preceding example with $m = 7$, this gives us 8,031,810,176 possible keys. This is large enough to make an exhaustive search by hand impossible. However, if Eve has a computer, this keyset is not too large to be searched by the machine. Furthermore, we shall see later that Eve has approaches other than the exhaustive key search she can employ.

EXAMPLE 8.1.3: The permutation cipher

In the permutation cipher, the plaintext characters in the message are unchanged, but their order is scrambled. Mathematically, this can be represented by using a permutation matrix, k_π. Let us assume that the plaintext is divided into blocks of m characters per block. k_π is then an m by m matrix in which every row and every column has precisely one "1". All other entries are "0". Representing the block of plaintext and the block of ciphertext as vectors, the encryption algorithm is

$$\bar{y} = E_{k_\pi}(\bar{x}) = \bar{x} k_\pi$$

and the decryption algorithm is

$$\bar{x} = D_{k_\pi}(\bar{y}) = \bar{y} k_\pi^{-1}.$$

To illustrate this, suppose $m = 6$ and

$$k_\pi = \begin{bmatrix} 0 & 0 & 1 & 0 & 0 & 0 \\ 0 & 0 & 0 & 0 & 0 & 1 \\ 1 & 0 & 0 & 0 & 0 & 0 \\ 0 & 0 & 0 & 0 & 1 & 0 \\ 0 & 1 & 0 & 0 & 0 & 0 \\ 0 & 0 & 0 & 1 & 0 & 0 \end{bmatrix}, \quad k_\pi^{-1} = \begin{bmatrix} 0 & 0 & 1 & 0 & 0 & 0 \\ 0 & 0 & 0 & 0 & 1 & 0 \\ 1 & 0 & 0 & 0 & 0 & 0 \\ 0 & 0 & 0 & 0 & 0 & 1 \\ 0 & 0 & 0 & 1 & 0 & 0 \\ 0 & 1 & 0 & 0 & 0 & 0 \end{bmatrix}.$$

Let our plaintext be "my dog has fleas" once again. This has 13 characters and will require three blocks to encrypt. This is done by adding a "pad" to the end of the message to "fill in" the last five "blank" positions. As an example, the padded plaintext message could be

mydogh asflea sdummy.

The resulting ciphertext becomes

$$(\text{mydogh}) \cdot k_\pi = (\text{DGMHOY}),$$
$$(\text{asflea}) \cdot k_\pi = (\text{FEAALS}),$$
$$(\text{sdummy}) \cdot k_\pi = (\text{UMSYMD}),$$

so

$$\bar{y} = \text{DGMHOYFEAALSUMSYMD}.$$

Decryption follows the same procedure, except k_π^{-1} is used in place of k_π.

The permutation cipher is easily implemented in hardware. In most modern ciphersystems, the plaintext message is represented as a sequence of binary digits, and blocking is performed every m bits. The permutation matrix is then simply a matter of wiring the incoming bit positions to specific output bit positions. Such a hardware implementation is commonly called a "P-box."

The preceding examples have all been examples of *linear* transformations from $\bar{x}$ to $\bar{y}$. That is, if

$$\bar{y}_1 = E_k(\bar{x}_1) \quad \text{and} \quad \bar{y}_2 = E_k(\bar{x}_2),$$

then

$$\bar{y}_3 = E_k(\bar{x}_1 + \bar{x}_2) = \bar{y}_1 + \bar{y}_2,$$

with the addition operation being defined as addition modulo 26. Linear ciphers are generally vulnerable to certain "attacks" by Eve (which we will say more about in the next section). The security of a ciphersystem can often be greatly increased by using a *nonlinear* transformation. We illustrate such a transformation in the following example.

EXAMPLE 8.1.4: The substitution cipher

In the substitution cipher, the ciphertext alphabet Y is a scrambled version of the plaintext alphabet X. In other words, each letter in X is replaced by some other letter in X. For instance, we could make the following assignments.

a = Q   b = W   c = E   d = R   e = T   f = Y   g = U   h = I   i = O   j = P
k = A   l = S   m = D   n = F   o = G   p = H   q = J   r = K   s = L   t = Z
u = X   v = C   w = V   x = B   y = N   z = M

With this substitution, our plaintext "my dog has fleas" becomes

$$\bar{y} = \text{DNRGUIQLYSTQL}.$$

It can be easily verified that this cipher is a nonlinear transformation by using Table 8.1 to convert the characters to numbers modulo 26 and adding two messages together such as

$$
\begin{array}{rccc}
& \text{mydog} & + \quad \text{mycat} & = \quad \text{ywfoz} \\
\Rightarrow & \text{DNRGUI} & \text{DNEQZ} & \text{NVYGM} \; (= \bar{y}_3), \\
\text{but} & \text{DNRGUI} & + \quad \text{DNEQZ} & = \quad \text{GAVOZ},
\end{array}
$$

and therefore, $\bar{y}_3 \neq \bar{y}_1 + \bar{y}_2$. The key for a substitution cipher is merely a table, like the one shown above, that defines the letter substitutions. If X has M letters in the alphabet, there are $M!$ keys. In the case of our 26 letter alphabet, this is $26! > 10^{26}$ keys.

In hardware implementations of the substitution cipher, it is common to convert the plaintext into a sequence of binary digits first. These digits are then blocked into groups of m bits, which are then encrypted using the substitution cipher. This is illustrated by the following example:

EXAMPLE 8.1.5: Binary-coded substitution cipher

Let us assume that our substitution cipher operates on blocks of three bits each. As our substitution, suppose we have

$$X: \quad 000 \quad 001 \quad 010 \quad 011 \quad 100 \quad 101 \quad 110 \quad 111$$
$$y: \quad 011 \quad 111 \quad 000 \quad 110 \quad 010 \quad 100 \quad 101 \quad 001$$

For the plaintext message "my dog has fleas", the first three letters are converted to binary as

$$m \ = \ 12 \ = \ 01100$$
$$y \ = \ 24 \ = \ 11000$$
$$d \ = \ 3 \ = \ 00011.$$

The substitution cipher therefore produces, as the first 15 bits of the ciphertext,

$$110 \quad 111 \quad 010 \quad 011 \quad 110$$

after "blocking" the binary plaintext into blocks of three bits each. Note how, in this example, individual letters of the original plaintext get "split" and distributed over different blocks. For example, "m" has three binary digits in the first block. The remaining two bits of "m" are combined with the first bit of "y" to form the second block. Therefore, in addition to "scrambling" the letters of the original English plaintext, this substitution cipher is capable of "splitting" a particular English character among different binary blocks before "scrambling."

In hardware implementations of the substitution cipher, the hardware that makes the substitution is usually called an "S-box."

8.2 ATTACKS ON CRYPTOSYSTEMS

The design of any cryptosystem faces the dual challenges of making the encryption and decryption operations easy for Alice and Bob, while, at the same time, making decryption as difficult and expensive for Eve as possible. Successful cryptosystems are usually classified as being either *unconditionally secure* (i.e., unbreakable) or else *computationally secure* (i.e., the amount of computation required to break the ciphersystem is such that the system could not be broken in fewer than some z number of years). In order to appreciate the issues that confront achievement of these goals, let us look at the methods by which a ciphersystem can be attacked. In analyzing attack methods, we assume that Eve knows what ciphersystem is being used, but does not possess the encryption key. (History has shown that this assumption is often justified.)

Attacks on ciphersystems are typically classified into four distinct levels. In order from least powerful to most powerful, these attacks are:

1. *Ciphertext-only attack*
   Eve possesses a string of ciphertext $\bar{y}$, some knowledge of the system being used, and knows the language used in the plaintext;
2. *Known-plaintext attack*
   Eve is in possession of a known plaintext and its associated ciphertext (in addition to the knowledge in (1) above);

3. *Chosen-plaintext attack*
   Eve has somehow obtained temporary access to the encryption machinery and has constructed a ciphertext $\bar{y}$ from a chosen plaintext $\bar{x}$ (this is often accomplished by tricking Alice into sending the chosen plaintext to Bob; this method was used by the Americans in World War II to learn that Midway Island was the target of a planned invasion by the Japanese; as a result, the U.S. Navy was able to ambush the Japanese task force on their way to Midway, resulting in the sinking of four Japanese aircraft carriers);

4. *Chosen-ciphertext attack*
   Eve has obtained temporary access to the decryption machinery and has reconstructed the plaintext $\bar{x}$ corresponding to the chosen ciphertext (this attack method is always possible in *public-key* cryptosystems, which thus guarantees that no public key cryptosystem can ever be unconditionally secure).

Generally, a cryptosystem is not considered secure unless it is secure from both a ciphertext-only attack and a known-plaintext attack.

Regardless of the attack level employed, Eve's task is always to determine the key that was used. Once Eve is in possession of the key, the cryptosystem is broken, and she can read Bob's mail as easily as he can. Therefore, we must consider the following questions. How secure is the system against cryptanalysis when Eve has unlimited time and computing resources? Does a cipher have a unique solution, and, if not, how many reasonable solutions does it have? How much ciphertext must be intercepted before the solution becomes unique? Are there cryptosystems that *never* yield a unique solution? Are there cryptosystems for which no information whatsoever is given to Eve, regardless of the amount of ciphertext intercepted?

Shannon considered these questions in his 1949 paper on secrecy systems. The theory presented in that paper, as well as much of the terminology, is widely considered to be the foundation of the mathematical theory of cryptography, which, until that time, was an esoteric art shrouded in such great secrecy that few people knew whether or not their cryptosystems were actually secure[1]. Not surprisingly, Shannon applied his new information theory to the questions outlined above. Most of the rest of this chapter is concerned with the results obtained from this theory.

8.3 PERFECT SECRECY

A cryptosystem is said to have *perfect secrecy* if knowledge of the ciphertext conveys no information about the plaintext. Under what conditions will this be true? Let us assume that the set of possible messages is

$$X = \{\bar{x}_1, \bar{x}_2, \cdots, \bar{x}_n\}. \qquad 8.3.1$$

We will further assume that the probability of message $\bar{x}_i$ being sent is p_i.

---

[1]The famous German commander, Erwin Rommel, sincerely believed that the German military codes were unbreakable ("a mathematical impossibility"); he never knew that the allied bombing attacks that sank most of his supplies en route to the Afrika Korps were the direct result of the allies breaking the German code.

The ciphertext is produced according to some encryption rule for which we have a finite set of possible keys

$$K = \{k_1, k_2, \cdots, k_m\}, \qquad 8.3.2$$

and we assume that the key is selected independently of the plaintext message. The resulting set of possible ciphertexts is then

$$Y = \{\bar{y} = E_k(\bar{x}) \forall k \in K, \forall \bar{x} \in X\}. \qquad 8.3.3$$

Note that we will have a total number of possible ciphertexts $|Y| \leq n \cdot m$, since it may be possible for the same $\bar{y}$ to be produced by different combinations of keys and plaintexts.

Now, perfect secrecy means that knowledge of $\bar{y}$ does not reduce our uncertainty of which has been transmitted. In terms of entropy, this means that

$$H(X|Y) = H(X). \qquad 8.3.4$$

The conditional entropy, which is also known as the *equivocation,* was defined in Chapter 1 as

$$H(X|Y) \equiv \sum_{\bar{y}_j \in Y} p_j \sum_{\bar{x}_i \in X} p_{i|j} \log_2(1/p_{i|j}), \qquad 8.3.5$$

where $p_{i|j} = \Pr(\bar{x}_i|\bar{y}_j)$. Applying the perfect secrecy condition of Equation 8.3.4 to Equation 8.3.5, we therefore have the equivalent condition

$$\Pr(\bar{x}_i|\bar{y}_j) = \Pr(\bar{x}_i), \qquad 8.3.6$$

for every possible plaintext message and every possible ciphertext.

In addition, since we can use the chain rule for entropy to write

$$H(X, Y) = H(X) + H(Y|X) = H(Y) + H(X|Y),$$

we can apply Equation 8.3.4 and obtain the result

$$H(Y|X) = H(Y), \qquad 8.3.7$$

which requires that

$$\Pr(\bar{y}_j|\bar{x}_i) = \Pr(\bar{y}_j), \qquad 8.3.8$$

for all possible plaintexts and ciphertexts. This result tells us that the probability of encrypting any particular $\bar{x}$ to obtain $\bar{y}$ must be the same for all possible $\bar{x}$. Consequently, there must be as many possible ciphertexts as there are plaintexts. In turn, this means that for all possible plaintext/ciphertext pairs $(\bar{x}, \bar{y})$, there must be some key k such that $\bar{y} = E_k(\bar{x})$. Furthermore, we must have the condition

$$\bar{y} = E_{k_1}(\bar{x}) = E_{k_2}(\bar{x}) \Rightarrow k_1 = k_2$$

if the number of possible keys equals the number of possible plaintexts.

If we have the condition $|K| = |X|$, then Equation 8.3.8 also requires that

$$\Pr(\bar{y}) = \Pr(\bar{y}|\bar{x}) = \Pr(E_k(\bar{x})|\bar{x}) = \Pr(k).$$

Since this condition must hold for all $(\bar{x}, \bar{y})$, and since we are assuming that each key is unique for a given encryption, this means that all keys must be used with *equal* probability, i.e., Pr($\bar{y}$). Since the number of keys is $|K|$, this means that Pr(k) = 1/$|K|$ for every key.

This result was first obtained by Shannon in his 1949 paper. We can summarize what we have done in this section as a theorem:

Theorem 8.3.1: A cryptosystem with $|K| = |X| = |Y|$ provides perfect secrecy if and only if every key is used with equal probability, and for every $\bar{x} \in X$ and $\bar{y} \in Y$ there is a unique $k \in K$ such that $E_k(\bar{x}) = \bar{y}$.

An obvious corollary to this theorem is: No cryptosystem in which the number of possible plaintexts is unlimited can be perfectly secret if the number of keys is finite.

EXAMPLE 8.3.1: The one-time pad

In the shift cipher of Example 8.1.1, we chose a single key, k, and used this key for encrypting every character in the plaintext. Suppose, instead, we pick a new key at random for each new character of plaintext. Our ciphertext is then given by

$$\bar{y} = (y_1, y_2, \cdots, y_n), \quad y_i = x_i + k_i \mod 26, \quad k_i \in [0, \cdots, 25].$$

This cryptosystem is called a "one-time pad." With n characters of plaintext, the number of possible keys is 26^n. It is obvious that we can satisfy the conditions of Theorem 8.3.1 using this system and, therefore, that the one-time pad has perfect secrecy.

When the plaintext, ciphertext, and keys are represented as binary vectors, and the addition operation is defined modulo 2, the one-time pad is easily implementable with digital hardware. The one-time pad was invented in 1917 by Gilbert Vernam. Mr. Vernam had claimed that this cryptosystem was unbreakable (a fact that was not rigorously proven until Shannon's 1949 paper), and he had hoped to make a lot of money from his invention. Unfortunately, there is one great practical difficulty with the one-time pad. Since we must chose a new key each time a new $\bar{x}$ is to be encrypted, and since this key must always be known to Bob, Alice must *securely* deliver to Bob a very large set of keys (equal to the number of messages to be sent). This is called the *key-management problem*.

We might ask ourselves what would be wrong with reusing the same (very long) key sequence over and over. That would certainly do much to solve the key management problem. Unfortunately, the one-time pad is quite vulnerable to a known-plaintext attack if the key sequence is reused for several different messages.

EXAMPLE 8.3.2

Eve intercepts the ciphertext GNGLOPSXFIGLXXQUAFBL. A few days later, she learns that this message was "my dog is sick with fleas". She strongly suspects that Alice and Bob are using a one-time pad but are reusing the key sequence. A week later, she intercepts the following message

NWDKSZFTOZDTUPNPEPXWWG.

Attempt to decode this message.

Solution: Using the known plaintext and the first message, we use Table 8.1 and solve for the following key sequence:

$$20, 15, 3, 23, 8, 7, 0, 5, 23, 6, 22, 15, 15, 4, 9, 15, 15, 1, 1, 19.$$

Applying this sequence to the new message, we have

$$\text{thanksforthefleapowd??}$$

where the last two "?" marks result from the fact that the new message is two characters longer than the first message. Looking at the context of the messages, "thanks for the flea powd??", it's a reasonable guess that the last word was "powder", and therefore, the next two keys in the sequence are 18 and 15.

8.4 LANGUAGE ENTROPY AND SUCCESSFUL CIPHERTEXT ATTACKS

8.4.1 The Key-Equivocation Theorem

In some situations, the need for secrecy is great enough to justify the use of the one-time pad to obtain perfect secrecy. However, in commercial and business applications it is often impractical or even impossible to solve the key-management problem. Imagine, for instance, the magnitude of this problem in, say, commercial banking or internet traffic. If the number of users is large, the practical aspects of securely distributing and managing the keys is formidable. We must, therefore, look at what can be achieved using smaller key sets and with reusing keys.

Now, recall that the basic goal of the cryptanalyst (Eve) is to find the key. Let us take a look at what the ciphertext reveals about the key. We will let X, K, and Y be the sets of possible plaintext, keys, and ciphertext, respectively. Using the chain rule for joint entropy, we have

$$H(X, Y, K) = H(Y) + H(K|Y) + H(X|Y, K)$$

and

$$H(X, Y, K) = H(K) + H(X|K) + H(Y|X, K).$$

Since X is completely determined given Y and K, and Y is completely determined given X and K, the last term in each of these two expressions is zero. Furthermore, since K is chosen independently of X, we have

$$H(X|K) = H(X).$$

Making this substitution and equating the expressions above, we get

$$H(K|Y) = H(K) + H(X) - H(Y). \qquad 8.4.1$$

Equation 8.4.1 is called the key-equivocation theorem. The key equivocation, $H(K|Y)$, is a measure of how much is revealed by the ciphertext about the key. From Eve's point of view, $H(K|Y) = 0$ means that the code has been broken. What, then, does $H(K|Y) > 0$ mean? Put simply, this means that, given an intercepted $\bar{y}$, there are two or more keys k that might have been used to produce the ciphertext. This, in turn, means there are two or more plaintexts that "make sense" given $\bar{y}$. The "extra" possible keys are called *spurious keys*.

8.4.2 Spurious Keys and Key Equivocation

Suppose we have intercepted a certain amount of ciphertext and are engaged in a ciphertext-only attack. How many spurious keys should we expect to have, given a certain amount of intercepted ciphertext?

Let $K(\bar{y})$ be the set of keys for which there are plaintexts which "make sense" for the encryption $\bar{y} = E_k(\bar{x})$. We know that one of the keys in $K(\bar{y})$ is the "real" key. The others are spurious. Therefore, given $\bar{y}$ the number of spurious keys is $|K(\bar{y})| - 1$. The expected number of spurious keys taken over Y is then given by the definition of the expected value as

$$s = \sum_{\bar{y} \in Y} \Pr(\bar{y}) \cdot [|K(\bar{y})| - 1] = \sum_{\bar{y} \in Y} \Pr(\bar{y})|K(\bar{y})| - 1. \qquad 8.4.2$$

EXAMPLE 8.4.1

Assume that Eve knows that Alice and Bob are using the English language for their plaintext and the shift cipher of Example 8.1.1. Given the ciphertext $\bar{y} =$ WNAJW, find the set of spurious keys.

Solution: We convert $\bar{y}$ into the numerical string 22, 13, 0, 9, 22 by using Table 8.1. We now apply the 25 nonzero keys of the shift cipher to obtain the following table:

Key	$D_k(\bar{y})$
1	v m z i v
2	u l y h u
3	t k x g t
4	s j w f s
5	r i v e r *
6	g h u d q
7	p g t c p
8	o f s b o
9	n e r a n
10	m d q z m
11	l c p y l
12	k b o x k
13	j a n w j
14	i z m v i
15	h y l u h
16	g x k t g
17	f w j s f
18	e v i r e
19	d u h q d
20	c t g p c
21	b s f o b
22	a r e n a *
23	z q d m z
24	y p c l y
25	x o b k x

From this, we see that "river" and "arena" are possible decryptions of the ciphertext. The possible keys are therefore 5 and 22.

We can relate Equation 8.4.2 to key equivocation in the following manner. Let $H(K|\bar{y})$ be the uncertainty in the key given ciphertext $\bar{y}$. We then have

$$H(K|Y) = \sum_{\bar{y} \in Y} \Pr(\bar{y}) H(K|\bar{y}). \qquad 8.4.3$$

Now, it is certainly true that

$$H(K|\bar{y}) \le \log_2 |K(\bar{y})|.$$

Applying this to Equation 8.4.3, we have

$$H(K|Y) \le \sum_{\bar{y} \in Y} \Pr(\bar{y}) \log_2 |K(\bar{y})|.$$

We now take advantage of an important property of logarithms, namely, that,

$$\sum_{\bar{y} \in Y} \Pr(\bar{y}) \log_2 |K(\bar{y})| \le \log_2 \left[\sum_{\bar{y} \in Y} \Pr(\bar{y}) \cdot |K(\bar{y})| \right].$$

This inequality is a consequence of a useful theorem known as Jensen's inequality. Substituting this into the previous expression and using Equation 8.4.2, we have

$$H(K|Y) \le \log_2(s + 1). \qquad 8.4.4$$

8.4.3 Language Redundancy and Unicity Distance

We are nearly ready to put all of this together and answer the question of how secure a cryptosystem that reuses its keys will be. First, however, we need to remind ourselves of the information per symbol in a natural language. In Chapter 2, we introduced the notion of the entropy rate of an n-character message. For a natural language, the "language entropy" is the entropy rate

$$H_L = \lim_{n \to \infty} \frac{H(A_1, A_2, \cdots, A_n)}{n}, \qquad 8.4.5$$

where A is the alphabet of that language. We may define the *redundancy* of the language as

$$R_L = 1 - \frac{H_L}{\log_2 |A|}. \qquad 8.4.6$$

For English, various measurements have empirically determined that $1.0 \le H_L \le 1.5$ bits per letter. With 26 characters in the English alphabet, this means that English is roughly 75% redundant!

If we have transmitted n plaintext characters, we may approximate the message entropy as

$$H(X) \approx nH_L = n(1 - R_L) \cdot \log_2 |A|. \qquad 8.4.7$$

Now suppose that Alice and Bob have been reusing one particular key. In this case, $|Y| = |X|$, and we have $H(Y) \le n \log_2 |A|$. Using this and Equation 8.4.7 in Equation 8.4.1, we have

$$H(K|Y) \ge H(K) - nR_L \log_2 |A| = \log_2 |K| - nR_L \log_2 |A|.$$

We now use Inequality 8.4.4 and a little algebra to obtain the result we are after,

$$s \geq \frac{|K|}{|A|^{nR_L}} - 1. \qquad 8.4.8$$

Inequality 8.4.8 relates the expected number of spurious keys to the size of the key set, the size of the plaintext alphabet, the redundancy of the natural language, and the number of intercepted ciphertext characters. *On the average,* Eve will be able to break the cryptosystem when the number of intercepted characters is sufficient to make the average number of spurious keys equal to zero. This number of characters is called the *unicity distance*. Setting $s = 0$ in Inequality 8.4.8 and solving for n, we have

$$n_0 = \frac{\log_2|K|}{R_L \log_2|A|}. \qquad 8.4.9$$

EXAMPLE 8.4.2

The substitution cipher has $|K| = 26!$ possible keys. Assuming that English is 75% redundant, estimate the expected number of spurious keys remaining after intercepting $n = 25$ characters of ciphertext, and find the unicity distance of the substitution cipher.

Solution: Applying inequality 8.4.8, we have

$$\log_{10}(26!) - 25 \cdot (.75) \cdot \log_{10}(26) = 0.07486876,$$

so

$$s \geq 10^{0.7486876} - 1 = 0.188,$$

and

$$n_0 = \frac{\log_{10}(26!)}{R_L \log_{10}(26)} = \frac{26.605619}{.75 \log_{10}(26)} \approx 25,$$

where we have used $\log_2(a)/\log_2(b) = \log_{10}(a)/\log_{10}(b)$. On the average, Eve can break the substitution cipher after only about 25 ciphertext characters have been intercepted.

EXAMPLE 8.4.3

Estimate the unicity distance of the Vigenàre cipher using a keyword length of five characters and assuming a language redundancy of 0.75.

Solution:

$$n_0 = \frac{5 \log_2(26)}{.75 \log_2(26)} = 6.67 \approx 7 \text{ characters.}$$

8.5 COMPUTATIONAL SECURITY

We might have expected that the relatively simple cryptosystems we have looked at so far could be broken without much trouble. However, we might not have expected it to be as easy as the examples at the end of Section 8.4 illustrate it to be! As Example 8.4.2

showed, the simple fact that we have an enormous set of possible keys is no guarantee that our cryptosystem will be hard to break. Equation 8.4.9 shows us the primary important factors affecting the security of the cryptosystem. If our system has perfect secrecy, then $\log|K| \to \infty$ as $n \to \infty$, and we are "safe." However, perfect secrecy comes at the expense of a very formidable problem in key management.

Any cryptosystem that does not have perfect secrecy can, in principle, be broken, given enough time and enough computing resources. A cryptosystem is said to have *computational security* if, even with the best known methods of cryptanalysis, the amount of computation time required to break the system is "unreasonably large." Let us look at some of the factors that contribute to the achievement of computational security.

In Section 8.4, we saw that the unicity distance is a measure of how secure a cryptosystem is against a ciphertext-only attack. The unicity distance is a function of the average number of spurious keys that, for a given ciphertext, result in "meaningful" plaintext. Now, the number of spurious keys is much smaller than the set of possible keys. This means that the great majority of keys result in "meaningless" text, and it is the great predominance of meaningless decryptions that enables Eve to break the system. This predominance, in turn, is a direct result of redundancy in the plaintext language.

In Chapter 1, we saw that redundancy in the representation of messages can be reduced by source compression. Since the entropy rate of a natural language is typically much less than the entropy of that language's alphabet, effective source coding for encryption purposes needs to be based on "block" coding rather than compression of the individual letters themselves. For instance, the entropy rate of English is known to lie between 1.0 and 1.5 bits per letter while the entropy of the English alphabet, taken letter by letter, lies between 3.9 and 4.1 bits per letter. This is a significant difference since merely compressing the letters themselves, without regard to the probabilistic structure introduced by words and phrases, does not do much to reduce the entropy rate of the resulting compressed text.

In principle, a source code for plaintext always exists that can approach H_L arbitrarily closely. (We will look at this in more detail in Chapter 9.) In practice, however, it may be quite difficult to achieve computational security through source coding alone. Shannon discussed some aspects of this in his 1951 paper on the entropy and prediction of printed English, listed in the chapter references. In this paper, he gave entropy estimates for digrams (letters taken two at a time) and trigrams (letters taken three at a time). He found entropies of 3.56 and 3.30 bits per letter, respectively. These results provide us with an indication of how strongly words and phrases influence the entropy rate of a natural language. Shannon found that blocks of 10 or more letters were needed to begin to closely approach the entropy rate of English. This would call for a rather formidable source code!

In any natural language, a word is (to quote Shannon), "a cohesive group of letters with strong statistical influence." If it is the structure of *words* in a language that mainly contributes to the language's entropy rate, might we not try source coding based on words rather than letters (or at least based on, say, the first 100 most commonly occurring words)? For instance, the four most commonly occurring words in English are "the," "of," "and," and "to." Since tables exist for the relative frequency of English words,[2] could we not build a kind of "hybrid" source encoder and achieve a great increase in the computational security of the system?

---

[2]See, for instance, G. Dewey, "Relative Frequency of English Speech Sounds," Harvard University Press, 1923.

In principle, the answer is yes; however, the practicality of such a system depends on the costs and complexity of implementing encoders and decoders for the system. Shannon's 1951 results show that the probability of the nth most frequent word in English is proportional to $1/n$, up to about $n = 8{,}727$. From this relationship, he found the "word entropy" of English to be about 11.8 bits per word. With an average word length of about 4.5 letters per word, this works out to be about 2.62 bits per letter—a big improvement, but it still presents us with a redundancy of about 43%, assuming an entropy rate of 1.5 bits per letter for English. This is not much of an improvement for the costs involved in a "word compressor" for English. Therefore, while compression can help to improve computational security, it is apparent that source coding alone is not enough; we need to look at some other techniques.

8.6 DIFFUSION AND CONFUSION

Cryptoanalysis often involves a great deal of "trial-and-error" work. We have seen that a large key set is a necessary ingredient for frustrating the efforts of the cryptanalyst. However, as we also have seen, a large key set in and of itself does not provide sufficient protection. The inherent structure of natural languages provides Eve with two important analysis tools. First, trials can be designed to progress from more probable to less probable hypotheses, based on the known statistical properties of the language. The relative frequencies of individual letters, digrams, trigrams, and specific words in the language are usually known to the cryptanalyst. These measures provide a basis for hypothesis construction during cryptoanalysis. Second, each "trial" in an effective ciphertext attack can be designed to eliminate a large *group* of possible keys, rather than simply eliminating one key at a time.

Consider, for instance, the Vigenère cipher with a very large keyword (say, 20 characters). The key set contains $26^{20} \approx 1.99 \cdot 10^{28}$ possible keys. However, the average length of an English word is about 4.5 letters. Therefore, a search for keyword *fragments,* say, 5 to 7 letters at a time, can reveal the most commonly occurring words within the ciphertext. Note that a fragment of 5 letters involves only about 10^7 key fragments. Such a "divide-and-conquer" approach can be carried out rather quickly by computer to rapidly narrow down the set of possible keyword fragments. An attack aimed at finding the most probable words within the ciphertext is, in fact, one of the cryptanalyst's most powerful weapons. There are few simple, classical cryptosystems that can resist the probable-word attack.

In his classic 1949 paper (see chapter references), Shannon suggested two methods for frustrating statistical ciphertext attacks: *diffusion* and *confusion*. Both methods are aimed at greatly increasing the amount of material that must be intercepted and analyzed in order to expose the underlying structure imposed on the plaintext by natural languages.

In diffusion, the statistical structure inherent in the plaintext is "diffused" into a statistical structure involving long *combinations* of letters in the text. One example, given by Shannon, is the "averaging" method. Suppose $\bar{x} = (x_1, x_2, \cdots x_n)$. From $\bar{x}$, we generate a new n-tuple $\bar{z}$ having elements

$$z_i = \sum_{\ell=0}^{\lambda} x_{i+\ell} \pmod{26}, \qquad 8.6.1$$

where λ is a parameter selected by Alice and Bob. The new "averaged" text string $\bar{z}$ is then encrypted to obtain the ciphertext.

In carrying out Equation 8.6.1, the final terms where $i + \ell > n$ may be obtained by padding the message with λ additional dummy characters. These pad characters are not encrypted and sent, but are known to Bob. "Unscrambling" $\bar{z}$ back into $\bar{x}$ can be carried out by Bob after decryption, starting with x_n and working backwards to x_1.

It can be shown that the redundancy in $\bar{z}$ is the same as that in $\bar{x}$, but that the "letter frequencies" in $\bar{z}$ are more nearly equal to each other than they are in $\bar{x}$. Thus, the symbol probabilities become "diffused" more uniformly throughout the plaintext. The method of diffusion is a counterattack against Eve's use of tables of known letter, digram, and trigram frequencies for the plaintext language.

The method of confusion is aimed at making the relation between the statistics of $\bar{y}$ and the description of the key set very complex and involved. Confusion is carried out by using a set of *nonlinear* enciphering equations and multiple key elements. We can look at this in general as a cryptosystem of the form

$$y_1 = f_1(x_1, x_2, \cdots, x_n, k_1, k_2, \cdots, k_m),$$
$$y_2 = f_2(x_1, x_2, \cdots, x_n, k_1, k_2, \cdots, k_m),$$
$$\vdots$$

It is best to have each f_i include several x_i and to have these x_i not be adjacent in the plaintext sequence. The general idea behind this strategy is to decrease the correlation between the x_i by selecting letters from different words within the plaintext.

The benefit of having nonlinear equations for the f_i is that such equations can produce multiple solutions for the k terms (given the y_i and the hypothetical plaintext). These solutions help to increase the number of spurious keys, making Eve's task involve much more work. It is especially important to have each f_i involve *multiple* key characters, in order to prevent Eve from solving the "simple" f_i and then attacking the others by substitution. (Remember, we must assume that Eve knows the encrypting functions f_i.)

The method of confusion is of great importance in combating the probable-word attack. While diffusion will complicate Eve's task, cryptosystems using only diffusion are still susceptible to the probable-word attack, especially when this attack is used as part of a known-plaintext attack. Confusion, on the other hand, is a powerful countermeasure even against probable-word cryptanalysis.

8.7 PRODUCT CIPHER SYSTEMS

A very common approach to the implementation of confusion and diffusion in a cryptosystem involves the use of what are known as product cipher systems. The basic approach was suggested by Shannon in his 1949 paper. The widely used Data Encryption Standard (DES) in the United States is an example of such a system. A detailed description of DES is given in several of the references at the end of the chapter. For our purposes here, a detailed discussion of DES is somewhat out of place, since such a discussion would have many tiny details, but relatively little overall pedagogical value. However, we will discuss the basics of product cipher systems; DES essentially implements these basics on a rather grand scale.

8.7.1 Commuting, Noncommuting, and Idempotent Product Ciphers

In its most basic form, a product cipher is a "cipher of a cipher." Suppose that plaintext $\bar{x}$ is first encrypted using cryptosystem $S_1(\bar{x}) = E^{(1)}_{k_1}(\bar{x})$. We then take the ciphertext resulting from this operation and encrypt it a second time, using cryptosystem $S_2(\bar{x}) = E^{(2)}_{k_2}(\bar{x})$. The overall resulting ciphertext is then given by

$$\bar{y} = E^{(2)}_{k_2}[E^{(1)}_{k_1}(\bar{x})] \equiv S_2 S_1(\bar{x}). \qquad 8.7.1$$

The system of Equation 8.7.1 is called a product cipher. This basic idea can be extended to as many "layers" of encryption as we want, i.e., $S_3 S_2 S_1(\bar{x})$, etc. Each cipher in the product can (and often does) have its own key, which is important for achieving confusion in the final ciphertext.

Shannon introduced the idea of the product cipher in his 1949 paper as a convenient way of implementing confusion or diffusion (or both) in a cryptosystem. In general, the *order* in which the individual cryptosystems are applied is important. A product cipher composed of two layers is said to be noncommuting if

$$S_2 S_1(\bar{x}) \ne S_1 S_2(\bar{x}).$$

Otherwise, the product cipher is said to commute.

Noncommuting systems are of great importance in obtaining what Shannon called a "good mixing transformation," a term which means that the system incorporates a significant level of confusion. To see why this is so, let us look at a simple example.

EXAMPLE 8.7.1

Let S_1 be the shift cipher. Let us implement a two-term product cipher using S_1 with a key of 5 for the inner cryptosystem and S_1 with a key of 23 for the outer cryptosystem. Let $\bar{x} =$ "n" $= 13$ (from Table 8.1). Then

$$S_1 S_1(\bar{x}) = [[(13 + 5) \bmod 26] + 23] \bmod 26 = 15 = \text{"P"}.$$

Note, however, that this is equivalent to encrypting $\bar{x}$ one time using S_1 with a key of 2:

$$13 + 2 \bmod 26 = 15 = \text{"P"}.$$

Therefore, the product cipher in this example really provides no additional security, since it is equivalent to a one-layer encryption using a different key.

A simple cryptosystem S_1 is said to be *idempotent* if $S_1 S_1(\bar{x}) = S_1(\bar{x})$ for some set of keys k_1, k_2, k_3. Most of the simple cryptosystems, including all of the ones in this chapter, are idempotent. Now, it would be a very desirable thing if we could use *simple* cryptosystems to construct a powerful product cipher cryptosystem. However, to do so, we would need to "build" our product cipher by iterating these simple systems several times. If our simple cryptosystems are idempotent, there is no improvement in security obtained by iterating them.

Suppose S_1 and S_2 are both simple, idempotent cryptosystems that commute. In this case, the product cipher $S_2 S_1$ will *also* be idempotent. To see this, we need only look at the following algebra:

$$(S_2S_1)(S_2S_1) = S_2(S_1S_2)S_1 \quad \text{(associative property)}$$
$$= S_2(S_2S_1)S_1 \quad \text{(commutative property)}$$
$$= (S_2S_2)(S_1S_1) \quad \text{(associative property)}$$
$$= S_2S_1 \quad \text{(idempotency of the two systems)}.$$

The critical step is the second one, in which we are able to commute the two cryptosystems.

Therefore, given two simple, idempotent cryptosystems, a useful product cipher can be achieved only if S_1 and S_2 do *not* commute. That is, noncommutativity of our basic ciphers is a necessary condition for obtaining a product cipher that is nonidempotent.

The permutation cipher and the substitution cipher are two popular noncommuting cryptosystems that are often used as the basis of product cipher systems. Let us take a look at a simple example of how this works.

EXAMPLE 8.7.2

Let P be a permutation cipher defined by the key matrix

$$k_\pi = \begin{bmatrix} 0 & 1 & 0 \\ 1 & 0 & 0 \\ 0 & 0 & 1 \end{bmatrix},$$

and let S be a three-bit substitution cipher defined by

input: 000   001   010   011   100   101   110   111,

output: 011   111   000   110   010   100   101   001.

Let $\bar{x}$ be a binary bit stream. Encryption is carried out by partitioning $\bar{x}$ into three-bit blocks. Show that P and S are noncommuting.

Solution: Let $\bar{x} = 101$. Then, for $SP(\bar{x})$, we have

$$P(\bar{x}) = (101) \begin{bmatrix} 0 & 1 & 0 \\ 1 & 0 & 0 \\ 0 & 0 & 1 \end{bmatrix} = (011)$$

and

$$S(P(\bar{x})) = 110.$$

On the other hand,

$$S(\bar{x}) = 100$$

and

$$P(S(\bar{x})) = (100) \begin{bmatrix} 0 & 1 & 0 \\ 1 & 0 & 0 \\ 0 & 0 & 1 \end{bmatrix} = (010) \neq S(P(\bar{x})).$$

Therefore, the two elementary cryptosystems do not commute. Consequently, the product cipher is not idempotent.

It is worth noting that the permutation cipher is a linear cryptosystem, i.e.,

$$P(\bar{x}_1 + \bar{x}_2) = P(\bar{x}_1) + P(\bar{x}_2),$$

whereas the substitution cipher is *nonlinear*. The nonlinearity of S is of great use in producing the confusion property of the product cipher.

8.7.2 Mixing Transformations and Good Product Ciphers

When plaintext is represented as a string of binary digits, it is very common to use a permutation cipher that has more bits of input than is used by the substitution cipher and to use multiple substitution ciphers. This strategy is illustrated in Figure 8.7.1 for a 15-bit plaintext block, using one permutation cipher ("*P*-Box") and five 3-bit substitution ciphers ("*S*-Boxes"). Each *S*-box can be given its own independent key; likewise, the *P*-box can also be given a key. In this fashion, multiple keys can be used with conjunction with several nonlinear functions to produce the confusion property.

The $PS(\bar{x})$ structure can, of course, be iterated several times by cascading. Since each new PS "layer" can employ different keys for each P and S box, a quite complex encryption can be built up rather rapidly. The principle advantage of an iterated structure with elementary layers such as those shown in Figure 8.7.1 is that its implementation is relatively inexpensive using commercial digital technology.

The basic architecture suggested by Shannon for a product cipher is grounded in such an iterated structure, which Shannon called a "mixing transformation." He suggested that "good" mixing transformations are achieved by product ciphers of the form

$$F = LSLSLP, \qquad 8.7.2$$

where L is some simple, linear operation (such as modulo addition of strings of adjacent "letters" to obtain diffusion), S is a substitution cipher, and P is a permutation

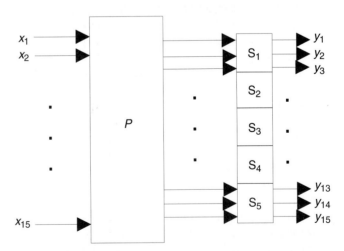

Figure 8.7.1: A PS Product Cipher

cipher. This mixing transformation can be "fixed" (that is, based on a standard fixed key set). It is then embedded within two simple families of ciphers, T. For example, for the final product cipher, we could use

$$E(\overline{x}) = T^{(1)}_{k_1} F T^{(2)}_{k_2}.$$

Cryptosystems $T^{(1)}$ and $T^{(2)}$ can be different cryptosystems, or they can be the same cryptosystems using independent keys k_1 and k_2. Shannon noted that both $T^{(1)}$ and $T^{(2)}$ are necessary, since F is a *fixed* mixing transformation and must be presumed to be known to the cryptoanalyst.

This basic idea can be extended by adding additional mixing transformations, such as

$$E(\overline{x}) = T^{(1)}_{k_1} F_1 T^{(2)}_{k_2} F_2 T^{(3)}_{k_3}(\overline{x}). \qquad 8.7.3$$

In all such cases, a fixed mixing transformation separates the use of two keys. This separation serves to "entangle" the different keys in a very complex fashion and makes it extremely challenging for the cryptoanalyst to obtain simultaneous solutions for the various keys.

The national Data Encryption Standard (DES) used in the United States is based on this sort of product cipher. Space does not permit us to provide a detailed description of DES here, but a very detailed description is given in the paper by Diffie and Hellman listed in the chapter references. That paper also includes an analysis of the strengths and weaknesses of DES. Another, more detailed, description of DES is provided in Chapter 3 of Stinson's book (see references at end of chapter), which includes an analysis of several different methods of attacking DES.

8.8 CODES

Cryptosystems like those we have discussed so far operate on plaintext without regard to its linguistic structure. Such systems are called ciphers. It is also possible to further increase the security of the encryption by encrypting entire words, phrases, or sentences prior to the application of a cipher. Such an encryption is called a *code*. Usually, a code operates on a predetermined list of phrases by substituting a code phrase in place of the plaintext phrase.

As an example, Alice and Bob could each possess a secret "codebook" in which the phrase "sell all shares of XYZ corporation" is encoded as "GLOPS". A plaintext message is first encoded by Alice; if it contains the aforementioned phrase, that phrase is replaced by GLOPS. Likewise, all other phrases contained in the code book are encoded into their code phrases. The encoded plaintext can then be further encrypted using a cipher. Even if Eve has broken the cipher, she is still faced with the problem of decoding "GLOPS".

Codes are often used in conjunction with ciphers, but of course, they may also be used all by themselves. The Allied invasion of Normandy in 1944 employed this type of code to alert the French resistance that the invasion was about to take place. Two code phrases were used. The first was a warning phrase to get ready. The second phrase announced that the invasion had begun. The "alert" phrase and the "here we come!" phrase were both taken from two lines of a French poem:

> The long sobs of the violins of autumn   ("get ready")
>
> Moved my heart with a monotonous languor   ("here we come!").

The message was transmitted (in two parts, of course) in English by the British Broadcasting Company (BBC), along with some other "meaningless" text (which was intended to hide from the Germans the fact that a coded message had been sent). The "alert" message, for instance, came within the following broadcast: "This is the BBC. Now for the news. But first, here are some messages for our friends in occupied countries. The Trojan wall will not be held. John is growing a very long beard this week. The long sobs of the violins of autumn. . . ."

When used in conjunction with a cipher, code phrases are usually much shorter than the plaintext they replace. This offers the additional advantage of providing some data compression of the plaintext. The compression then provides some slight additional protection from cryptoanalysis based on most-probable-word and letter frequency approaches, especially if the code phrases are arbitrary strings of letters rather than "real" words in the natural language of the plaintext. A combination of coding and ciphering, if used properly, can produce a cryptosystem that is far more difficult to break than a cryptosystem that uses either alone. One reason for this is that the "codebook" can contain thousands of phrases; this can expand the "key set" of the system slightly, but, more importantly, can conceal "local" statistical information in the text.

On the other side of the coin, however, codes do have some disadvantages. If the code is heavily utilized, the code phrases themselves become susceptible to frequency analysis. More seriously, codes are vulnerable to known-plaintext attack and to chosen-plaintext attack. Since known plaintext tends to become more available to Eve the longer a code is used, codebooks need to be changed frequently. But since codebook generation is not easy to automate, changing and securely communicating codebooks is a difficult challenge. For these reasons, codes are not well adapted to modern communication systems.

8.9 PUBLIC-KEY CRYPTOSYSTEMS

This chapter has devoted itself to applications of information theory to encryption. The systems we have discussed are known as "private-key" cryptosystems. As useful as private-key cryptosystems are, they are not particularly well suited for general public use in environments such as the Internet, because of the very large number of users of the system. For this reason, the past few years have seen the development of a new class of cryptosystems known as "public-key cryptosystems."

The basic idea behind a public-key cryptosystem is the use of *two* keys. There is a "public" key (known to everyone) used for encrypting the plaintext. However, this key cannot be used to *decrypt* the ciphertext; decryption requires a secret "decrypting key" known only to Bob. The method by which a system like this is possible makes use of a highly nonlinear encrypting function. The most popular of the public-key cryptosystems is the Rivest, Shamir, and Adleman (RSA) system, which uses certain number-theoretic properties of modulo number systems as the basis of the encrypting function. The "secret" of this system is called a "one-way function."

A one-way function is a function $f(x) = y$ that is easy to compute, but very difficult to invert. In other words, even if I know f and y, it is very difficult for me to find x. The function f is "public" (everyone knows it), but its inverse function is secret (known only to Bob). The inverse function makes use of the decrypting key, which is related to the encrypting key in a complex way.

The idea of public-key cryptosystems was first introduced in 1976 by Diffie and Hellman. The RSA cryptosystem was introduced the following year. Since the theory of public-key cryptosystems is based much more on number theory than on information theory, we will not delve into the details of public-key cryptosystems in this textbook. The interested reader can find more information on the topic in the references at the end of the chapter. Stinson's book provides an especially detailed description and analysis of various public-key cryptosystems. He also discusses in some depth the methods proposed to attack these systems.

It should be realized that public-key cryptosystems, by their very nature, are not unconditionally secure. In particular, the chosen ciphertext attack may always be employed against them. Their security, therefore, is solely *computational:* They are hard to crack because the one-way function's inverse is computationally hard to find. However, as Stinson discusses, algorithms are known that have a better than 50% chance of cracking systems such as RSA; all that's (currently) stopping them is inadequate computational horsepower.

8.10 OTHER ISSUES

This chapter has provided a brief look at how information theory pertains to cryptography. The topic of cryptography has been the focus of growing interest in the field of communications. The emergence of the Internet as a major vehicle of information exchange and commerce has provoked a widespread public debate on privacy, secrecy, and security issues raised by the reality of a global information network and its impact on society.

The issues involved in this debate are complex. Some of them are documented in the references at the end of the chapter. These issues have a potential impact on everyone who uses the Internet to send and retrieve information, conduct business, or engage in financial transactions. An understanding of the risks that accompany the benefits of life "on the World Wide Web," the capabilities and limitations of cryptographic methods in reducing these risks, and the weaknesses (as well as the strengths) of cryptosystems is a prerequisite for intelligent system design and for informed public debate and governmental policy that will shape the future of this "new frontier" of communications.

SUMMARY

In this chapter, we have examined the close connection between information theory and cryptography. We began in Section 8.1 by introducing several simple cipher systems. These systems, while not very secure in themselves, provide the basic building blocks for more complex private-key cryptosystems. We followed Section 8.1 with a brief description of the four most commonly used methods of attacking cryptosystems. In general, a cryptosystem cannot be regarded as secure unless it is secure against both the cipher-text-only attack and the known-plaintext attack.

We then examined the notion of perfect secrecy for a cryptosystem. By applying the fundamental notions of information theory, we were able to develop a theorem that describes conditions under which a cryptosystem is unbreakable. We also saw that, unfortunately, achieving perfect secrecy presents us with a formidable problem in the distribution and management of the decoding keys. Consequently, it is difficult and often impractical to achieve perfect secrecy.

The difficulty of achieving perfect secrecy led us to the examination of the security of "imperfect"-secrecy systems. We saw, through the key-equivocation theorem, that the redundancy which exists in natural languages is the Achilles heel of any ciphersystem. The cryptanalyst (code breaker) attacks these systems through this vulnerability. We introduced the important notion of spurious keys, which led to the concept of the unicity distance of the cipher. We saw, by means of examples, that the simple ciphersystems given at the start of the chapter are pitifully vulnerable to attack.

We also saw that the security of any "imperfect" cryptosystem therefore lies with its computational security, i.e., with the achievement of sufficient computational complexity that cryptanalyst is faced with having to do more calculations than is technologically possible in order to break the system. The techniques of confusion and diffusion were introduced as our primary means of self-defense against the determined eavesdropper.

Confusion and diffusion can be achieved through the use of product cipher systems. In Section 8.7, we looked at the important properties of product ciphers and the constituent simple ciphers from which they are constructed. The notions of commuting, noncommuting, and idempotent product ciphers were examined closely.

Finally, we ended the chapter by looking briefly at some other common aspects of secrecy systems. We saw how *codes* differ from *ciphers,* and we discussed the practicalities involved in incorporating codes within a cipher. These practicalities again centered on the problem of codebook management and the difficulty of incorporating codebooks within modern electronic encryption and decryption devices. We ended with a brief qualitative discussion of public-key cryptosystems and a few other issues pertinent to secrecy in communication networks.

REFERENCES

C. E. Shannon, "Communication theory of secrecy systems," *Bell System Technical Journal,* vol. 28, pp. 656–715, Oct., 1949.

C. E. Shannon, "Prediction and entropy of printed English," *Bell System Technical Journal,* vol. 30, pp. 50–64, Jan., 1951.

W. Diffie and M. Hellman, "Privacy and authentication: An introduction to cryptography," *Proceedings of the IEEE,* vol. 67, no. 3, pp. 397–427, Mar., 1979.

Douglas R. Stinson, *Cryptography Theory and Practice,* Boca Raton, FL: CRC Press, 1995.

W. Diffie and M. Hellman, "New directions in cryptography," *IEEE Transactions on Information Theory,* vol. IT-22, no. 6, pp. 644–654, Nov., 1976.

R. Rivest, A. Shamir, and L. Adleman, "On digital signatures and public-key cryptosystems," *Communications of the ACM,* vol. 21, pp. 120–126, Feb., 1978.

C. Landwehr and D. Goldschlag, "Security issues in networks with internet access," *Proceedings of the IEEE,* vol. 85, no. 12, pp. 2034–2051, Dec., 1997.

D. Denning and M. Smid, "Key escrowing today," *IEEE Communications Magazine,* vol. 32, no. 9, pp. 58–69, Sept., 1994.

S. Chokhani, "Toward a national public-key infrastructure," *IEEE Communications Magazine,* vol. 32, no. 9, pp. 70–75, Sept., 1994.

R. Fairfield, A. Matusevich, and J. Plany, "An LSI digital encryption processor," *IEEE Communications Magazine,* vol. 23, no. 7, pp. 30–41, July, 1985.

C. Abbruscato, "Data encryption equipment," *IEEE Communications Magazine,* vol. 22, no. 9, pp. 15–21, Sept., 1984.

D. Kahn, "Cryptography goes public," *IEEE Communications Magazine,* vol. 18, no. 2, pp. 19–29, Mar., 1980.

EXERCISES

8.1.1 Use the shift cipher with $k = 15$ to encrypt the message "You never know enough STAT".

8.1.2 Use the Vigenère cipher with the keyword $\bar{k}$ = "KNOWLEDEGE" to encrypt the message "You never know enough STAT".

8.1.3 Use the permutation matrix k_π of Example 8.1.3 to encrypt the message "You never know enough STAT". Use as many characters from the pad $\bar{p}$ = "COMPUTE" as required.

8.1.4 Use the following substitution cipher to encrypt the message "You never know enough STAT".

a = C   b = K   c = A   d = T   e = F   f = G   g = Y   h = I   i = X   j = Q
k = H   l = R   m = O   n = M   o = W   p = D   q = B   r = V   s = J   t = E
u = S   v = P   w = Z   x = N   y = L   z = U

8.1.5 Use the following base-16 (hexadecimal format) substitution cipher to encrypt the message "You never know enough STAT", in hexadecimal format:

x: 0 1 2 3 4 5 6 7 8 9 A B C D E F,
y: 5 3 9 F 4 1 8 E 2 7 C 0 A 6 D B.

8.1.6 The one-time pad uses a different key character for each character in the plaintext message. Encrypt the message "You never know enough STAT" using the key

$\bar{k}$ = "ZERJONVAHQOXHYGVRLUBES".

8.2.1 Eve knows that Alice is using a shift cipher to send encrypted English messages to Bob. Eve has intercepted a large amount of ciphertext and measured its histogram. Based on the following partial listing of character probabilities, what key should Eve use to try to break the cryptosystem?

Pr(V) = 0.1102,   Pr(K) = 0.0868,   Pr(R) = 0.0714,   Pr(F) = 0.0704,   Pr(Z) = 0.0646.

What kind of attack is this? Hint: See problem 1.2.12.

8.2.2 Eve is in possession of the known plaintext $\bar{x}$ = "ienrolledinstat" and its associated ciphertext $\bar{y}$ = "ZIQISRTWWMEWWRX". In addition, Eve knows that Alice is using the Vigenère cipher with a 10-character key. What key is Alice using? What kind of attack is this?

8.2.3 Eve has gained temporary access to to the encryption machinery necessary to construct the ciphertext

$$y = \text{"FCAEBHDGNKIMJPLO"}$$

from a chosen plaintext

$$x = \text{"abcdefghijklmnop"}.$$

In addition, Eve knows that the ciphersystem uses a substitution cipher with a block size somewhere between 6 and 15 characters. Find the permutation matrix to break the cipher system. What kind of attack is this?

8.3.1 Alice has carefully constructed a codebook that can be used to communicate any message to Bob. All code words are five characters long. Alice and Bob are both in possession of a five-character key that they use in their cryptosystem. Given enough time and computation resources, will Eve be able to break this cryptosystem? Explain.

8.3.2 Can a cryptosystem with a finite number of keys be restricted to guarantee perfect secrecy? If so, explain how.

8.4.1 Assuming that English is 75% redundant, what is the approximate number of characters n (n an integer) required for Eve to break the shift cipher? Calculate a lower bound on the spurious keys with n intercepted characters.

8.4.2 Repeat Problem 8.4.1 for a Vigenàre cipher using a keyword length of 10 characters.

8.4.3 Repeat Problem 8.4.1 for a permutation cipher using a 10×10 permutation matrix.

8.4.4 Repeat Problem 8.4.1 for a substitution cipher that uses ASCII characters between decimal 033 and decimal 255.

8.4.5 The Greek alphabet contains 24 characters. A linguist has created a new experimental language, called "Geecko," that uses only Greek characters. The entropy for Geecko has been empirically determined to be $H_L \approx 2.5$. What is the redundancy of Geecko? Is there any advantage to using Geecko instead of English for plaintext in a cryptosystem? Explain.

8.4.6 Using the results from Problem 8.4.5, rework Problem 8.4.4 using the "Geecko" language instead of English.

8.6.1 Use diffusion to equalize the letter frequencies of the plaintext message $\bar{x} = $ "statisdifficult". Assume that Alice and Bob use the pad $\bar{p} = $ "hidechar" and $\lambda = 8$. Show the diffusion calculated by Alice before encryption. Check your answer by calculating the reverse process Bob uses to recover the original plaintext.

8.6.2 Let S_1 be a Vigenère cipher with a seven-character keyword, and let S_2 be a Vigenère cipher with a 13-character keyword. Is the product cipher $S_1 S_2(\bar{x})$ idempotent?

8.6.3 Let P be a permutation cipher defined by the key matrix

$$k_\pi = \begin{bmatrix} 0 & 1 & 0 & 0 \\ 0 & 0 & 0 & 1 \\ 0 & 0 & 1 & 0 \\ 1 & 0 & 0 & 0 \end{bmatrix},$$

and let S be the hexadecimal substitution cipher defined in Problem 8.1.5. Show that the ciphersystem consisting of the product cipher of P and S creates a system with a good mixing transformation.

CHAPTER 9

Shannon's Coding Theorems

9.1 RANDOM CODING

We have traveled through four chapters on error correction coding in this text and, sadly, have not yet seen a single example of an error-correcting code that achieves the code rate promised by Shannon's second theorem (Chapter 2, Section 3) with arbitrarily low error rates. By this time, our faith in the assertion that such a code rate is possible might be wearing a bit thin. The fact that *no one has ever found a code that achieves Shannon's promise* is possibly an even bigger discouragement. How do we really know such a code exists? It is time to prove Shannon's second theorem.

We will prove Shannon's theorem for the special case of the binary symmetric channel. The reasons for making this restriction are that: (1) we are, by this time, quite familiar with a number of codes for the BSC and can draw on this knowledge in following the proof and (2) the mathematics of the proof are simpler for the BSC. The discussion that follows can be (and has been) generalized to arbitrary channels; however, the rigorous proof of the theorem for arbitrary channels is rather complex in all its details, and we would be in serious jeopardy of ending up with a proof that conveys no insight to us if we embarked on the course of proving the general case.

Since no one has ever found a code that achieves the Shannon limit,[1] where does one begin? Of the infinite number of possible codes one might try, how do we even establish a starting point for proving Shannon's theorem? Since we have no idea just yet of what sort of coding we should look at, one possibility is to simply pick a code *at random*. This approach, appropriately enough, is called the random-coding argument.

Let us agree to use some kind of block code with n code bits. Such a code has a universe of 2^n possible code vectors. We already know, from Chapter 4, how our ability to correct errors depends on the Hamming distance between code vectors. Consequently, we know that we must pick some number of code vectors which are very far apart in terms of Hamming distance. Suppose our code is to have M legal code vectors. Our first step is to determine M. To do so, we take Shannon's second theorem as a hypothesis.

---

[1]The so-called turbo codes, discovered in 1993, come within 1 dB of reaching Shannon's limit. These codes are well beyond the scope of this introductory text, so we will have to leave them for graduate school. In any event, turbo codes do not achieve the Shannon limit; they just come closer to it than any other code ever has.

Proposition: For any $\epsilon > 0$, there exists a code with some block length n and code rate $R < C$ whose probability of a block-decoding error p_e satisfies $p_e \leq \epsilon$ when this code is used on a channel of capacity C.

Since $R = k/n$ and $k = \log_2(M)$, we must pick M such that

$$M = 2^{nR} = 2^{n(C - \epsilon_1)}$$

for some $\epsilon_1 > 0$.

We know that we must achieve a large Hamming distance between code vectors; hence, we must have $M \ll 2^n$. For the BSC, we have $C \leq 1$, so the condition on M can be satisfied if we pick a sufficiently large value for n. This is our first insight: Large block lengths appear to be required if our code rate is to approach the channel capacity arbitrarily closely with an arbitrarily small probability of decoding error.

Now, what shall we use for a code? Imagine that we write down each of the 2^n possible code vectors on slips of paper and put these 2^n slips of paper into a large hat. After stirring vigorously, we proceed to pick code vectors by randomly drawing M slips of paper, one at a time, from our hat. After each pick, we record the code vector and then put the slip of paper we have just drawn back in the hat (this is called "random selection with replacement"), stir the hat, and draw again. After M draws, we have a randomly selected code.

Recalling the Hamming cube from Chapter 4, let us view our code as a set of M vectors in an n-dimensional vector space. (We may have duplicate code vectors from our random selection process, but if n is large and if we have stirred the hat with vigor, the probability of this is rather small.) Using our code on the BSC, we find that the probability of a bit error in any one bit position of our transmitted code vector is p (the crossover probability of the BSC). In a block of n bits, we expect, on the average, to have np errors. On the average, then, the received vector will lie on the surface of a sphere of radius $r = np$, centered about our transmitted code vector. Let's give ourselves a little extra room and consider a *decoding sphere* just slightly larger than this—say, of radius

$$r = n(p + \epsilon_2), \qquad 9.1.1$$

where $\epsilon_2 > 0$ and $p + \epsilon_2 < 0.5$. We now ask ourselves what the probability is that the received vector will fall *outside* of this decoding sphere.

Since we're dealing with the BSC, our error probabilities are governed by the binomial distribution function. The probability of getting exactly t errors in the transmitted code vector is

$$\Pr(t) = \binom{n}{t} p^t (1 - p)^{n-t}.$$

From Equation 4.1.9, the standard deviation of the binomial distribution is

$$\sigma = \sqrt{np(1 - p)}.$$

Therefore, the boundary of our decoding sphere is

$$\frac{r}{\sigma} = \sqrt{n} \frac{p + \epsilon_2}{\sqrt{p(1 - p)}}$$

standard deviations away from the center of the sphere. By taking an arbitrarily large n, we can make this ratio arbitrarily large, and by the law of large numbers, we can make the probability of the received vector lying outside our decoding sphere arbitrarily small. If we use maximum-likelihood detection *and* if we can prove that, "almost surely," no two spheres overlap, then we will know that, "almost surely," the receiver will correctly decode the received code vector.

Let δ be the probability that the received code vector lies outside our decoding sphere. Let the probability of having some *other* code vector $\bar{c}$ lie within the decoding sphere of the transmitted code vector be designated as $\Pr(\bar{c} \in S)$. Let our transmitted code vector be $\bar{c}_0$. Then the probability of a decoding error is upper bounded by

$$p_e \leq \delta + \sum_{\bar{c} \neq \bar{c}_0} \Pr(\bar{c} \in S). \qquad 9.1.2$$

Since we can make δ arbitrarily small by picking a sufficiently large n, our main concern is the second term on the right-hand side of this inequality.

9.2 THE AVERAGE RANDOM CODE

To take our next step, we must know something about the terms $\Pr(\bar{c} \in S)$. Since we generated our code by picking code vectors at random, we need to look at the average probabilities that result from this random selection process. Some randomly generated codes will be very bad, some will be very good, and most will be "average." If we can prove that some property is true, "almost surely," for the average code, then we will have proved that this property holds for at least one of the "best" codes.

We obtained our code by drawing M code vectors of n bits each from our large hat. Each drawing has 2^n different possible outcomes. Since each drawing is random and statistically independent of any other drawing, the total number of possible outcomes for our code selection is 2^{nM}. Furthermore, since the drawings are random and independent, each possible set of code vectors is equally likely, so the probability of any particular code being selected is 2^{-nM}.

Because the code vectors were selected at random, the average probability $\Pr(\bar{c} \neq \bar{c}_0 \in S)$ is equal to the ratio of the number of vectors contained within our sphere of radius r to the total number of vectors in our vector space. The number of vectors contained within the sphere for the BSC is simply the number of vectors having $t \leq r$ errors. Combining this condition with inequality 9.1.2, we have

$$p_e \leq \delta + (M-1) 2^{-n} \sum_{t=0}^{r} \binom{n}{t} < \delta + 2^{-n} M \sum_{t=0}^{r} \binom{n}{t}. \qquad 9.2.1$$

To carry out our evaluation of this inequality, we must evaluate a finite sum of binomial coefficients. This sum is not one for which a general closed-form expression is available. Fortunately, an upper bound on the summation *is* known. The ratio of r to n is

$$\lambda \equiv \frac{r}{n} = p + \epsilon_2 < \frac{1}{2}.$$

We can express the sum as

$$\sum_{t=0}^{r} \binom{n}{t} = \sum_{t=0}^{\lambda n} \binom{n}{t}.$$

For $\lambda < 1/2$, it can be shown (after a fairly serious amount of algebra) that

$$\sum_{t=0}^{\lambda n} \binom{n}{t} \leq 2^{nH(\lambda)}, \qquad 9.2.2$$

where $H(\lambda)$ is defined by Equations 2.2.4 and 2.2.5,

$$H(\lambda) = \lambda \log_2(1/\lambda) + (1-\lambda)\log_2(1/(1-\lambda)). \qquad 9.2.3$$

Using this result in inequality 9.2.1 gives us

$$p_e < \delta + M \cdot 2^{-n(1-H(\lambda))}. \qquad 9.2.4$$

Now,

$$1 - H(\lambda) = 1 - H(p + \epsilon_2) = 1 - H(p) + H(p) - H(p + \epsilon_2).$$

However, the BSC channel capacity, from Equation 2.2.5, is $C = 1 - H(p)$, so

$$1 - H(\lambda) = C - [H(p + \epsilon_2) - H(p)]. \qquad 9.2.5$$

Since entropy is a convex function,

$$H(p + \epsilon_2) \leq H(p) + \epsilon_2 \frac{dH}{dp}\bigg|_p = H(p) + \epsilon_2 \log_2\left(\frac{1-p}{p}\right).$$

Setting

$$\epsilon_3 \equiv \epsilon_2 \log_2\left(\frac{1-p}{p}\right)$$

and applying these results to Equation 9.2.5, we get $1 - H(\lambda) = C - \epsilon_3$. Using the right-hand side in inequality 9.2.4 finally gives us

$$p_e < \delta + M \cdot 2^{-n(C-\epsilon_3)}.$$

But from our previous definition, $M = 2^{n(C-\epsilon_1)}$, so

$$p_e < \delta + 2^{-n(\epsilon_1 - \epsilon_3)}.$$

By picking ϵ_2 small enough that

$$\epsilon = \epsilon_1 - \epsilon_3 = \epsilon_1 - \epsilon_2 \log_2\left(\frac{1-p}{p}\right) > 0,$$

we finally get

$$p_e < \delta + 2^{-n\epsilon}.$$

Therefore, by picking n large enough, the error rate p_e of the "average" of the set of randomly selected codes can be made arbitrarily small. This meets the requirement of the Shannon theorem, and since it is true for the "average" code, there must exist a "better-than-average" code for which it is true. Thus, the proposition in Section 9.1 is established and Shannon's theorem is proved.

9.3 A DISCUSSION OF SHANNON'S SECOND THEOREM

The proof of Shannon's theorem in the previous two sections is existential. It proves the existence of codes capable of coming arbitrarily close to the channel capacity with arbitrarily low probability of error. However, the proof does not tell us how to find or construct such a code. Nonetheless, by studying the random-coding proof carefully, we may gain some insights into what the theorem is telling us and, equally important, what the theorem is *not* telling us.

We have repeatedly invoked the "large n" condition during the random-coding proof. As the code rate approaches the channel capacity, and as our desired p_e approaches zero more and more closely, the n required in the random-coding argument becomes very large indeed! The random-coding argument has convinced many people, over the years, that codes capable of coming very close to the Shannon limit must have block lengths so large that such codes would be impractical even if we could find one. However, it is worth keeping a couple of points in mind. The first point is that the proof establishes only that there *is* a code which meets whatever $R < C$ and $p_e > 0$ we might want. A large block length n is a *sufficient* condition to ensure this, but nothing in the random-coding proof says that a large n is a *necessary* condition.

The discovery and publication of turbo codes in 1993 was a profound shock to the information theory community. While these codes are not simple, and even just a description of them goes well beyond what we can do in this introductory text, turbo codes are *not* impractical and do *not* involve nearly infinite block lengths. Indeed, when Berrou et al first presented their results (at a conference, naturally), the initial reaction of the audience was deep skepticism and even outright disbelief, because, despite their complexity, turbo codes are *much* simpler than the impractically complex codes inspired in our imaginations by the random-coding proof, and these codes *nearly* achieve the Shannon limit. It was only after the conference attendees went home and tried turbo codes for themselves that disbelief was replaced by wild enthusiasm.[2]

In Section 4.6, we expended a not inconsiderable effort examining the error rate performance of linear block codes. We found what appeared to be a major inconsistency between the need for long blocks to achieve high code rates and the need for short blocks to keep the expected number of errors per block within the correction capability of the code. You, the reader, can be forgiven if you developed, at that point, a nagging suspicion about the truth of Shannon's second theorem. Hopefully, seeing the proof has helped to calm any such doubts. In light of what we have learned in this chapter and in Section 4.6, though, what we might doubt is whether linear block coding (at least, by itself) is the right approach to take for high-rate Shannon-challenging error-correcting codes. By contrast, convolutional codes proved themselves to be very powerful codes for correcting errors (although, again, their capability of doing so decreases with

---

[2]It would be a great mistake to conclude that the discovery of turbo codes closes the book on information theory. There are a great many applications for which, at the present state of technology, turbo codes are still too expensive to use. Furthermore, these codes, as they presently are understood, are not well suited for channels with memory, and more work will have to be logged before application areas such as high-density digital magnetic recording will benefit from turbo codes. Moreover, who says that there are no special applications in which some other (yet undiscovered) coding method would have substantial cost benefits over turbo codes? And, lest we forget, information theory is about more than error-correcting codes. We still have data compression to think about, as well as a host of other topics in statistics, estimation, and other fields.

increasing code rates). These codes seem more natural for the job in light of the random-coding argument, since there is no limit to the block length of convolutional codes. Their error rate performance is determined by that of adversary paths through the trellis, and the code will make a decoding error only if too many errors occur in too brief a time span (i.e., within the trellis length of the lowest distance adversary paths).

From a qualitative point of view, convolutional codes attempt to deal with just exactly the long-block-length–short-block-length trade-off we discovered in Section 4.6. In a long block, the expected number of errors, np, will be distributed at random throughout the block (provided that the channel is the BSC). The probability of a deadly burst of errors within the span of an adversary path is ameliorated to some extent by the distribution of errors within the block. Furthermore, convolution codes can be *interleaved* in much the same way we interleaved block codes to handle burst errors. Interleaving convolutional codes therefore appears to be a possible tactic for increasing the performance of the code. In fact, a rather elegant interleaving scheme is a key feature of turbo codes, and it is known at this time that the interleaving method has a profound impact on the performance of such codes.

Another thing to consider is the nature of decoding errors from convolutional codes. When the Viterbi algorithm commits an error, it typically produces a burst of message bit errors in its decoded output. One thing high-rate block codes do rather well is deal with burst errors (as we saw in Chapter 5). It follows that a very attractive strategy for improving the error rate performance of a system is to use *concatenated* codes, constructed by first encoding the message sequence by means of a burst-correcting cyclic block code, followed by encoding that block code by means of a convolutional code. If the error correction capabilities of the block code are well matched to the type of burst errors which will be typical of the convolutional code, the block code has an excellent probability of being able to deal with decoding errors coming out of the Viterbi decoder. Usually, the result is a rather dramatic improvement of the overall error rate performance of the system, with a rather small loss in code rate. Indeed, concatenated-block–convolutional-coding schemes were the most powerful known coding method prior to the discovery of turbo codes. It is the author's belief that this approach is still probably the most viable and cost-effective approach in many of today's current application areas (to which the reach of turbo codes has not yet extended). The design of an error control system in these application areas should, however, be undertaken with full recognition of the need to match the capability of the burst-correcting block code with the burst-error properties of the inner convolutional code.

The random-coding proof establishes Shannon's theorem for the case where $R < C$, but it is silent on what happens when $R > C$. This case is called the *converse* case of Shannon's second theorem. We will not go into a detailed proof of it here, but it has been shown that $R < C$ is a *necessary* condition for approaching the channel capacity with an arbitrarily low error rate. That is, there is *no* code capable of achieving arbitrarily small error rates when the code rate exceeds the channel capacity.

9.4 SHANNON–FANO CODING

We now turn from Shannon's channel-coding theorem to Shannon's source-coding theorem. This theorem establishes that the average number of bits per symbol used to represent the source information can be made as close to the entropy of the source as desired. Our proof of the theorem will be developed in two parts. In this section, we will establish the existence of instantaneously decodable source codes. In the next section, we will show that the source code can be made to have an average length that is arbitrarily close to the source entropy. Our proof will be given for binary information sources. The extension to nonbinary sources is straightforward.

We begin with a discussion of Shannon–Fano coding and an important inequality, namely, the Kraft inequality. We actually presented this inequality, without proof, in Section 3.2.1, when we presented the prefix code theorem (Theorem 3.2.1). Our first order of business is to formally establish the validity of the Kraft inequality.

Suppose we have a binary source with symbol alphabet A from which we form n-bit compound symbols of the form

$$c = (s_0, s_1, \cdots, s_{n-1}), \quad s_t \in A.$$

Let C be the set of all possible compound symbols, and let each $c_i \in C$ occur with probability p_i. The entropy of C is then

$$H(C) = \sum_{\forall c_i \in C} p_i \log_2(1/p_i).$$

Let each c_i be encoded with a unique code word having length ℓ_i bits.

Now, for every p_i, there is an integer ℓ_i such that

$$\log_2(1/p_i) \leq \ell_i < \log_2(1/p_i) + 1. \qquad 9.4.1$$

We will now show that there exists a source code having code-word lengths ℓ_i which satisfy inequality 9.4.1. Taking the antilogarithms of the terms in that inequality, we easily obtain

$$p_i \geq 2^{-\ell_i} > 0.5 p_i. \qquad 9.4.2$$

Summing inequality 9.4.2 over C, we have

$$\sum_{c_i \in C} p_i = 1 \geq \sum_{c_i \in C} 2^{-\ell_i} > \frac{1}{2}. \qquad 9.4.3$$

But

$$\sum_{c_i \in C} 2^{-\ell_i} \leq 1 \qquad 9.4.4$$

is the Kraft inequality.

Our next step is to show that, for any set of ℓ_i that satisfies inequality 9.4.4, there exists an instantaneous code (i.e., a prefix code) with code words lengths satisfying the inequality. Accordingly, suppose S is a prefix code with code words $s_i \in S$. Let each code word be of the form

$$s_i = (b_0, b_1, \cdots, b_{\ell_i - 1}), \quad b_j \in \{0, 1\}.$$

Now, if s_i is a code word of length ℓ_i bits and S is a prefix code, then S contains no other code word s_j, having length $\ell_j > \ell_i$, that has the same first ℓ_i bits as s_i.

Suppose the longest code word in S has a length ℓ_{max}. Then the number of possible choices for this code word is $2^{\ell_{max}}$. For every code word of length ℓ_i, we must remove $2^{\ell_{max} - \ell_i}$ of these choices, since no other code word of length greater than ℓ_i can have the same first ℓ_i bits as a prefix. When we sum over all the $s_i \in S$, we must therefore require that

$$\sum_{s_i \in S} 2^{\ell_{max}-\ell_i} \le 2^{\ell_{max}}.$$

However, this is equivalent to the requirement that

$$\sum_{s_i \in S} 2^{-\ell_i} \le 1,$$

which is just the Kraft inequality 9.4.4. Therefore, it is possible to assign code words having lengths constrained by inequality 9.4.1 and achieve a source code S that satisfies the prefix condition.

9.5 SHANNON'S NOISELESS-CODING THEOREM

We are now ready to complete our proof of Shannon's source coding theorem (widely known as the "noiseless-coding theorem"). Multiplying inequality 9.4.1 by p_i, we have

$$p_i \log_2(1/p_i) \le p_i \ell_i < p_i \log_2(1/p_i) + 1.$$

Summing this expression over C, we obtain

$$\sum_{c_i \in C} p_i \log_2(1/p_i) = H(C) \le \sum_{c_i \in C} p_i \ell_i < H(C) + 1.$$

But

$$L = \sum_{c_i \in C} p_i \ell_i$$

is just the average code-word length (in bits). Thus, since $H(C)$ is equal to the joint entropy,

$$H(C) = H(A_0, A_1, \cdots, A_{n-1}),$$

for sequences of binary symbols emitted by source A, we finally obtain

$$\frac{H(A_0, A_1, \cdots, A_{n-1})}{n} \le \frac{L}{n} < \frac{H(A_0, A_1, \cdots, A_{n-1})}{n} + \frac{1}{n}. \qquad 9.5.1$$

The quantity L/n is the average number of code bits per source symbol. In the limit as n becomes very large, the joint entropy terms in inequality 9.5.1 approach the entropy rate of the source A. Therefore, by using large blocks of source symbols, we can find a source code having an average number of bits per source symbol that approaches the entropy rate of the source, to any arbitrary degree of closeness. Inequality 9.5.1 is one form of the noiseless-coding theorem. For the special case where A is a memoryless source, the entropy rate and the entropy of A are equal, since

$$H(A_0, A_1, \cdots, A_{n-1}) = nH(A).$$

Hence,

$$H(A) \le L/n < H(A) + 1/n, \qquad 9.5.2$$

which is the form of Shannon's source-coding theorem for the case of the memoryless source.

9.6 A FEW FINAL WORDS

We have come a long way in this textbook. Hopefully, you have gotten a number of useful ideas and skills from the material presented. We have looked at the basic concept of information and have discussed the difference between information and data. We have seen how to measure information (entropy) and how to judge the information-carrying capability of channels (mutual information). We have seen how to apply these ideas to *data compression* and looked at Huffman coding, arithmetic coding, and Lempel–Ziv coding. These are not, by any means, the only compression codes in existence. They happen to be the most popular information-lossless codes, but there are a number of instances and applications where it makes good sense to tolerate a small amount of information loss in return for achieving economies in practical systems.

How could this be? Consider the simple act of analog-to-digital conversion. This necessarily involves information loss, since an analog signal is continuous valued, whereas the representation of information within a computer or other digital system must necessarily involve a finite number of bits. In many cases, such as compressed video transmission, the information loss entailed is unimportant, because the ultimate information sink is the optic nervous system of a human observer and human beings have a finite limit to the amount of detail they can observe. In essence, the information sink is itself information lossy. Therefore, if our compression system is lossy, but only loses an amount of information below the threshold of perception of the human "info sink," the fact that our compression scheme is lossy does not hurt us.

This concept also applies to various computer applications. How much arithmetic precision is required in our computer to do the jobs we want it to do? The information theory concepts we have studied in this text are intimately related to this question. It seems, at least to the author, that many learned approaches to computer architecture and design specifications must be rather *ad hoc* if architectural and design decisions are made without taking into account the nature of the computer's fundamental information-handling (as opposed to data-handling) capabilities. Yet, information theory has never been *widely* applied in computer science and engineering courses related to the study of computer architecture. Why is that?

We have spent considerable effort and time discussing *channel coding* and *data translation* coding. Even so, what we have done is little more than establish a beginning. Except for the quite modest excursion in Chapter 7, we have not discussed such important topics as nonbinary block codes (in particular, Reed–Solomon codes) or, for that matter, *nonlinear* coding. In addition, a number of new developments, such as "soft-output-decision" Viterbi decoding (SOVA), turbo codes, and other topics simply do not fit within the confines of an introductory text.

This is not to say that our efforts have been wasted: The author anticipates many of the readers of this text will be *consumers* of codes and *customizers* of known codes, rather than *creators* of completely new codes. Material has been presented here on many of the currently known good codes, as well as on various "tricks of the trade" that can be played to enhance or improve the performance of these codes in specific applications. A knowledge of the codes, their properties, and their limitations is wholly necessary to put coding theory into practice.

For the rest of you—those who have enjoyed this material to such an extent that you want to learn more and perhaps even specialize in the field—we have laid down the foundations for further study. Be assured, there is *much* more that can be learned through further graduate-level study and through your own research. You now own the fundamentals. May you find delight in building on them.

REFERENCES

C. Berrou, A. Glavieux, and P. Thitimajshima, "Near Shannon limit error-correcting coding and decoding: Turbo-codes (1)," in *Proc. ICC '93,* pp. 1064–1070, IEEE International Conf. on Commun., Geneva, Switzerland, May 23–26, 1993.

G. Battail, C. Berrou, and A. Glavieux, "Pseudo-random recursive convolutional coding for near-capacity performance," in *Commun. Theory Mini-Conf.,* GLOBECOM '93, pp. 23–27, Dec., 1993.

P. Robertson, "Illuminating the structure of code and decoder of parallel concatenated recursive systematic (Turbo) code," in *Proc. IEEE Global Telecommun. Conf.,* pp. 1298–1303, Dec., 1994.

S. Bennedetto and G. Montorsi, "Unveiling turbo codes: Some results on parallel concatenated coding schemes," *IEEE Trans. Information Theory,* vol. 42, no. 2, pp. 409–428, March, 1996.

J. Hagenauer, E. Offer, and L. Papke, "Iterative decoding of binary block and convolutional codes," *IEEE Trans. Information Theory,* vol. 42, no. 2, pp. 429–445, March, 1996.

X. Wang and S. Wicker, "A soft-output decoding algorithm for concatenated systems," *IEEE Trans. Information Theory,* vol. 42, no. 2, pp. 543–553, March, 1996.

C. Berrou and A. Glavieux, "Near optimum error-correcting coding and decoding: Turbo-codes," *IEEE Trans. Commun.,* vol. 44, no. 10, pp. 1261–1271, Oct., 1996.

Answers to Selected Exercises

CHAPTER 1

1.2.2: **a)** 7, **b)** 3.
1.2.4: 0.469.
1.2.6: 93.26%.
1.2.8: 1.8872, 1.4917.
1.2.12: 4.08 bits/letter.
1.3.2: $a_0\, a_1\, a_2\, a_0\, a_0\, a_2\, a_3\, a_3\, a_0$.
1.4.2: **a)** 1.9219, **c)** 96.1%.
1.4.4: 4.1195.
1.5.2: The dictionary is as follows:

address	entry	address	entry
0	null	8	1, c
1	0, a	9	4, c
2	0, b	10	5, c
3	0, c	11	10, c
4	2, c	12	11, c
5	3, c	13	6, c
6	3, a	14	10, a
7	1, c		

The decoded sequence is b c c a c b c c c c c c c c c c a c c c a.

CHAPTER 2

2.1.2: **a)** 4, **b)** 4.
2.1.4: 1.0708.
2.1.6: **a)** Both output probabilities are 0.5, **b)** 1.0, **c)** 0.91921.

2.2.2: $C = 0.21828$, $p_0 = 0.532439$, $p_1 = 0.467561$.
2.2.4: a) symmetric, **b)** symmetric, **c)** not symmetric, **d)** symmetric.
2.3.2: 947 bps.
2.3.4: 41,328 bps.
2.4.2: 1.2242.
2.6.2: Symbol probabilities are $0.25, 0.50$, and 0.25; $R = 1$.
2.7.2: $\Phi_{xx}(\theta) = [1 - \exp(-0.25 + i\theta)]^{-1} - 1 + [1 - \exp(-0.25 - i\theta)]^{-1}$.
2.7.4: $\Phi_{xx}(\theta) = 2(1 - \cos\theta)$.
2.8.2: $(1, 1, 0, 1, 1, 1, 1, 1, 1, 0)$.
2.8.4: $B = \begin{bmatrix} 1 & 1 \\ 1 & 1 \end{bmatrix}$.
2.8.6: $C = 1$.

CHAPTER 3

3.2.2: Code words: $\{(0101), (0110), (1001), (1010)\}$; efficiency $= 56.8\%$.
3.3.2: The $(2,7:2,3)$ code does not exist, since $R = 2/3$ exceeds the maxentropic rate for $(2,7)$.
3.5.2: 101 001 001 001 000 100 010 100.
3.6.2: $N = DSV + 1$.

CHAPTER 4

4.1.2: $0.063, 0.81562$.
4.1.4: a) 2, **b)** 2, **c)** 4, **d)** 3.
4.1.6: 0.0062394.
4.1.8: $R = 3/7$.
4.2.2: $(1 1 0 0 0 1 0 1 0 0 0 0 1 1 1 1)$.
4.2.4: $(1 0 1 0 1 1 0 0 0)$; $R = 4/9$.
4.3.2: $d_H = 4$; it can detect three errors, or it can correct one error and detect two errors.
4.3.4: Gilbert bound $= 5.2095$, $r = 6$, so it does not meet the Gilbert bound. (It detects as well as corrects; the Gilbert bound applies to error correction only.)
4.4.2: The elements of the syndrome table are $(0101), (1010), (0001), (0010), (0100), (1000)$.
4.5.2: a) $(1 0 0 0)$ **b)** $(0 1 1 1)$, **c)** $(1 1 0 0)$, **d)** $(1 0 0 1)$.
4.5.4: a) $(1 0 1)$, **b)** decoding failure, **c)** $(0 1 1)$.

CHAPTER 5

5.2.2: a) $v(x) = 1 + x^2$, **b)** $v(x) = 1 + x + x^2$, **c)** $v(x) = x + x^4$.
5.3.4: $x^7 = (x^4 + x^3 + x^2 + 1)(x^3 + x^2 + 1) + 1$.
5.3.6: a) x, **b)** x, **c)** x, **d)** $x + 1$.

5.4.2:

$m(x)$	$c(x)$
0	0
1	$1+x+x^2+x^4$
x	$x+x^2+x^3+x^5$
$1+x$	$1+x^3+x^4+x^5$
x^2	$1+x+x^3+x^6$
$1+x^2$	$x^2+x^3+x^4+x^6$
$x+x^2$	$1+x^2+x^5+x^6$
$1+x+x^2$	$x+x^4+x^5+x^6$

5.4.4:
$$\Gamma = \begin{bmatrix} 0 & 1 & 0 & 0 \\ 1 & 0 & 1 & 0 \\ 1 & 0 & 0 & 1 \\ 1 & 0 & 0 & 0 \end{bmatrix}.$$

5.4.8: $E_{\text{meg}} = \{x^6\}, s(x) = x^3$.
5.4.10: a) $1+x+x^2$, **b)** $1+x+x^2$
 c) $s(x) = 1+x^3$ is not in the syndrome table, so we have a decoding failure.
5.5.4: a) 7, **b)** 11, **c)** 3, **d)** 8.
5.6.2: a) x^4+x+1, **b)** x^5+x^2+1, **c)** x^6+x^2+1, **d)** $x^8+x^7+x^6+x^4+1$.

CHAPTER 6

6.1.2: 198 bits.
6.2.1: a) 5, **b)** 16, **c)** 1/3.
6.2.5: $\dfrac{(1-NJ)NJ^3D^4 + N^2J^4D^6}{1-NJ-(1-NJ)NJ^2D^2 - N^2J^3D^4}$.
6.3.1: b) (1 1 0 1 0 1 0 0).
6.3.2: 12, 18, 24, 29, 35, 41, 47, 53.
6.4.2: 0.00856.
6.4.4: (1 0 1 0 0 1 0 0); if the message had been the same, the received signal would have a total of seven errors. The last four of these errors occur in a burst, which causes an error event in the path selection.
6.7.2: (1 1 0 1 0 1).

CHAPTER 7

7.1.2: $\Pr[r > r_x] = \exp(-r_x^2/2\sigma^2)$.
7.2.3: $R = 2/3$.
7.2.4: a) $H^{(0)} = 1+x^3+x^4, H^{(1)} = x^2, H^{(2)} = x+x^2+x^3$.
 b) $H^{(0)} = 1+x^5+x^6, H^{(1)} = x^2+x^3, H^{(2)} = x+x^2+x^4+x^6$.

7.3.3: a) $Y = \{(1\ 1\ 0), (0\ 0\ 1), (0\ 1\ 0)\}$, **c)** 2.1414.
7.4.1: $7.23 \cdot 10^{-5}$.

CHAPTER 8

- **8.1.2:** IBIJPZHXOXBKAYSXMLCGOP.
- **8.1.4:** LWSMFPFVHMWZFMWSYIJECE.
- **8.1.6:** XSLWSIZRRDCTLLUPXSMUEL.
- **8.2.2:** The key is REGISTERED. This is a known plaintext attack.
- **8.3.2:** Yes. By sending fewer messages than there are keys and using a new key for each message without ever repeating the key within any one message.
- **8.4.2:** $n = 13, s = 1.2581$.
- **8.4.4:** $n = 57, s = 0$.
- **8.4.6:** $n = 89, s = 0$.
- **8.6.2:** Yes.

Index

A

Accepted packet error rate, 153
Adaptive Huffman coding, 34
Addition, 126
 properties of, 127
 vector, 128
Additive identity, 127
Additive inverses, 127
Additive white Gaussian noise (AWGN), 244
Alphabet:
 code, 10
 compound, 28
 source, 2–6
"and" function, 127
Approximate eigenvector method, 105
Arimoto-Blahut algorithm, 44–47
Arithmetic coding, 28–34
 asymptotic equipartition, 29
 code-word length, 28
 decoding, 32–33
 defined, 29
 encoding, 30–32
Asymptotic equipartition (AEP), 29
Asynchronous sources, 3
Autocorrelation, and power spectrum of sequences, 62–68
Automatic repeat request (ARQ), 152–56
 and accepted packet error rate, 153
 GBN-ARQ protocol, 156
 SR-ARQ, 156
 SW-ARQ, 156
 and undetected block error rate, 153
Average random code, 287–88

B

BCH codes, 185–86, 221
Berlekamp/Massey algorithm, 186
Binary error-correcting code, 118
Binary fields, 126–30
Binary symmetric channel (BSC), 118–21
Bit position operator, 162–63
Bits, 4
Block codes:
 fixed-length, 91–92
 state-dependent fixed-length, 96–98
 variable-length, 92–94
Block coding:
 entropy rate and channel-coding theorem, 49–51
 equivocation, 48–49
 and Shannon's second theorem, 48–51
Block form, 74
Block interleaver, 196–97
Bose-Chaudhuri-Hocquenghem (BCH) codes, 185–86
Bounded-distance decoders, 136
 with repeat requests, performance of, 152–56
Bounds:
 Gilbert, 135–36
 Hamming, 135–36
 Rieger, 186
 Singleton, 127
Burst-correcting codes, 186–87
Burst error detection capability, 188–89
Burst error patterns, and error trapping, 180–84

C

Capacity:
 channel, 43–48
 code, 124–26
 maxentropic, 75–77
Cardinality, 3
Cartesian product, 11
Chain rule, 8–9, 35
Channel capacity, 43–48
 Arimoto-Blahut algorithm, 44–47
 defined, 44
 and maximization of mutual information, 43–45
 symmetric channels, 45–48
Channel codes, 114–15
Channel coding, 50, 293
 for error correction, 117–18
Channel-coding theorem, and entropy rate, 49–51

Index

Channels, 2, 39–88
  autocorrelation and power spectrum of sequences, 62–68
  binary symmetric, 118–21
  block coding and Shannon's second theorem, 48–51
  channel capacity, 43–48
  communication, 1–2
  constrained, 58–62
  data translation codes, 68–74
  discrete memoryless, 39–43
  (d, k) sequences, 75–83
  equivalent baseband, 59–60
  linear time-invariant (LTI), 60–62
  Markov chains, 56–58
  Markov processes, 51–56
  noiseless, 118
  symmetric, 45–48
Chosen-ciphertext attack, 266
Chosen-plaintext attack, 266
Ciphers, 279, 282
Ciphersystems:
  basic elements of, 259–61
  binary-coded substitution cipher, 265
  ciphertext, 260
  cryptoanalysis, 262
  decryption rule, 260
  permutation cipher, 263–64
  product ciphersystems, 275–79
  shift cipher, 261–62
  substitution cipher, 264
  Vigenere cipher, 262–63
Ciphertext, 260
Ciphertext-only attack, 265
CJB code, 104
Clock recovery block, 59
Code:
  alphabet, 10
  average random, 287–88
  BCH, 185–86, 221
  complete, 95
  concatenated, 187, 290
  convolutional, 201–41
  cyclic, 160–200
  DTC, 68–74, 89–90
  error-correcting, 117–59
  expanded, 144–47, 189–90
  extending, 189–90
  Huffman, 16–21
  incomplete, 95
  look-ahead, 102–7
  Matched Spectral Null (MSN), 109–14
  MFM, 96, 98
  perfect, 136
  prefix, 16–17, 94–96
  shortened, 188
  source, 5
  state-dependent, 96–98
  state-independent, 96–98
  variable-length, 92–94
  variable-length fixed-rate, 98–102
Code alphabet B, 10
Code capacity, and Hamming distance, 124–26
Code rate, 118
Codes, 279–80
Code-word concatenation, look-ahead codes, 102–4
Code words, 10, 118, 131
Coding:
  arithmetic, 28–34
  block, 48–51
  data translation, 68–74, 89–90
  Huffman, 16–21
  Lempel-Ziv (LZ) coding, *See* LZ codes
  random, 285–87
Coding gain, of TCM code, 254–55
Communication system, 1–2
  examples of, 1
Commuting ciphers, 276–78
Complete decoder, 136
Compound symbol, 6, 8
Compress command, UNIX, 22
Computational security, 272–74
Concatenated codes, 187, 290
Conditional entropy, 7, 48
  See also Equivocation
Conditional mutual information, 57
Conditional probability, 6
Confusion, 274–75
Constrained channels, 58–62
  and linear time-invariant channels, 60–62
  and modulation theory, 58–60
Content-addressable memory (CAM), and Huffman coding, 20
Convolutional codes, 201–41
  definition of, 201–5
  known-good, 221–23
  punctured, 234–38
    known-good, 236–38
  state diagram and trellis representations, 205–7
  structural properties of, 205–10
  transfer functions of, 207–10
  Viterbi algorithm, 210–21
    hard-decision decoding, 215–21
    soft-decision decoding, 223–29
  Viterbi decoding, traceback method, 229–34
Convolutional interleaver, 196–97
Convolution sum, 162
Coset leaders, 138
Cosets, 138
CRC, *See* Cyclic redundancy check (CRC)
Cross interleaver, 196
Crosstalk, 90
Cryptography, 259–84
Cryptosystems, 259–65
  attacks on, 265–66
    chosen-ciphertext attack, 266
    chosen-plaintext attack, 266
    ciphertext-only attack, 265
    known-plaintext attack, 265
  ciphersystems:
    basic elements of, 259–61
    binary-coded substitution cipher, 265
    ciphertext, 260
    cryptoanalysis, 262
    decryption rule, 260
    permutation cipher, 263–64
    product ciphersystems, 275–79
    shift cipher, 261–62

Index

substitution cipher, 264
 Vigenere cipher, 262–63
ciphertext attacks, and language entropy, 269–72
codes, 279–80
key-management problem, 268
perfect secrecy, 266–69
public-key cryptosystems, 280–81
Cutoff rate, 51
Cyclic:
 code, 160–200
 shift, 160
Cyclic codes, 160–200
 cyclic shift property, 160
 definition of, 160
 error-trapping decoders, 178–84
 expanding, 189–90
 extending, 189–90
 generation/decoding of, 169–78
 generator polynomial of, 164
 interleaving, 186, 194–97
 and linear block code properties, 161
 Meggitt decoder, 175, 186
 polynomial modulo arithmetic, 164–68
 polynomial representation of, 162–64
 properties of, 160–62
 shortening, 188, 190–93
  noncyclicity of shortened codes, 193–94
 simple modifications to, 189–97
 standard, 184–88
  BCH codes, 185–86
  burst-correcting codes, 186–87
  cyclic redundancy check (CRC) codes, 187–89
  Hamming codes, 184–85
 systematic, 164
Cyclic redundancy check (CRC):
 burst detection capability, 188–89
 codes, 187–89
 error pattern coverage, 188
 undetected error probability, 188

D

Data, information compared to, 4
Data compression, 5
 lossless, 35–36
 lossy, 35
 See also Arithmetic coding; Huffman codes; Lempel-Ziv codes; Source coding
Data Encryption Standard (DES), 275–79
Data processing:
 inequality, 58
 and Markov chains, 56–58
 source encoder, 5, 10
Data storage system, 2
Data translation codes, 68–74, 89–90, 293
 capacity of, 73–74
 data sequences, constraints on, 68–70
 and run-length-limited codes, 89–91
 state space and trellis description of code, 70–72
Dc-free codes, 107–14
 running digital sum/digital sum variation, 107–9
 state-splitting/matched spectral null codes, 109–14

Decision feedback equalization (DFE), 90
Decoder, 2
 bounded distance, 152–56
 building, 14
 burst-correcting, 186–87
 complete, 136
 error-trapping, 178–84
 failure, 136
 forward error correction (FEC), 147
 hardware implementation of, 171–74
 maximum likelihood, 123–24
 minimum-distance, 136
 shortened code, 188
 soft-decision, 223–29
 standard-array, 136, 138
Decoder failure, 136
Decoding:
 depth, 215
 hard-decision, 215–21
 instantaneous, 92
 soft-decision, 223–29
 Viterbi, 229–34
Decoding depth, 215
Decoding sphere, 286
Decryption rule, 260
De-interleaver, 195, 197
Detection:
 maximum likelihood sequence, 89
 symbol-by-symbol (SBS), 89
Dictionary codes:
 and Lempel-Ziv (LZ) coding, 21–28
 rationale behind, 21–22
Diffusion and confusion, 274–75
Digital communication and storage system, 1–2
Digital frequency, 64
Digital sum variation (DSV), 108
Discrete Fourier transform, 65
Discrete memoryless channel model, 39–43
 output entropy and mutual information, 41–43
 transition probability matrix, 39–41
Discrete memoryless channels (DMC), 242
Discrete sources and entropy, 1–68
 arithmetic coding, 28–34
 dictionary codes and Lempel-Ziv (LZ) coding, 21–28
 digital communication and storage systems, 1–2
 entropy of symbol blocks and the chain rule, 8–9
 Huffman coding, 16–21
 joint and conditional entropy, 6–7
 source alphabets and entropy, 2–6
 source coding, 10–16
Discrete-time sources, examples of, 2
Distance:
 Euclidean, 224
 free, 252
 Hamming, 122–26, 133–35, 224
 unicity, 272
(d, k) sequences, 75–83
 maxentropic sequences:
  defined, 77
  power spectrum of, 77–81
  and run-length-limited codes, 75–77

Dual code:
  defined, 140
  of a Hamming code, 143–44
Dual space, 140

E

Encoder, 2
  encryption, 14
  FIR, 202, 248
  hardware implementation of, 171–74
  IIR, 202
  source, 5, 10
  Ungerboeck, 248–49
Encryption, 14–15
Encryption encoders, 14
Entropy, 1–68
  chain rule for, 8–9
  conditional, 6–7, 48
  defined, 4
  joint, 6–7
  language, 269–72
  output, 41–43
  rate, 27, 29, 49–51, 54–56
  and source alphabets, 2–6
  of symbol blocks, 8–9
Entropy rate, 27, 29, 50
  and channel-coding theorem, 49–51
  and steady-state probability, 54–56
Equivalent baseband channel, 59–60
Equivocation, 48–49, 267
Ergodic Markov process, 54–56
Error:
  burst, 180–84
  event, 217–19
  undetectable, 153, 188
Error-correcting codes:
  binary, 118
  burst, 186–87
  convolutional, 201–41
  cyclic, 160–200
  Hamming, 136, 140–47, 184–85, 221
  linear, 126–31
  trellis-coded modulation (TCM), 202fn, 242–58
Error correction, 121–23
Error detection, 121–23
Error distributions, for the binary symmetric channel, 118–21
Error pattern coverage, 188
Error polynomial, 169
Error rate:
  for the binary symmetric channel, 118–21
  performance bounds, 147–52
  PSK, 245–47
  QAM, 247–48
  TCM, 242
  undetected block, 153
  unrecoverable, 120
Error rate performance bounds, for linear block error-correcting codes, 147–52

Error-trapping decoders, 178–84
  burst error patterns and error trapping, 180–84
  updating the syndrome during correction, 178–79
Euclidean distance metric, 224
Euclid's algorithm, 186
Exclusive-or operation, 127
Expanded Hamming codes, 144–47

F

Fast-adaptive Huffman coding, 34
Fields, 126
Finite impulse response (FIR) encoder, 202, 248
Fire codes, 186, 221
Fixed-length block codes, 91–92
Forward error correction (FEC), 147
Fractional rate loss, 204
Franaszek code, 94–95, 98, 101
Free distance, of a code, 252
Frequency domain analysis, 62

G

Generator matrix, 138
Generator polynomial, 164, 169
Gilbert bound, 135–36
GLOPS, 279
Go-back-N ARQ (GBN-ARQ) protocol, 156
Golay code, 136

H

Hamming bound, 135–36
Hamming codes, 136, 140–47, 184–85, 221
  design of, 140–43
  dual code of, 143–44
  expanded, 144–47
Hamming cube, 133–34
Hamming distance, 122–26, 133–35, 224
  and code capacity, 124–26
Hamming sphere, 135–36
  volume of, 135
Hamming weight, 133, 207–8
Hard-decision decoding, 215–21
  bit error rate, bounds on, 219–21
  error event probability, 217–19
  maximum likelihood under, 215–17
Huffman codes, 16–21
  coding efficiency, robustness of, 20–21
  construction of, 17–19
  content-addressable memory (CAM), 20
  decoding of, 19–20
  hardware implementation approaches, 19–20
  instantaneous decoding, 16–17
Huffman coding:
  adaptive, 34
  fast-adaptive, 34
  prefix codes, 16–17

I

Idempotent cryptosystem, 276–77
Impulse response, 60
Incomplete prefix codes, 95
Indeterminate, 162
Infinite impulse response (IIR) channel, 60
Infinite impulse response (IIR) encoder, 202
Information:
  compared to data, 4
  knowledge compared to, 3
Information-bearing waveform signal, 243
Information content, 4
Information efficiency, of source, 5
Information-lossy process, 228
Information theory, 2–4
  ciphersystems, 259–64
  computational security, 272–74
  and cryptography, 259–84
  crytosystems, 250–65
  diffusion and confusion, 274–75
  product cipher systems, 275–79
Informative, use of term, 3
In-phase signals, 242–43
Instantaneous decoding, 17, 92
Interleaving, 186, 194–97

J

Joint entropy, 6–7
Joint probability, 6
Jordan form, 74

K

Key:
  public, 280–81
  spurious, 269
Key-equivocation theorem, 269
Key-management problem, 268
Known-plaintext attack, 265
Kraft equality, and prefix codes, 95–96
Kronecker delta function, 63
Kuhn-Tucker theorem, 47

L

Language entropy, and successful ciphertext attacks, 269–72
Language redundancy, 271–72
Lempel-Ziv (LZ) coding, *See* LZ codes
Lempel-Ziv-Welch (LZW) algorithm, 22
Linear block codes, 131–36
  decoding, 136–40
  Hamming cube, 133–34
  Hamming distance, 133–35
  Hamming sphere, 135–36
  Hamming weight, 133
  syndrome decoders and the parity-check theorem, 138–39
  vector spaces, elementary properties of, 131–33

Linear block error-correcting codes, 117–59
  binary fields/binary vector spaces, 126–31
  bounded distance decoders with repeat requests, performance of, 152–56
  channel coding for error correction, 117–18
  error detection/correction, 121–23
  error rate performance bounds for, 147–52
    bit error rate, 148–52
    block error rates, 147–48
  error rates/error distributions for the binary symmetric channel, 118–21
  Hamming codes, 140–47
  Hamming distance, 122–26
    and code capacity, 124–26
  linear block codes, 131–36
  maximum likelihood decoding principle, 123–24
Linear code, definition of, 131
Linear time-invariant channels (LTI), 60–62
Linear transforms, 162
Linked lists, 22–25
Log likelihoods, 123
Look-ahead codes, 102–7
  code-word concatenation, 102–4
  informal/formal design methods, 105–7
  k constraint, 104–5
Lossless compression codes, 35–36
  arithmetic coding, 28–34
  Huffman coding, 16–21, 35–36
  LZ codes, 21–28, 36
LZ codes, 21–28
  decoding process, 25–26
  linked lists, 22–25
  LZ compression, large-block requirement of, 26–28

M

Markov chains, and data processing, 56–58
Markov processes, 51–56
m-ary phase shift keying (m-PSK), 242
Matched Spectral Null (MSN) codes, 109–14
Matrix:
  generator, 138
  parity-check, 138
  state transition, 71
  transition probability, 39–41
Maxentropic sequences:
  defined, 77
  power spectrum of, 77–81
  and run-length-limited codes, 75–77
Maximization of mutual information, and channel capacity, 43–45
Maximum likelihood decoding principle, 123–24
Maximum likelihood sequence estimation (MLSE), 89
Meggitt decoder, 175, 186
Meggitt's theorem, 168, 170, 175–77
Memoryless source, 8
Merging rule, 96
Message word, 118
MFM state-dependent code, 96, 98
Minimum distance decoder, 136
Minimum Hamming distance of the code, 124–26

Modulation theory, 89
  and constrained channels, 58–60
Modulo arithmetic, 164–68
  algebraic identities for, 166–68
  polynomial rings, 164–66
Monic polynomial, 170
Multiamplitude/multiphase discrete memoryless channels, 242–48
  I-Q modulation, 242–43
  n-ary PSK signal constellation, 243–45
  PSK error rate, 245–47
  quadrature amplitude modulation, 247–48
Multiplication, 126
  properties of, 127
  scalar, 128
Multiplicative identity, 127
Multiplicative inverses, 127
Multi-stage pipelining, 179
Mutual information, 12–14, 35
  conditional, 57
  maximization of, and channel capacity, 43–45

N

Natural units (nats), 4
Noise:
  additive white Gaussian noise (AWGN), 244
  channel, 42
  power spectral density, 245–46
  quantizing, 58
Noiseless channels, 118
Noise power spectral density, 245–46
Noncommuting ciphers, 276–78
Nonlinear error-correcting codes, 242
Null character, 3–4
Null symbol, 23

O

Operator, 162
Ordered n-tuple, 6
Ordered pair, 6

P

Parity-bit generator, 138
Parity bits, 138
Parity-check polynomials, 169, 249–50
Parity-check theorem, and syndrome decoders, 138–39
Partial response maximum-likelihood (PRML), 90
Partial response signaling, 90
Path survivor register, 229
Perfect code, 136
Periodic interleaver, 196–97
Permutation cipher, 263–64
Phrases, 27
Playback system, 2
Polynomial:
  error, 169
  generator, 164, 169
  monic, 170
  parity-check, 169, 249–50
  primitive, 188
  rings, 164–66
  syndrome, 169
Polynomial modulo arithmetic, 164–68
  algebraic identities for, 166–68
  polynomial rings, 164–66
Power spectrum of sequences, 64–68
Prefix codes, 16–17, 94–96
  complete, definition of, 95
  and Kraft equality, 95–96
Prefix condition, 92
  satisfying, 17
Primitive binary polynomial, 188
Principal states, 98, 105
Product cipher systems, 275–79
  commuting/noncommuting/idempotent product ciphers, 276–78
  mixing transformations and good product ciphers, 278–79
Programmable logic arrays (PLAs), and Huffman coding, 20
Public-key cryptosystems, 280–81
Punctured convolutional codes, 234–38
  good codes, 236–38
Puncturing a code, 203, 235

Q

Quadrature amplitude modulation (QAM), 247–48
Quadrature signals, 242–43
Quantizing noise, 58

R

Random coding, 285–87
Reader, 2
Recorder, 1
Redundancy, 271
Redundant symbols, 50
Reed-Muller codes, 221
Reed-Solomon codes, 131, 221
Rieger bound, 186
Rivest, Shamir, and Adleman (RSA) system, 280–81
Run-length-limited (RLL) codes, 89–116
  data translation coding, general considerations for, 89–91
  dc-free codes, 107–14
  look-ahead codes, 102–7
  and maxentropic sequences, 75–77
    average runlength for, 77
  prefix codes/block codes, 91–96
  variable-length fixed-rate codes, 98–102
Running digital sum (RDS), 108

S

Scalar multiplication, 128
Selective-repeat ARQ (SR-ARQ) protocol, 156
Self-punctuating code, 16–17, 92
Set partitioning, 251–54
Shannon-Fano coding, 291–92
Shannon-McMillan-Breiman theorem, 29

Index

Shannon's coding theorems, 285–94
  average random code, 287–88
  random coding, 285–87
  Shannon-Fano coding, 291–2
  Shannon's noiseless-coding theorem, 292
  Shannon's second theorem, 48–51, 118, 289–90
Shannon's noiseless-coding theorem, 292
Shannon's second theorem, 48–51, 118, 289–90
  converse case of, 290
Shortening cyclic code, 188
Signal constellation, 243
Signal mapping, 251–54
Signal space redundancy, 257
Signal to noise ratio (SNR), 245, 248, 255
Singleton bound, 125–26
Soft-decision decoding, 223–29
  elimination of ties/information loss, 226–28
  Euclidean distance and maximum likelihood, 223–29
  likelihood metric, calculation of, 228–29
Soft-output-decision Viterbi decoding (SOVA), 293
Source:
  alphabet, 2–6
  asynchronous, 3
  coding, 10–16, 35
  efficiency, 5
  information, 1
  information efficiency of, 5
  memoryless, 8
  synchronous, 3
Source alphabets:
  defined, 3
  and entropy, 2–6
Source code, 5
Source coding, 10–16, 35
  encryption, 14–15
  mapping functions and efficiency, 10–11
  mutual information, 12–14
Source data, 4
Source encoder, 5, 10
Source words, 94–95
Spurious keys, 269
Standard-array decoder, 136, 138
State-dependent fixed-length block codes, 96–98
State-splitting, 109–10
State transition matrix, 71
Stationary source, 3
Steady-state probability, and entropy rate, 54–56
Stop-and-wait ARQ (SW-ARQ) protocol, 156
Storage systems, 1–2
Strings of source symbols, and LZ coding, 23
Substitution cipher, 264
Superposition, linear block code, 131–32
Supremum, 27–28
Symbol blocks, entropy of, 8–9
Symbol-by-symbol (SBS) detection, 89
Symbols, time sequence of, 2
Symmetric channels, 45–48
Synchronous sources, 3
Syndrome:
  bits, 129
  decoders, 138–39
  polynomial, 169
  updating, 178–79
  vector, 130
Systematic block codes, 169–71
Systematic cyclic codes, hardware implementation of encoders for, 171–74
Systematic recursive convolutional encoders, 248–50

T

3PM code, 104
Time redundancy, 257
Time sequences, statistics of, 62–64
Traceback method, 212, 229–34
Transfer function operators, 208
Transfer functions, of convolutional codes, 207–10
Transition probability matrix, 39–41
Transmitter, 1
Trellis-coded modulation (TCM), 202fn, 242–58
  known good trellis codes for PSK and QAM, 254–57
  multiamplitude/multiphase discrete memoryless channels, 242–48
  signal mapping and set partitioning, 251–54
  systematic recursive convolutional encoders, 248–50
Trellis diagram, 71
Truncation error, 215
Turbo codes, 51
Typical set, 30

U

Undetected block error rate, 153
Ungerboeck encoder, 248–49
Unicity distance, 272
Unrecoverable error rate, 120

V

Variable-length block codes, 92–94
Variable-length fixed-rate codes, 98–102
Vector addition, 128
Vector space, 128
  dimension of, 132–33
  elementary properties of, 131–33
Vector subspace, 133
Vigenere cipher, 262–63
Viterbi algorithm, 210–21, 290
Viterbi decoding, traceback method, 229–34

W

White noise, 244
White spectrum, 66
Writer, 1

Z

Zero modulation (ZM) code, 104